To my mother Madhuri, my wife Meena, and children Nupur and Kunal

"I think I now know . . . that what matters in our work—whether you act on the stage or write stories—what really matters is not fame, or glamour, not the things I used to dream about—but knowing how to endure things. How to bear one's cross and have faith."
                        Anton Chekhov (*The Seagull*)

Prac
and

Parag K

Prentice Hall
Upper Saddle River, NJ Columbus, Ohio

Library of Congress Cataloging-in-Publication Data

Lala, Parag K.
   Practical digital logic design and testing / Parag K. Lala.
     p.   cm.
   Includes bibliographical references and index.
   ISBN 0-02-367171-8
     1. Logic circuits—Design—Data processing. 2. Logic design—Data processing. 3. Computer-aided design. I. Title.
TK7868,L6L348   1996   95-20486
621.39'5—dc20   CIP

Cover Art: © Gregory Macnicol/ Photo Researchers
Editor: Dave Garza
Production Supervision: Julie Anderson Peters
Cover Design Coordinator: Jill E. Bonar
Text Design, Production, and Illustration Coordination: York Production Services/ Susan Free
Cover Design: Brian Deep
Production Manager: Patricia A. Tonneman
Marketing Manager: Debbie Yarnell
Illustrations: Dartmouth Publishing, Inc.

This book was set in Times Roman by Alphabet Graphics and was printed and bound by Quebecor Printing/Book Press. The cover was printed by Phoenix Color Corp.

© 1996 by Prentice-Hall, Inc.
A Pearson Education Company
Upper Saddle River, NJ 07458

All rights reserved. No part of this book may be reproduced, in any form or by any means, without permission in writing from the publisher.

Transferred to Digital Print on Demand 2002
Printed and bound by Antony Rowe Ltd, Eastbourne

10 9 8 7 6 5 4 3 2 1

ISBN: 0-02-367171-8

Prentice-Hall International (UK) Limited,London
Prentice-Hall of Australia Pty. Limited, Sydney
Prentice-Hall Canada Inc., Toronto
Prentice-Hall Hispanoamericana, S.A., Mexico
Prentice-Hall of India Private Limited, New Delhi
Prentice-Hall of Japan, Inc., Tokyo
Pearson Education Asia Pte. Ltd., Singapore
Editora Prentice-Hall do Brasil, Ltda., Rio de Janeiro

Preface

This book covers all major topics that are needed in a modern logic design course. It presents in-depth discussion of many topics that are relevant to current logic design practices but either are inadequately covered or not covered at all in introductory texts. Major features of this book are computer-aided minimization techniques, multilevel logic design, PLD (Programmable Logic Device)–based digital design, design description using ABEL, state assignment using computer-aided techniques, test generation techniques, and designing for testability. The material has been discussed in an informal manner, although the nature of certain topics makes an abstract discussion unavoidable. The objective is not to achieve understanding at the expense of avoiding necessary theory but to clarify the theory with illustrative examples in order to establish the theoretical basis for practical implementations.

The book is divided into eleven chapters. Since this is an introductory book, a complete bibliography is not provided. References at the end of a chapter are limited mainly to textbooks that a student may wish to read for intensive study of a particular topic.

Chapter 1 provides standard coverage of number representations and considers various number formats. It also discusses binary arithmetic operations such as addition, subtraction, multiplication, and division.

Chapter 2 provides a comprehensive coverage of the most important error detecting codes (e.g., the Berger, m-out-of-n, and residue codes). Conventional binary codes used to represent decimal numbers are also covered, as well as the Hamming error correcting code.

Chapter 3 contains a miscellany of basic topics in discrete mathematics required for understanding material presented in later chapters. Also, the operations of various gates that are used to construct logic circuits are discussed.

Chapter 4 provides an in-depth coverage of traditional combinational logic analysis, minimization, and synthesis techniques. The basic concepts of multilevel logic synthesis have been clearly explained. The use of the Berkeley CAD tool SIS[1] in the minimization and synthesis of multilevel logic has been illustrated. This chapter also provides a detailed coverage of combinational logic implementation using PLDs. The basics of the design entry software program ABEL, which is used specifically for describing logic circuits for PLD-based implementation, have been presented. Several examples of creating design descriptions using ABEL have been included. Also, a technique for implementing combinational logic implementation using EX-OR gates is presented.

Chapter 5 presents the basic concepts of sequential circuits. The operation of memory elements is analyzed. The use of state diagrams and state tables to represent the behavior of sequential circuits is discussed. Also, the distinction between *synchronous* and *asynchronous* operation of sequential circuits is clarified.

Chapter 6 provides a clear picture of how sequential circuits are designed using fundamental building blocks (e.g., latches and flip-flops) rather than presenting a rigorous mathematical structure of such circuits. A good coverage of partition algebra for deriving state assignment has been included. Also, computer-aided techniques for state assignment—e.g., NOVA and JEDI (part of the SIS tool) are discussed. A detailed discussion on sequential circuit implementation using PLDs is presented. Several examples of ABEL-based design descriptions of sequential circuits are included.

Chapter 7 covers design principles for traditional fundamental mode nonsynchronous sequential circuits. The concepts of race and hazard are clarified with examples, and state assignment techniques to avoid these are discussed.

Chapter 8 provides comprehensive coverage of shift registers and counters, which are important in many digital applications. Several design examples and illustrations are provided to clarify the use of these devices. The realization of sequential circuits using shift registers is thoroughly discussed.

Chapter 9 contains discussions on all types of arithmetic circuits (unlike many other texts in which only adders and subtractors are discussed in the context of combinational logic design). BCD addition/subtraction algorithms and carry-save addition techniques are included. Multiplication and division are thoroughly discussed. Also, a technique for implementing symmetric functions by full adders is incorporated.

Chapter 10 provides detailed coverage of integrated-circuit families used in the design of custom and semicustom chips. This chapter will be especially useful for students with a computer science background.

Chapter 11 Testing logic circuits has become one of the most important issues in VLSI design community. This chapter discusses fundamental concepts of test generation and design for testability. It provides sufficient background for anybody interested in this area to enable him or her to understand and apply currently available techniques in practical design of testable circuits.

---

[1] SIS can be obtained by contacting `anonymous ftp: ilpsoft.berkeley.edu` on the Internet. A `readme` file provides instructions for copying the tools.

This book is primarily intended as a college text for a two-semester course in logic design for students in electrical/computer technology or electrical/computer engineering and computer science programs. It does not require any previous knowledge of electronics; only some general mathematical ability is assumed. In the first semester (introductory course), the following sequence of chapters may be covered: Chapter 1, Chapter 2 (Sections 2.1–2.2), Chapter 3, Chapter 4 (Sections 4.1–4.7), Chapter 5, Chapter 6 (Sections 6.1–6.5), and Chapter 10. In the second semester (advanced-level course), the suggested sequence of chapters is: Chapter 2 (Sections 2.2–2.4), Chapter 4 (Sections 4.5–4.11), Chapter 6 (Sections 6.5, 6.6, 6.7, 6.8), Chapter 7, Chapter 8, Chapter 9, and Chapter 11.

Although the book is meant for a two-semester course sequence, certain sections can be omitted to fit the material in a typical one-semester course. Sections that can be omitted without loss of continuity have been marked with asterisks. However, individual instructors may select chapters at their discretion to suit the needs of a particular class they are teaching.

This book should also be extremely useful for practicing engineers who took logic design courses five or more years ago, to update their knowledge. Electrical engineers who are not logic designers by training but wish to become so can use this book for self-study. The author can be contacted by e-mail (Lala @garfield.ncat.edu) or by phone (910) 334-7760 ext. 215.

Acknowledgments

I am grateful to several of my students—Xiaoying Ma, Jian-Qing Wang, Kowen Lai, Nitin Patel, and especially Shihyu Yang—who read various drafts of the manuscript, suggested ways to improve the presentation, and corrected mistakes. My sincere thanks go to Dave Garza, my editor at Prentice Hall, for his encouragement and cooperation, and to the reviewers of the manuscript for their critical and constructive review. These reviewers are:

William E. Barnes, New Jersey Institute of Technology

Dr. Walter Buchanan, University of Central Florida

David M. Hata, Portland Community College

Samuel Kraemer, Oklahoma State University–Stillwater

Roberto Uribe, Kent State University

Steve Yelton, Cincinnati State Technical and Community College

I am also greatly indebted to my wife, Meena, for her help in typing and editing the manuscript. She has been a constant source of support throughout the writing of the book. Lastly, I acknowledge the patience and understanding of my daughter Nupur and son Kunal, who are not yet old enough to appreciate what writing a book involves.

Parag K. Lala

Contents

1
Number Systems — 1

1.1	Decimal numbers	1
1.2	Binary numbers	2
1.3	Octal numbers	8
1.4	Hexadecimal numbers	12
1.5	Signed numbers	14
1.6	Floating-point numbers	19
	Exercises	20
	References	21
	Further Reading	21

2
Error Detecting and Correcting Codes — 23

2.1	Weighted codes	23
2.2	Non-weighted codes	25
*2.3	Error detecting codes	28
*2.4	Error correcting codes	32
	Exercises	37
	Further Reading	38

3

Sets, Relations, Graphs, and Boolean Algebra — 39

3.1	Set theory	39
3.2	Relations	41
3.3	Partitions	43
3.4	Functions	45
3.5	Graphs	46
3.6	Boolean algebra	47
3.7	Boolean functions	52
3.8	Derivation and classification of Boolean functions	54
3.9	Canonical forms of Boolean functions	56
3.10	Logic gates	59
	Exercises	64
	Further Reading	68

4

Combinational Logic Design — 69

4.1	Introduction	69
4.2	Minimization of Boolean expressions	70
4.3	Karnaugh maps	73
4.4	The Quine-McCluskey method	82
*4.5	ESPRESSO	88
*4.6	Minimization of multiple-output functions	90
4.7	NAND-NAND and NOR-NOR logic	92
*4.8	Multilevel logic design	97
*4.9	Combinational logic implementation using EX-OR gates	102
4.10	Logic design using MSI chips	108
4.11	Combinational circuit design using PLDs	121
	Exercises	137
	References	140
	Further Reading	140

5

Fundamental Concepts of Sequential Logic — 141

5.1	Introduction	141
5.2	Synchronous and asynchronous operation	141
5.3	Latches	143
5.4	Flip-flops	145
5.5	Timing in synchronous sequential circuits	151
5.6	State tables and state diagrams	154
5.7	Mealy and Moore models	157
5.8	Analysis of synchronous sequential circuits	160
	Exercises	162
	References	164

6

Synchronous Sequential Circuits — 165

6.1	Introduction	165
6.2	Problem definition of sequential circuits	166
6.3	State reduction	169
6.4	Derivation of design equations	174
6.5	State assignment	182
6.6	Incompletely specified sequential circuits	195
6.7	Algorithmic State Machine (ASM)	198
6.8	Sequential logic design using PLDs	201
	Exercises	230
	References	234
	Further Reading	235

7

Asynchronous Sequential Circuits — 237

7.1	Flow table	238
7.2	Reduction of primitive flow tables	241
7.3	State assignment	243
7.4	Excitation and output functions	251
7.5	Hazards	253
	Exercises	262
	References	265
	Further Reading	265

8

Registers and Counters — 267

8.1	Ripple (asynchronous) counters	267
8.2	Asynchronous up-down counters	272
8.3	Synchronous counters	272
8.4	Gray code counters	276
8.5	Integrated-circuit counters	278
8.6	Shift registers	284
*8.7	Shift register counters	287
8.8	Ring counters	296
*8.9	Johnson counters	299
*8.10	Pseudo-random sequence generation using shift registers	303
*8.11	Sequential circuit realization using shift registers	306
8.12	Random access memory	309
	Exercises	310
	References	312
	Further Reading	312

9
Arithmetic Circuits — 313

9.1	Half-adders	313
9.2	Full adders	313
9.3	Serial and parallel adders	316
9.4	Carry-lookahead adders	317
9.5	Carry-save addition	319
*9.6	Implementation of symmetric functions using full adders	320
9.7	BCD adders	325
9.8	Half-subtractors	325
9.9	Full subtractors	326
9.10	Two's complement subtractors	329
9.11	BCD subtractors	330
9.12	Multiplication	331
9.13	Division	336
9.14	Comparators	338
	Exercises	340
	Further Reading	341

10
Digital Integrated Circuits — 343

10.1	NMOS transistors	343
10.2	NMOS inverters	344
10.3	Logic design using NMOS gates	347
10.4	CMOS logic	352
10.5	Bipolar logic families	358
10.6	BiCMOS logic	371
	Exercises	373
	References	377

11
Testing and Testability — 379

11.1	Fault models	379
11.2	Fault detection in logic circuits	385
11.3	Test generation for combinational circuits	386
11.4	Design for testability	395
11.5	Built-in self-test (BIST)	403
11.6	Autonomous self-testing	407
	Exercises	408
	References	413

Index — 415

1
Number Systems

Number systems provide the basis for all operations in information processing systems. In conventional arithmetic, a number system based upon ten units (0 to 9) is used. However, arithmetic and logic circuits used in computers and other digital systems operate with only 0's and 1's because it is very difficult to design circuits that require ten distinct states. The number system with the basic symbols 0 and 1 is called binary. In this chapter we discuss number systems in general and the binary system in particular. In addition, we consider the octal and hexadecimal number systems and fixed- and floating-point representation of numbers.

1.1 DECIMAL NUMBERS

The invention of decimal number systems has been the most important factor in the development of science and technology. The term *decimal* comes from the Latin word for "ten." The decimal number system uses positional number representation, which means that the value of each digit is determined by its position in a number.

The base (also called radix) of a number system is the number of symbols that the system contains. The decimal system has ten symbols: 0, 1, 2, 3, 4, 5, 6, 7, 8, 9; in other words it has a base of 10. Each position in the decimal system is 10 times more significant than the previous position. For example, consider the four-digit number 2725:

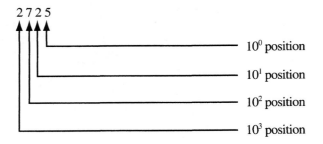

Notice that the 2 in the 10^3 position has a different value than the 2 in the 10^1 position. The value of a decimal number is determined by multiplying each digit of the number by the value of the position in which the digit appears and then adding the products. Thus, the number 2725 is interpreted as

$$2 \times 1000 + 7 \times 100 + 2 \times 10 + 5 \times 1 = 2000 + 700 + 20 + 5$$

i.e., two thousand seven hundred twenty-five. In this case, 5 is the least significant digit (LSD) and the leftmost 2 is the most significant digit (MSD).

In general in a number system with a base or radix r, the digits used are from 0 to $r - 1$. The number can be represented as

$$N = a_n r^n + a_{n-1} r^{n-1} + \cdots + a_1 r^1 + a_0 r^0 \tag{1.1}$$

where, for $n = 0, 1, 2, 3, \ldots$
r = base or radix of the number system
a = number of digits having values between 0 and $r - 1$.

Thus, for the number 2725, $a_3 = 2$, $a_2 = 7$, $a_1 = 2$, and $a_0 = 5$. Equation (1.1) is valid for all integers. For numbers between 0 and 1, i.e., fractions, the following equation holds

$$N = a_{-1} r^{-1} + a_{-2} r^{-2} + \cdots + a_{-n+1} r^{-n+1} + a_{-n} r^{-n} \tag{1.2}$$

Thus for the decimal fraction 0.8125,

$$\begin{aligned} N &= 0.8000 + 0.0100 + 0.0020 + 0.0005 \\ &= 8 \times 10^{-1} + 2 \times 10^{-2} + 1 \times 10^{-3} + 8 \times 10^{-4} \\ &= a_{-1} \times 10^{-1} + a_{-2} \times 10^{-2} + a_{-3} \times 10^{-3} + a_{-4} \times 10^{-4} \end{aligned}$$

where
$$\begin{aligned} a_{-1} &= 8 \\ a_{-2} &= 1 \\ a_{-3} &= 2 \\ a_{-1} &= 5 \end{aligned}$$

1.2 BINARY NUMBERS

The binary number has a radix of 2. As $r = 2$, only two digits are needed, and these are 0 and 1. A binary digit, 0 or 1, is called a bit. Like the decimal system, binary is a positional system, except that each bit position corresponds to a power of 2 instead of a power of 10. In digital systems, the binary number system and other number systems closely related to it are used almost exclusively. However, people are accustomed to using decimal number system; hence, digital systems must often provide conversion between decimal and binary numbers. The decimal value of a binary number can be formed by multiplying each power of 2 by either 1 or 0, and adding the values together.

■ **EXAMPLE 1.1**

Let us find the decimal equivalent of the binary number 101010.

$$\begin{aligned} N &= 101010 \\ &= 1 \times 2^5 + 0 \times 2^4 + 1 \times 2^3 + 0 \times 2^2 + 1 \times 2^1 + 0 \times 2^0 \quad \text{(using Eq. 1.1)} \\ &= 32 + 0 + 8 + 0 + 2 + 1 + 0 \\ &= 43 \end{aligned}$$

■

An alternative method of converting from binary to decimal begins with the leftmost bit and works down to the rightmost bit. It starts with a sum of 0. At each step the current sum is multiplied by 2, and the next digit to the right is added to it.

EXAMPLE 1.2

The conversion of 11010101 to decimal would use the following steps:

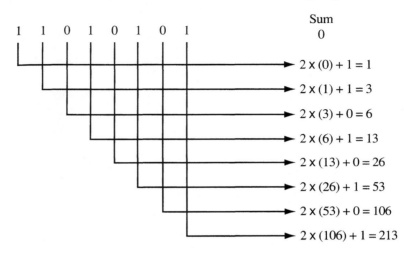

The reverse process, the conversion of decimal to binary, may be made by first decomposing the given decimal number into two numbers—one corresponding to the positional value just lower than the original decimal number and a remainder. Then the remainder is decomposed into two numbers: a positional value just equal to or lower than itself and a new remainder. The process is repeated until the remainder is 0. The binary number is derived by recording 1 in the positions corresponding to the numbers whose summation equals the decimal value.

EXAMPLE 1.3

Let us consider the conversion of decimal number 426 to binary:

$$426 = 256 + 170$$
$$= 256 + 128 + 42$$
$$= 256 + 128 + 32 + 10$$
$$= 256 + 128 + 32 + 8 + 2$$

with arrows pointing to 2^8, 2^7, 2^5, 2^3, 2^1

Thus, $426_{10} = 110101010_2$ (the subscript indicates the value of the radix).

An alternative method for converting a decimal number to binary is based on successive division of the decimal number by the radix number 2. The remainders of the divisions, when written in reverse order (with the first remainder on the right), yield the binary equivalent to the decimal number. The process is illustrated below by converting 353_{10} to binary.

$$\frac{353}{2} = 176, \text{remainder } 1$$

$$\frac{176}{2} = 88, \text{remainder } 0$$

$$\frac{88}{2} = 44, \text{remainder } 0$$

$$\frac{44}{2} = 22, \text{remainder } 0$$

$$\frac{22}{2} = 11, \text{remainder } 0$$

$$\frac{11}{2} = 5, \text{remainder } 1$$

$$\frac{5}{2} = 2, \text{remainder } 1$$

$$\frac{2}{2} = 1, \text{remainder } 0$$

$$\frac{1}{2} = 0, \text{remainder } 1$$

Thus, $353_{10} = 101100001_2$.

So far we have only considered whole numbers. Fractional numbers may be converted in a similar manner.

■ **EXAMPLE 1.4**

Let us convert the fractional binary number 0.101011 to decimal. Using Equation (1.2),

$$N = 0.101011$$
$$= 1 \times 2^{-1} + 0 \times 2^{-2} + 1 \times 2^{-3} + 0 \times 2^{-4} + 1 \times 2^{-5} + 1 \times 2^{-6}$$

where $a_{-1} = 1, a_{-2} = 0, a_{-3} = 1, a_{-4} = 0, a_{-5} = 1, a_{-6} = 1$.
Thus,

$$N = 0.101011$$
$$= \frac{1}{2} + \frac{1}{8} + \frac{1}{32} + \frac{1}{64} = 0.671875$$

■

A decimal fraction can be converted to binary by successively multiplying it by 2; the integral (whole number) part of each product, 0 or 1, is retained as the binary fraction.

■ **EXAMPLE 1.5**

Derive the binary equivalent of the decimal fraction 0.203125. Successive multiplication of the fraction by 2 results in

1.2 BINARY NUMBERS

	0.203125
	× 2
$a_{-1} = 0$	0.406250
	× 2
$a_{-2} = 0$	0.812500
	× 2
$a_{-3} = 1$	0.625000
	× 2
$a_{-4} = 1$	0.250000
	× 2
$a_{-5} = 0$	0.500000
	× 2
$a_{-6} = 1$	0.000000

Thus, the binary equivalent of 0.203125_{10} is 0.001101_2. The multiplication by 2 is continued until the decimal number is exhausted (as in the example) or the desired accuracy is achieved. Accuracy suffers considerably if the conversion process is stopped too soon. For example, if we stop after the fourth step, then we are assuming 0.0011 is approximately equal to 0.20315, whereas it is actually equal to 0.1875, an error of about 7.7%. ∎

Basic Binary Arithmetic

Arithmetic operations using binary numbers are far simpler than the corresponding operations using decimal numbers due to the very elementary rules of addition and multiplication. The rules of binary addition are

$$0 + 0 = 0$$
$$0 + 1 = 1$$
$$1 + 0 = 1$$
$$1 + 1 = 0 \text{ (carry 1)}$$

As in decimal addition, the least significant bits of the addend and the augend are added first. The result is the sum, possibly including a carry. The carry bit is added to the sum of the digits of the next column. The process continues until the bits of the most significant column are summed.

■ **EXAMPLE 1.6**

Let us consider the addition of the decimal numbers 27 and 28 in binary.

Decimal	Binary	
27	11011	Addend
+ 28	+ 11100	Augend
55	110111	Sum
	11000	Carry

To verify that the sum is correct, we convert 110111 to decimal;

$$1 \times 2^5 + 1 \times 2^4 + 0 \times 2^3 + 1 \times 2^2 + 1 \times 2^1 + 1 \times 2^0$$
$$= 32 + 16 + 0 + 4 + 2 + 1$$
$$= 55$$

EXAMPLE 1.7

Let us add -11 to -19 in binary. Since the addend and the augend are negative, the sum will be negative.

Decimal	Binary	
19	10011	
11	01011	
30	11110	Sum
	00011	Carry

In all digital systems, the circuitry used for performing binary addition handles two numbers at a time. When more than two numbers have to be added, the first two are added, then the resulting sum is added to the third number, and so on.

Binary subtraction is carried out by following the same method as in the decimal system. Each digit in the **subtrahend** is deducted from the corresponding digit in the **minuend** to obtain the difference. When the minuend digit is less than the subtrahend digit, then the radix number, i.e., 2, is added to the minuend, and a **borrow** 1 is added to the next subtrahend digit. The rules applied to the binary subtraction are

$$0 - 0 = 0$$
$$0 - 1 = 1 \quad \text{(borrow 1)}$$
$$1 - 0 = 1$$
$$1 - 1 = 0$$

EXAMPLE 1.8

Let us consider the subtraction of 21_{10} from 27_{10} in binary

Decimal	Binary	
27	11011	Minuend
$-$ 21	$-$ 10101	Subtrahend
6	00110	Difference
	00100	Borrow

It can easily be verified that the difference 00110_2 corresponds to decimal 6.

EXAMPLE 1.9

Let us subtract 22_{10} from 17_{10}. In this case, the subtrahend is greater than the minuend. Therefore, the result will be negative.

Decimal	Binary	
17	10001	
$-$ 22	$-$ 10110	
$-$ 5	$-$ 00101	Difference
	00001	Borrow

Binary multiplication is performed in the same way as decimal multiplication, by multiplying, then shifting one place to the left, and finally adding the partial products. Since the multiplier can only be 0 or 1, the partial product is either zero or equal to the multiplicand. The rules of multiplication are

$$0 \cdot 0 = 0$$
$$0 \cdot 1 = 0$$
$$1 \cdot 0 = 0$$
$$1 \cdot 1 = 1$$

■ **EXAMPLE 1.10**

Let us consider the multiplication of the decimal numbers 67 by 13 in binary

Decimal	Binary	
67	1000011	Multiplicand
× 13	1101	Multiplier
871	1000011	1st partial product
	0000000	2nd partial product
	1000011	3rd partial product
	1000011	4th partial procuct
	1101100111	Final product

■

■ **EXAMPLE 1.11**

Let us multiply 13.5 by 3.25.

Decimal	Binary	
13.5	1101.10	Multiplicand
× 3.25	11.01	Multiplier
43.875	110110	1st partial product
	000000	2nd partial product
	110110	3rd partial product
	110110	4th partial product
	101011.1110	Final product

The decimal equivalent of the final product is $43 + 0.50 + 0.25 + 0.125 = 43.875$. ■

The process of binary division is very similar to standard decimal division. However, division is simpler in binary because when one checks to see how many times the divisor fits into the dividend, there are only two possibilities, 0 or 1.

■ **EXAMPLE 1.12**

Let us consider the division of 101110 (46_{10}) by 111 (7_{10}).

```
                    0001      Quotient
         Divisor 111 )101110   Dividend
                      0111
                       100
```

Since the divisor, 111, is greater than the first three bits of the dividend, the first three quotient bits are 0. The divisor is less than the first four bits of the dividend; therefore, the division is possible, and the fourth quotient bit is 1. The difference is less than the divisor, so we bring down the next bit of the dividend:

$$
\begin{array}{r}
00011 \\
111\overline{)101110} \\
\underline{0111} \\
1001 \\
\underline{111} \\
10
\end{array}
$$

The difference is less than the divisor, so the next bit of the dividend is brought down

$$
\begin{array}{r}
000110 \\
111\overline{)101110} \\
\underline{0111} \\
1001 \\
\underline{111} \\
100 \quad \text{Remainder}
\end{array}
$$

In this case the dividend is less than the divisor; hence, the next quotient bit is 0 and the division is complete. The decimal conversion yields 46/7 = 6 with remainder 4, which is correct. ∎

The methods we discussed to perform addition, subtraction, multiplication, and division are equivalents of the same operations in decimal. In digital systems, all arithmetic operations are carried out in modified forms; in fact, they use only addition as their basic operation. Chapter 9 discusses in detail how binary arithmetic operations are performed in digital systems.

1.3 OCTAL NUMBERS

Digital systems operate only on binary numbers. Since binary numbers are often very long, two shorthand notations, **octal** and **hexadecimal**, are used for representing large binary numbers. The octal number system uses a base or radix of 8; thus, it has digits from 0 to $r - 1$, or $8 - 1$, or 7. As in the decimal and binary systems, the positional value of each digit in a sequence of numbers is definitely fixed. Each position in an octal number is a power of 8, and each position is 8 times more significant than the previous position. The number 375_8 in the octal system therefore means

$$
\begin{aligned}
&3 \times 8^2 + 7 \times 8^1 + 5 \times 8^0 \\
&= 192 + 56 + 5 \\
&= 253_{10}
\end{aligned}
$$

■ **EXAMPLE 1.13**

Let us determine the decimal equivalent of the octal number 14.3.

$$
\begin{aligned}
14.3_8 &= 1 \times 8^1 + 4 \times 8^0 + 3 \times 8^{-1} \\
&= 8 + 4 + 0.375 \\
&= 12.375
\end{aligned}
$$

∎

The method for converting a decimal number to an octal number is similar to that used for converting a decimal number to binary (Sec. 1.2), except that the decimal number is successively divided by 8 rather than 2.

EXAMPLE 1.14

Let us determine the octal equivalent of the decimal number 278.

$$\frac{278}{8} = 34, \text{ remainder } 6$$

$$\frac{34}{8} = 4, \text{ remainder } 2$$

$$\frac{4}{8} = 0, \text{ remainder } 4$$

Thus, $278_{10} = 426_8$

Decimal fractions can be converted to octal by progressively multiplying by 8; the integral part of each product is retained as the octal fraction. For example, 0.651_{10} is converted to octal as follows:

```
              0.651
                  8
      5       0.208
                  8
      1       0.664
                  8
      5       0.312
                  8
      2       0.496
                  8
      3       0.968
             etc.
```

According to Equation (1.2), $a_{-1} = 5$, $a_{-2} = 1$, $a_{-3} = 5$, $a_{-4} = 2$, and $a_{-5} = 3$; hence, $0.651_{10} = 0.51523_8$. More octal digits will result in more accuracy.

A useful relationship exists between binary and octal numbers. The number of bits required to represent an octal digit is three. For example, octal 7 can be represented by binary 111. Thus, if each octal digit is written as a group of three bits, the octal number is converted into a binary number.

EXAMPLE 1.15

The octal number 324_8 can be converted to a binary number as follows

```
  3     2     4
  ↓     ↓     ↓
 011   010   100
```

Hence, $324_8 = 11010100_2$; the most significant 0 is dropped because it is meaningless, just as 0123_{10} is the same as 123_{10}.

The conversion from binary to octal is also straightforward. The binary number is partitioned into groups of three starting with the least significant digit. Each group of three binary digits is then replaced by an appropriate decimal digit between 0 and 7 (Table 1.1).

EXAMPLE 1.16

Let us convert 110011101001_2 to octal

$$\underbrace{110}_{6} \; \underbrace{011}_{3} \; \underbrace{101}_{5} \; \underbrace{001}_{1}$$

The octal representation of the binary number is 6351_8. If the leftmost group of a partitioned binary number does not have three digits, it is padded on the left with 0's. For example, 1101010 would be divided as

$$\underbrace{001}_{1} \; \underbrace{101}_{5} \; \underbrace{010}_{2}$$

The octal equivalent of the binary number is 152_8. In case of a binary fraction, if the bits cannot be grouped into 3-bit segments, then 0's are added on the right to complete groups of three. Thus, 110111.1011 can be written as

$$\underbrace{110}_{6} \; \underbrace{111}_{7} \; . \; \underbrace{101}_{5} \; \underbrace{100}_{4}$$

As shown in the previous section, binary equivalent of a decimal number can be obtained by successively dividing the number by 2 and using the remainders as the answer, the first remainder being the lowest significant bit, and so on. A large number of divisions by 2 are required to convert from decimal to binary if the decimal number is large. It is

TABLE 1.1
Binary to octal conversion

Binary	Octal
000	0
001	1
010	2
011	3
100	4
101	5
110	6
111	7

often more convenient to convert from decimal to octal and then replace each digit in octal in terms of three digits in binary. For example let us convert 523_{10} to binary by going through octal.

$$\frac{523}{8} = 65, \text{ remainder } 3$$

$$\frac{65}{8} = 8, \text{ remainder } 1$$

$$\frac{8}{8} = 1, \text{ remainder } 0$$

$$\frac{1}{8} = 0, \text{ remainder } 1$$

Thus,

$$(523)_{10} = (\ 1 \quad 0 \quad 1 \quad 3\ )_8$$
$$= (001 \ 000 \ 001 \ 011)_2$$

It can be verified that the decimal equivalent of 001000001011_2 is 523_{10}:

$$1 \times 2^9 + 1 \times 2^3 + 1 \times 2^1 + 1 \times 2^0$$
$$= 512 + 8 + 2 + 1$$
$$= 523_{10}$$

Addition and subtraction operations using octal numbers are very much similar to that used in decimal systems. In octal addition, a carry is generated when the sum exceeds 7_{10}. For example,

$$\begin{array}{r} 153_8 \\ +327_8 \\ \hline 502_8 \end{array}$$

$3 + 7 = 10_{10} = 2 + 1$ carry ⟵ 1st column
$5 + 2 + 1$ carry $= 0 + 1$ carry ⟵ 2nd column
$1 + 3 + 1$ carry $= 5$ ⟵ 3rd column

In octal subtraction, a borrow requires that 8_{10} be added to the minuend digit and a 1_{10} be added to the left adjacent subtrahend digit.

$$\begin{array}{r} 670_8 \\ -125_8 \\ \hline 543_8 \end{array}$$

$0 - 5 = (8 - 5 + 1 \text{ borrow})_{10} = 3 + 1$ borrow ⟵ 1st column
$7 - (2 + 1 \text{ borrow}) = 7 - 3 = 4$ ⟵ 2nd column
$6 - 1 = 5$ ⟵ 3rd column

1.4 HEXADECIMAL NUMBERS

The hexadecimal numbering system has a base 16; i.e., there are 16 symbols. The decimal digits 0 to 9 are used as the first ten digits as in the decimal system, followed by the letters A, B, C, D, E, and F, which represent the values 10, 11, 12, 13, 14, and 15 respectively. Table 1.2 shows the relationship between decimal, binary, octal, and hexadecimal number systems. The conversion of a binary number to a hexadecimal number consists of partitioning the binary numbers into groups of 4 bits, and representing each group with its hexadecimal equivalent.

EXAMPLE 1.17

The binary number 1010011011110001 is grouped as

$$1010 \quad 0110 \quad 1111 \quad 0001$$

which is shown here in hexadecimal:

$$A6F1_H$$

The conversion from hexadecimal to binary is straightforward. Each hexadecimal digit is replaced by the corresponding 4-bit binary equivalent from Table 1.2. For example, the binary equivalent of $4AC2_H$ is

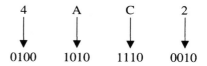

Thus, $4AC2_H = 0100101011000010_2$.

Sometimes it is necessary to convert a hexadecimal number to decimal. Each position in a hexadecimal number is 16 times more significant than the previous position. Thus, the decimal equivalent for $1A2D_H$ is

$$1 \times 16^3 + A \times 16^2 + 2 \times 16^1 + D \times 16^0$$
$$= 1 \times 16^3 + 10 \times 16^2 + 2 \times 16^1 + 13 \times 16^0$$
$$= 6701$$

Hexadecimal numbers are often used in describing the data in a computer memory. A computer memory stores a large number of words, each of which is a standard size collection of bits. An 8-bit word is known as a **byte**. A hexadecimal digit may be considered as half of a byte. Two hexadecimal digits constitute one byte, the rightmost 4 bits corresponding to half a byte, and the leftmost 4 bits corresponding to the other half of the byte. Often a half-byte is called a **nibble**.

Hexadecimal addition and subtraction are performed as for any other positional number system.

EXAMPLE 1.18

Let us find the sum of 688_H and 679_H.

1.4 HEXADECIMAL NUMBERS   13

TABLE 1.2
Number equivalents

Decimal	Binary	Octal	Hexadecimal
0	0000	0	0
1	0001	1	1
2	0010	2	2
3	0011	3	3
4	0100	4	4
5	0101	5	5
6	0110	6	6
7	0111	7	7
8	1000	10	8
9	1001	11	9
10	1010	12	A
11	1011	13	B
12	1100	14	C
13	1101	15	D
14	1110	16	E
15	1111	17	F

$$688_H$$
$$679_H$$
$$\overline{D01_H}$$

$8 + 9 = 17_{10} = 1 + 1$ carry ← 1st column
$8 + 7 + 1$ carry $= 16_{10} = 0 + 1$ carry ← 2nd column
$6 + 6 + 1$ carry $= 13_{10} = D$ ← 3rd column

Hexadecimal subtraction requires the same need to carry digits from left to right as in octal and decimal.

EXAMPLE 1.19

Let us compute $2A5_H - 11B_H$ as shown:

$$2A5_H$$
$$11B_H$$
$$\overline{18A_H}$$

$5 - B = (21 - 11 + 1 \text{ borrow})_{10} = 10 + 1$ borrow
$\qquad\qquad\qquad\qquad\qquad = A + 1$ borrow ← 1st column
$A - (1 + 1 \text{ borrow}) = (10 - 2)_{10} = 8$ ← 2nd column
$\qquad\qquad\qquad 2 - 1 = 1$ ← 3rd column

1.5 SIGNED NUMBERS

So far, the number representations we considered have not carried sign information. Such unsigned numbers have a magnitude significance only. Normally a prefix + or − may be placed to the left of the magnitude to indicate whether a number is positive or negative. This type of representation, known as **sign-magnitude representation,** is used in the decimal system. In binary systems, an additional bit known as sign bit, is added to the left of the most significant bit to define the sign of a number. A 1 is used to represent − and a 0 to represent +. Table 1.3 shows 3-bit numbers in terms of signed and unsigned equivalents. Notice that there are two representations of the number 0, namely +0 and −0. The range of integers that can be expressed in a group of three bits is from $-(2^2 - 1) = -3$ to $+(2^2 - 1) = +3$, with one bit being reserved to denote the sign.

Although the sign-magnitude representation is convenient for computing the negative of a number, a problem occurs when two numbers of opposite signs have to be added. To illustrate, let us find the sum of $+2_{10}$ and -6_{10}.

$$\begin{aligned} +2_{10} &= 0010 \\ -6_{10} &= 1110 \\ \hline &10000 \end{aligned}$$

The addition produced a sum that has 5 bits, exceeding the capability of the number system (sign + 3 bits); this results in an **overflow.** Also, the sum is wrong, it should be −4 (i.e., 1100) instead of 0.

An alternative representation of negative numbers, known as the **complement** form, simplifies the computations involving signed numbers. The complement representation enjoys the advantage that no sign computation is necessary. There are two types of complement representations:

Diminished radix complement

Radix complement

Diminished Radix Complement

In the decimal system $(r = 10)$ the complement of a number is determined by subtracting the number from $(r - 1)$, i.e., 9. Hence the process is called finding the 9's complement. For example,

$$\begin{aligned} &\text{9's complement of 5 } (9 - 5) = 4 \\ &\text{9's complement of 63 } (99 - 63) = 36 \\ &\text{9's complement of 110 } (999 - 110) = 889 \end{aligned}$$

In binary notation $(r = 2)$, the diminished radix complement is known as the **1's complement.** A positive number in 1's complement is represented in the same way as in sign-magnitude representation. The 1's complement representation of a negative number x is derived by subtracting the binary value of x from the binary representation of $(2^n - 1)$, where n is the number of bits in the binary value of x.

■ **EXAMPLE 1.20**

Let us compute the 1's complement form of −43.
The binary value of 43 = 00101011. Since $n = 8$, the binary representation of $2^8 - 1$ is

TABLE 1.3
Signed and unsigned binary numbers

Binary	Decimal Equivalent	
	Signed	Unsigned
000	+0	0
001	+1	1
010	+2	2
011	+3	3
100	−0	4
101	−1	5
110	−2	6
111	−3	7

$$2^8 = 100000000$$
$$\underline{-1}$$
$$2^8 - 1 = 11111111$$

Hence the 1's complement form of −43 is

$$2^8 - 1 = 11111111$$
$$-43 = \underline{-00101011}$$
$$11010100$$

Notice that the 1's complement form of −43 is a number that has a 0 in every position that +43 has a 1, and vice versa. Thus, the 1's complement of any binary number can be obtained by complementing each bit in the number. ■

A 0 in the most significant bit indicates a positive number. The sign bit is not complemented when negative numbers are represented in 1's complement form. For example, the 1's complement form of −25 will be represented as follows

$$-25 = 111001 \text{ (sign-magnitude form)}$$
$$ = 100110 \text{ (1's complement form)}$$

Table 1.4 shows the comparison of 3-bit unsigned, signed, and 1's complement values. The advantage of using 1's complement numbers is that they permit us to perform subtraction by actually using the addition operation. It means that in digital systems, addition and subtraction can be carried out by using the same circuitry.

The addition operation for two 1's complement numbers consists of the following steps:

i. Add the two numbers including the sign bits.
ii. If a carry bit is produced by the leftmost bits, i.e., the sign bits, add it to the result. This is called **end-around carry**.

■ **EXAMPLE 1.21**

Let us add −7 to −5

TABLE 1.4
Comparison of 3-bit signed, unsigned, and 1's complement values

Binary	Unsigned	Decimal Equivalent Sign-Magnitude	1's Complement
000	0	+0	+0
001	1	+1	+1
010	2	+2	+2
011	3	+3	+3
100	4	−0	−3
101	5	−1	−2
110	6	−2	−1
111	7	−3	−0

```
       1's complement form of −7 = 11000
       1's complement form of −5 = 11010
                         Sum     110010
                         Carry        1
                1's complement sum  10011
```

The result is a negative number. By complementing the magnitude bits we get

$$11100, \text{ i.e., } -12$$

Thus, the sum of −7 and −5 is −12, which is correct. ∎

The subtraction operation for two 1's complement numbers can be carried out as follows:

i. Derive the 1's complement of the subtrahend and add it to the minuend.
ii. If a carry bit is produced by the leftmost bits, i.e., the sign bits, add it to the result.

EXAMPLE 1.22

Let us subtract +21 from +35

```
     Minuend    = +35 = 0100011
     Subtrahend = +21 = 1101010    (in 1's complement form)
                 Sum  = 10001101
                 Carry         1
         1's complement sum = 0001110
```

The sign is correct, and the decimal equivalent +14 is also correct. ∎

The 1's complement number system has the advantage that addition and subtraction are actually one operation. Unfortunately, there is still a dual representation for 0. With 3-bit numbers, for example, 000 is **positive zero** and 111 is **negative zero.**

Radix Complement

In the decimal system, the radix complement is the 10's complement. The 10's complement of a number is the difference between 10 and the number. For example,

10's complement of 5, $10 - 5 = 5$
10's complement of 27, $100 - 27 = 73$
10's complement of 48, $100 - 48 = 52$

In binary number system, the radix complement is called the 2's complement. The **2's complement** representation of a positive number is the same as in sign-magnitude form. The 2's complement representation of a negative number is obtained by complementing the sign-magnitude representation of the corresponding positive number, and adding a 1 to the least significant position. In other words, the 2's complement form of a negative number can be obtained by adding 1 to the 1's complement representation of the number.

EXAMPLE 1.23

Let us compute the 2's complement representation of -43.

$$
\begin{aligned}
+43 &= 0101011 \quad \text{(sign-magnitude form)} \\
&= 1010100 \quad \text{(1's complement form)} \\
& \underline{+1} \quad \text{(add 1)} \\
& 1010101 \quad \text{2's complement form}
\end{aligned}
$$

Table 1.5 shows the comparisons of four representations of 3-bit binary numbers. The bit positions in a 2's complement number have the same weight as in a conventional binary number except that the weight of the sign bit is negative. For example, the 2's complement number 1000011 can be converted to decimal in the same manner as a binary number:

$$
\begin{aligned}
&-2^6 + 2^1 + 2^0 \\
={}& -64 + 2 + 1 \\
={}& -61
\end{aligned}
$$

TABLE 1.5 Various representations of 3-bit binary numbers

Binary	Unsigned	Decimal Equivalent		
		Signed	1's Complement	2's Complement
000	0	+0	+0	+0
001	1	+1	+1	+1
010	2	+2	+2	+2
011	3	+3	+3	+3
100	4	-0	-3	-4
101	5	-1	-2	-3
110	6	-2	-1	-2
111	7	-3	-0	-1

A distinct advantage of 2's complement form is that unlike 1's complement form, there is a unique representation of 0 as can be seen in Table 1.5. Moreover, addition and subtraction can be treated as the same operation as in 1's complement; however, the carry bit can be ignored and the result is always in correct 2's complement notation. Thus, addition and subtraction are easier in 2's complement than in 1's complement.

An **overflow** occurs when two 2's complement numbers are added, if the carry-in bit into the sign bit is different from the carry-out bit from the sign bit. For example, the following addition will result in an overflow:

```
         0  ← Carry-in
    101010        (−22)
    101001        (−23)
   1010011
   ↑
Carry-out
```

Hence the result is invalid.

EXAMPLE 1.24

Let us derive the following using 2's complement representation

(i) $+13$ (ii) -15 (iii) -9 (iv) -5
     -7 -6 $+6$ -1

```
                   1  ← Carry-in
(i)  +13        01101
     −7         11001
     +6        100110
              ↑
        Carry-out
```

Since the carry-in is equal to the carry-out, there is no overflow; the carry-out bit can be ignored. The sign bit is positive. Thus, the result is +6.

```
                   0  ← Carry-in
(ii)  −15       10001
      −6        11010
               101011
              ↑
        Carry-out
```

The carry-out bit is not equal to the carry-in bit. Thus, there is an overflow and the result is invalid.

```
                   0  ← Carry-in
(iii)  −9       10111
       +6       00110
       −3      011101
              ↑
        Carry-out
```

There is no overflow, and the sign bit is negative. The decimal equivalent of the result is $-2^4 + 2^3 + 2^2 + 2^0 = -3$.

```
                              1  ← Carry-in
    (iv)       −5      1011
               −1      1111
               ___     _____
               −6     11010
          Carry-out ─────┘
```

There is no overflow, and the sign bit is negative. The result, as expected, is $-2^3 + 2^1 = -6$. ∎

An important advantage of 2's complement numbers is that they can be **sign-extended** without changing their values. For example, if the 2's complement number 101101 is shifted right (i.e., the number becomes 1101101), the decimal value of the original and the shifted number remains the same (−19 in this case).

1.6 FLOATING-POINT NUMBERS

Thus far, we have been dealing mainly with fixed-point numbers in our discussion. The word *fixed* refers to the fact that the radix point is placed at a fixed place in each number, usually either to the left of the most significant digit or to the right of the least significant digit. With such a representation, the number is always either a fraction or an integer. The main difficulty of fixed-point arithmetic is that the range of numbers that could be represented is limited. Figure 1.1 illustrates the fixed-point representation of a signed four-digit decimal number; the range of numbers that can be represented using this configuration is 9999. In order to satisfy this limited range, scaling has to be used.

For example, to add +50.73 to +40.24 we have to multiply both numbers by 100 before addition and then adjust the sum, keeping in mind that there should be a decimal point two places from the right. The scale factor in this example is 100.

An alternative representation of numbers, known as the **floating-point** format, may be employed to eliminate the scaling factor problem. In a floating-point number, the radix point is not placed at a fixed place; instead, it "floats" to various places in a number so that more digits can be assigned to the left or to the right of the point. More digits on the left of the radix point permit the representation of larger numbers, whereas more digits on the right of the radix point result in more digits for the fraction. Numbers in floating-point format consist of two parts: a fraction and an exponent; they can be expressed in the form

$$\text{fraction} \times \text{radix}^{\text{exponent}}$$

The fraction is often referred to as the **mantissa** and can be represented in sign-magnitude, diminished radix complement, or radix complement form. For example, the decimal number 236,000 can be written as 0.236×10^6. In a similar manner, very small numbers may be represented using negative exponents. For example, 0.00000012 may be written as 0.12×10^{-6}. By adjusting the magnitude of the exponent, the range of numbers covered can be considerably enlarged. Leading 0's in a floating-point number may be removed by shifting the man-

FIGURE 1.1
Fixed-point number representation

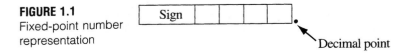

tissa to the left and decreasing the exponent accordingly; this process is known as **normalization** and floating-point numbers without leading 0's are called normalized. For example, the normalized form of the floating-point number 0.00312×10^5 is 0.312×10^3. Similarly, a binary fraction such as 0.001×2^4 would be normalized to 0.1×2^2.

Arithmetic operations with floating-point numbers are more complicated than the same operations using fixed-point numbers. An excellent discussion on floating-point arithmetic can be found in [1].

EXERCISES

1. Convert the following decimal numbers to binary.
   - **a.** 623
   - **b.** 73.17
   - **c.** 53.45
   - **d.** 2.575
2. Convert the following binary numbers to decimal.
   - **a.** 10110110
   - **b.** 110000101
   - **c.** 100.1101
   - **d.** 1.001101
3. Convert the following binary numbers to hexadecimal and octal.
   - **a.** 100101010011
   - **b.** 001011101111
   - **c.** 1011.111010101101
   - **d.** 1111.100000011110
4. Convert the following octal numbers to binary and hexadecimal.
   - **a.** 1026
   - **b.** 7456
   - **c.** 5566
   - **d.** 236.2345
5. Convert the following hexadecimal numbers to binary and octal.
   - **a.** EF69
   - **b.** 98AB5
   - **c.** DAC.1BA
   - **d.** FF.EE
6. Perform the addition of the following binary numbers.
   - **a.** 100011 + 1101
   - **b.** 10110110 + 11100011
   - **c.** 10110011 + 1101010
7. Perform the following subtractions, where each of the numbers is in binary form.
   - **a.** 101101 − 111110
   - **b.** 1010001 − 1001111
   - **c.** 10000110 − 1110001
8. Add the following pairs of numbers, where each number is in hexadecimal form.
   - **a.** ABCD + 75EF
   - **b.** 129A + AB22
   - **c.** EF23 + C89
9. Repeat Exercise 1.8 using subtraction instead of addition.
10. Add the following pairs of numbers, where each number is in octal form.
    - **a.** 7521 + 4370
    - **b.** 62354 + 3256
    - **c.** 3567 + 2750

11. Repeat Exercise 1.10 using subtraction instead of addition.
12. Derive the 6-bit sign-magnitude, 1's complement, and 2's complement for the following decimal numbers.
    a. $+22$
    b. -31
    c. $+17$
    d. -1
13. Find the sum of the following pairs of decimal numbers assuming 8-bit 1's complement representation of the numbers.
    a. $+61 + (-23)$
    b. $-56 + (-55)$
    c. $+28 + (+27)$
    d. $-48 + (+35)$
14. Repeat Exercise 1.13 assuming 2's complement representation of the decimal numbers.
15. Assume that X is the 2's complement of an n-bit binary number Y. Prove that the 2's complement of X is Y.
16. Find the floating-point representation of the following numbers.
    a. $(326.245)_{10}$
    b. $(101100.100110)_2$
    c. $(-64.462)_8$
17. Normalize the following floating-point numbers.
    a. 0.000612×10^6
    b. 0.0000101×2^4

References

1. J. F. Cavanaugh, *Digital Computer Arithmetic*, McGraw-Hill, New York, 1984.

Further Reading

1. K. Hwang, *Computer Arithmetic*, John Wiley, New York, 1979.

2
Error Detecting and Correcting Codes

The last chapter showed how decimal numbers can be represented by equivalent binary numbers. Although digital systems use binary numbers for their internal operations, communication with the external world has to be done in decimal systems. In order to simplify the communication, every decimal number may be represented by a unique sequence of binary digits; this is known as **binary encoding**. Since there are ten decimal digits (0, 1, ..., 9), the minimum number of binary digits required is four, giving 16 ($=2^4$) possible combinations, of which only ten are used. Binary codes are divided into two groups—**weighted** and **non-weighted**.

2.1 WEIGHTED CODES

In a weighted code each binary digit is assigned a **weight** w; the sum of the weights of the 1 bits is equal to the decimal number represented by the four-bit combination. In other words, if d_i ($i = 0,..., 3$) are the digit values and w_i ($i = 0,..., 3$) are the corresponding weights, then the decimal equivalent of a 4-bit binary number is given by

$$d_3w_3 + d_2w_2 + d_1w_1 + d_0w_0$$

If the weight assigned to each binary digit is exactly the same as that associated with each digit of a binary number (i.e., $w_0 = 2^0 = 1$, $w_1 = 2^1 = 2$, $w_2 = 2^2 = 4$, and $w_3 = 2^3 = 8$), then the code is called the **BCD (Binary-Coded-Decimal) code**. The BCD code differs from the conventional binary number representation in that in the BCD code each decimal digit is binary-coded. For example, the decimal number 15 in conventional binary number representation is

$$1111$$

whereas in the BCD code, 15 is represented by

$$\underbrace{0001}_{1} \quad \underbrace{0101}_{5}$$

the decimal digits 1 and 5 each being binary-coded.

Several forms of weighted codes are possible, since the codes depend on the weight assigned to the binary digits. Table 2.1 shows the decimal digits and their weighted code equivalents. The 7421 code has fourteen 1's in its representation, which is the minimum number of 1's possible. However, if we represent decimal 7 by 0111 instead of 1000, the 7421 code will have sixteen 1's instead of fourteen. In the 4221 code, the sum of the weights is exactly 9 ($= 4 + 2 + 2 + 1$). Codes whose weights add up to 9 have the property that the 9's complement of a number (i.e., $9 - N$, where N is the number) represented in the code can be obtained simply by taking the 1's complement of its coded representation. For example, in the 4221 code shown in Table 2.1, the decimal number 7 is equivalent to the code word 1101; the 9's complement of 7 is 2 ($= 9 - 7$), and the corresponding code word is 0010, which is the 1's complement of 1101. Codes having this property are known as **self-complementing** codes. Similarly, the $84\overline{2}\overline{1}$ is also a self-complementing code.

Among the weighted codes, the BCD code is by far the most widely used. It is useful in applications where output information has to be displayed in decimal. The addition process in BCD is the same as in simple binary as long as the sum is decimal 9 or less. For example,

```
Decimal        BCD
   6           0110
  +3          +0011
   9           1001
```

However, if the sum exceeds decimal 9, the result has to be adjusted by adding decimal 6 (0110) to it. For example, let us add 5 to 7:

```
Decimal        BCD
   7           0111
  +5          +0101
  12           1100     12 (not a legal BCD number)
              +0110     Add 6
             00010010
              ‿  ‿
              1  2
```

TABLE 2.1
Weighted binary codes

Decimal number	8421	7421	4221	8$4\overline{2}\overline{1}$
0	0000	0000	0000	0000
1	0001	0001	0001	0111
2	0010	0010	0010	0110
3	0011	0011	0011	0101
4	0100	0100	1000	0100
5	0101	0101	0111	1011
6	0110	0110	1100	1010
7	0111	1000	1101	1001
8	1000	1001	1110	1000
9	1001	1010	1111	1111

As another example, let us add 9 to 7:

$$
\begin{array}{cc}
\text{Decimal} & \text{BCD} \\
9 & 1001 \\
+7 & +0111 \\
\hline
16 & \underbrace{0001}_{1}\ \underbrace{0000}_{0}
\end{array}
$$

Although the result consists of two valid BCD numbers, the sum is incorrect. It has to be corrected by adding 6 (0110). This is required when there is a carry from the most significant bit of a BCD number to the next higher BCD number. Thus, the correct result is

$$
\begin{array}{cc}
0001 & 0000 \\
+ & 0110 \\
\hline
\underbrace{0001}_{1} & \underbrace{0110}_{6}
\end{array}
$$

Other arithmetic operations in BCD can also be performed.

2.2 NON-WEIGHTED CODES

In the non-weighted codes there are no specific weights associated with the digits, as was the case with weighted codes. A typical non-weighted code is the **excess-3** code. It is generated by adding 3 to a decimal number and then converting the result to a 4-bit binary number. For example, to encode the decimal number 6 to excess-3 code, we first add 3 to 6. The resulting sum, 9, is then represented in binary, i.e., 1001. Table 2.2 lists the excess-3 code representations of the decimal digits.

Excess-3 code is a self-complementing code and is useful in arithmetic operations. For example, consider the addition of two decimal digits whose sum will result in a number greater than 9. If we use BCD code, no carry bit will be generated. On the other hand, if excess-3 code is used, there will be a natural carry to the next higher digit; however, the sum has to be adjusted by adding 3 to it. For example, let us add 6 to 7:

TABLE 2.2
Excess-3 code

Decimal	Excess-3
0	0011
1	0100
2	0101
3	0110
4	0111
5	1000
6	1001
7	1010
8	1011
9	1100

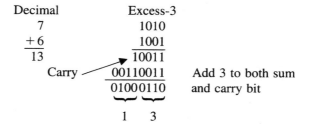

```
        Decimal        Excess-3
           7             1010
          +6             1001
          ---           -----
          13            10011
          Carry ──→   00110011      Add 3 to both sum
                       01000110     and carry bit
                       ‿‿‿‿ ‿‿‿‿
                         1    3
```

In excess-3 code, if we add two decimal digits whose sum is 9 or less, then the sum should be adjusted by subtracting 3 from it. For example,

```
        Decimal        Excess-3
           6             1001
          +2            +0101
          ---            ----
           8             1110
                         0011       Subtract 3
                         ----
                         1011
                         ‿‿‿‿
                           8
```

For subtraction in excess-3 code, the difference should be adjusted by adding 3 to it. For example,

```
     Decimal        Excess-3
       17         0100   1010
      -11         0100   0100
      ---         ----   ----
        6         0000   0110
                         0011       Add 3
                         ----
                         1001
                         ‿‿‿‿
                           6
```

Another code that uses four unweighted binary digits to represent decimal numbers is the **cyclic code**. Cyclic codes have the unique feature that the successive codewords differ only in one bit position. Table 2.3 shows an example of such a code.

TABLE 2.3
Cyclic code

Decimal	Cyclic
0	0000
1	0001
2	0011
3	0010
4	0110
5	0100
6	1100
7	1110
8	1010
9	1000

TABLE 2.4
Gray code

Decimal	Binary	Gray
0	0000	0000
1	0001	0001
2	0010	0011
3	0011	0010
4	0100	0110
5	0101	0111
6	0110	0101
7	0111	0100
8	1000	1100
9	1001	1101
10	1010	1111
11	1011	1110
12	1100	1010
13	1101	1011
14	1110	1001
15	1111	1000

One type of cyclic code is the **reflected code**, also known as the **Gray code**. A 4-bit Gray code is shown in Table 2.4. Notice that in Table 2.4, except for the most significant bit position, all columns are "reflected" about the midpoint; in the most significant bit position, the top half is all 0's and the bottom half all 1's.

A decimal number can be converted to Gray code by first converting it to binary. The binary number is converted to the Gray code by performing a modulo-2 sum of each digit (starting with the least significant digit) with its adjacent digit. For example, if the binary representation of a decimal number is

$$b_3 \, b_2 \, b_1 \, b_0$$

then the corresponding Gray code word, $G_3 G_2 G_1 G_0$, is

$$G_3 = b_3$$
$$G_2 = b_3 \oplus b_2$$
$$G_1 = b_2 \oplus b_1$$
$$G_0 = b_1 \oplus b_0$$

where $\oplus$ indicates exclusive-OR operation (i.e., modulo-2 addition), according to the following rules

$$0 \oplus 0 = 0$$
$$0 \oplus 1 = 1$$
$$1 \oplus 0 = 1$$
$$1 \oplus 1 = 0$$

As an example let us convert decimal 14 to Gray code.

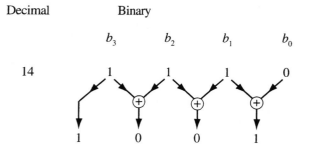

Thus the Gray code word for decimal 14 is

$$G_3\ G_2\ G_1\ G_0$$
$$1\ \ 0\ \ 0\ \ 1$$

The conversion of a Gray code word to its decimal equivalent is done by following this sequence in reverse. In other words, the Gray code word is converted to binary and the resulting binary number is then converted to decimal. To illustrate, let us convert 1110 from Gray to decimal:

$$G_3\ G_2\ G_1\ G_0$$
$$1\ \ 1\ \ 1\ \ 0$$

$$b_3 = G_3 = 1$$
$$G_2 = b_3 \oplus b_2 = 1 = 1 \oplus b_2 \quad \text{since } b_3 = 1$$
$$\therefore\ b_2 = 0$$
$$G_1 = b_2 \oplus b_1 = 1 = 0 \oplus b_1 \quad \text{since } b_2 = 0$$
$$\therefore\ b_1 = 1$$
$$G_0 = b_1 \oplus b_0 = 0 = 1 \oplus b_0 \quad \text{since } b_1 = 1$$
$$\therefore\ b_0 = 1$$

Thus, the binary equivalent of the Gray code word 1110 is 1011, which is equal to decimal 11. The first ten code words of the Gray code shown in Table 2.4 can be utilized as reflected BCD code if desired. The middle ten code words can be used as reflected excess-3 code.

*2.3 ERROR DETECTING CODES

Digital data may be corrupted during transmission due to noise. Besides, physical defects in a digital system may result in alteration of stored data. Hence, if data is to be transmitted or stored, some means of detecting error/s that might have been introduced should be incorporated. In general, only single-bit errors are considered in most practical systems, since the probability of two or more errors occurring simultaneously is very small.

Let us suppose that in BCD code the second bit of the code word 0110 (=6) is erroneously changed from 1 to 0, resulting in 0010 (=2). Since 0010 is a valid code word in BCD, it is not possible to say whether or not this code word was the intended one. Thus, one can detect the occurrence of a single-bit error in a code if the occurrence of the error transforms a valid code word into a non–code word.

Parity-Checked Codes

One way of introducing error-detecting ability into a code is to add an additional bit, this bit being known as a parity bit, to each code word. This parity bit makes the total number of 1's in a code word either odd or even, respectively giving rise to an "odd parity" and "even parity" code. Table 2.5 shows odd-parity-checked and even-parity-checked BCD codes. Even though the odd and even parity checks are mathematically equivalent, odd parity is generally preferred, since it ensures at least one 1 in any code word. From Table 2.5 it can be seen that the code word for 6 with odd parity is 01101. If the second bit is changed to 0, an error will be indicated, since the total number of 1's in 00101 is even. It should be apparent that parity checking can detect only single-bit errors. Double errors in the data bits cannot be detected, since the parity remains unchanged. For example, if the second and the third bit in the code word 01101 are changed, the code word 00001 will result, which is still correct.

M-out-of-N Codes

Another class of codes, called m-out-of-n codes, also has the single-error-detecting capability. In an m-out-of-n code, all valid code words have n bits, of which m bits are 1's. For representing the ten decimal symbols, the 2-out-of-5 code is suitable, since there are exactly ten combinations of five things taken two at a time, i.e., (C_2^5).

For a 2-out-of-5 code it is not possible to correctly "weight" all the ten code words; however, it is possible to weight nine of them. Table 2.6 shows a 2-out-of-5 code with the weights assigned as shown. Any error that causes a 0 → 1 or 1 → 0 will result in a non–code word, i.e., not a 2-out-of-5 code word, and hence will be detected. Double errors (e.g., two 1 → 0 or two 0 → 1) will also be detected; however, a double error consisting of a 0 → 1 and a 1 → 0 error will remain undetected.

TABLE 2.5 Parity-checked BCD code

Decimal Digit	BCD Code	Odd Parity	Even Parity
0	0000	1	0
1	0001	0	1
2	0010	0	1
3	0011	1	0
4	0100	0	1
5	0101	1	0
6	0110	1	0
7	0111	0	1
8	1000	0	1
9	1001	1	0

TABLE 2.6
A 2-out-of-5 code

Decimal	2-out-of-5 Code
	weight 63210
0	01001 ← not weighted
1	00011
2	00101
3	00110
4	01010
5	01100
6	10001
7	10010
8	10100
9	11000

Biquinary Codes

This is a seven-bit weighted code. It consists of two parts: One part is a weighted 1-out-of-2 code; the other part is a weighted 1-out-of-5 code. Table 2.7 shows a biquinary code. Any single error in such a code can be detected easily. Multiple errors are also detected, provided such errors are not compensating errors (i.e., $0 \to 1$ and $1 \to 0$).

Residue Codes

Residue codes are very useful in detecting errors in the results produced by arithmetic operations. A **residue** is defined as the remainder after a division. The residue representation of an integer N may be obtained from

$$N = I \cdot m + r$$

where m is a check base and I is an integer, so that $0 \leq r < m$. The quantity r is called

TABLE 2.7
Biquinary code

	50	43210
0	01	00001
1	01	00010
2	01	00100
3	01	01000
4	01	10000
5	10	00001
6	10	00010
7	10	00100
8	10	01000
9	10	10000

the residue N modulo m or

$$r = N \text{ modulo } m$$

The following example shows how the residue of decimal number (mod 9) can be calculated

$$I = 2658$$
$$r = (2 + 6 + 5 + 8) \text{mod } 9$$
$$= (21) \text{mod } 9$$
$$= 3$$

The calculation of the residue can be performed step by step by adding the digits one at a time and subtracting 9 from the sum if it exceeds 9. For example, in the given number, $2 + 6 = 8$; $8 + 5 = 13$, which is greater than 9, so 9 is subtracted from it ($13 - 9 = 4$). Proceeding, we get

$$4 + 8 = 12$$
$$12 - 9 = 3 \quad \text{the residue } r$$

In residue codes the data bits define a number N and the residue of the number; i.e., r is appended to the data bits as **check bits**. The number of check bits is $\lceil \log_2 m \rceil$[1]. For example, $N = 10111 (=23_{10})$, $m = 3$, $r = 23 \mod 3 = 2$ (i.e., 10 in binary), code word = 10111, concatenate 10 = 1011110. An alternative way to form residue codes is to multiply the number to be coded, i.e., N, by the desired check base, i.e., m. For example, $N = 10111$ $m = 3$, code word = 1000101.

In the first coding technique, the data bits are separate from the check bits and hence can be handled separately. In the second technique, the data bits can be extracted from the encoded word only after dividing it by the check base m. Notice that in residue codes the check bits are derived from the weighted value of the data bits, whereas in parity checking the parity bit is derived from the unweighted data bits.

Berger Codes

Berger codes are used in designing logic circuits that can detect their own errors, called **self-checking circuits**. They are formed as follows:

   i. A binary number corresponding to the number of 1's in the data bits is formed.
   ii. The complement of each bit in this number is taken.
   iii. The resulting binary number constitutes the check bits and is appended to the data bits to form a code word.

If there were n data bits, then $k = \lceil \log_2(n + 1) \rceil$ = number of check bits required in a Berger code word.

For example, if $n = 101011$, $k = \lceil \log_2(6 + 1) \rceil = 3$; hence, the Berger code word must have 9 (= 6 + 3) bits. The 3 check bits are derived as follows:

$$\text{Number of 1's in data bits} = 4$$
$$\text{Binary equivalent of } 4 = 100$$

---

[1]The ceiling of x, $\lceil x \rceil$, is the smallest integer greater than x.

The bit-by-bit complement of 100 is 011, which provides the three check bits. Hence,

$$\text{code word} = 101011011$$

Alternatively, the k check bits may be the binary number representing the number of 0's in n data bits, provided the total number of bits in n is equal to $2^k - 1$. For example, if $n = 1000110$, $k = \lceil \log_2(7 + 1) \rceil = 3$. Since the total number of bits in n is 7 ($= 2^k - 1$), the check bits are given by the binary equivalent of the number of 0's in data bits; in other words, the check bits are 100 ($=4$). Hence,

$$\text{code word} = 1000110100$$

To illustrate the error detection capability of the Berger code, let us assume that there are two erroneous bits in the data part of the code word; thus, the code word becomes

$$1011110100$$

There are now two 0's in the data bits; therefore, the check bits will be 010. Since the new check bits differ from the original check bits ($=100$), an error will be indicated. Notice that if in the original code word, one bit is changed from 1 to 0 and another is changed from 0 to 1, no error will be indicated. For example, if the code word 1000110100 has two erroneous bits,

$$1010010100$$

there will be no change in the check bits; hence, error detection is not possible. Thus, Berger codes are capable of detecting only single-bit errors and errors in which bits change from 0 to 1 or 1 to 0 but not both.

*2.4 ERROR CORRECTING CODES

As discussed in the last section, the basic idea of coding is to add check bits to the data bits such that if an error occurs in the data bits it can be detected. In certain applications the automatic correction of a detected error becomes indispensable (e.g., in the main memory of computer systems). None of the codes discussed so far can correct any errors. An error correcting code should not only have the ability to detect bit error(s), it should also be able to indicate their locations.

The error detecting and correcting capability of a code can be defined in terms of the **Hamming distance** of a code. This is the number of bits in which two distinct code words differ. The minimum Hamming distance is the minimum number of bits that must be changed to convert one code word to another. The relationship between the Hamming distance of a code and its error detecting and correcting capabilities can be defined as

$$d = C + D + 1 \quad \text{with } D \geq C$$

where
d = minimum Hamming distance of a code
D = number of bit errors that can be detected
C = number of bit errors that can be corrected

Since no error can be corrected without being detected, C is less than D. The relationship between C and D for values of d up to five are tabulated in Table 2.8.

In codes with a minimum distance of 1, two valid code words differ only in one bit position. Therefore, a single error in a code word will result in another code word; consequently, single errors cannot be detected in such codes.

EXAMPLE 2.1

Let us consider the following two-bit code words

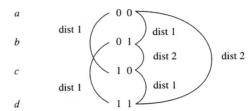

The minimum distance of the code words is 1. An error in a code word will result in another code word, thereby preventing the detection of an error. To illustrate, let us assume that the first bit of the code word a is erroneous (i.e., the bit has changed from 0 to 1). In other words, code word a is converted to code word c.

In codes with a minimum distance of 2, a single bit error in a code word will convert it into a non–code word because it will not match any other code word. However, two or more bit errors in a code word may convert it into another valid code word, and therefore these errors will not be detected. Thus, codes with a minimum Hamming distance of 2 are single-error-detecting codes.

EXAMPLE 2.2

Let us consider the following four-bit code words

a 0001
b 0110
c 1010
d 1111

TABLE 2.8 Relation between Hamming distance and error detecting/correcting capability

d	D	C
1	0	0
2	1	0
3	1	1
	2	0
4	3	0
	2	1
5	4	0
	3	1
	2	2

The minimum distance of 2 in the code words has been obtained by adding two extra bits to the code words of Example 2.1. Suppose the rightmost bit in code word *b* is changed to 1. The resulting word, 0111, is different from any of the other code words. Thus, the presence of the error is detected. On the other hand, if the first and the third bit in code word *c* are erroneous, the resulting word will be confused as code word *d*; hence, the double-bit error cannot be detected. ∎

In codes with a minimum distance of 3, all code words differ in at least three bit positions. A single error in a code word will result in a non–code word, i.e., a pattern that will have a distance of 1 from one code word, and a distance of 2 or more from all others. Let us consider the following code words with minimum distance 3:

$$
\begin{array}{cc}
a & 00111 \\
b & 11010 \\
c & 01001 \\
d & 10100
\end{array}
$$

Suppose the second bit of code word *a* is erroneous; i.e., code word *a* is changed to 00101. Since 00101 is a non–code word, the presence of the error is indicated. Similarly, 2-bit errors will also result in a non–code word. If it is assumed that only one error is likely to occur in a distance-3 code, then the error can be corrected as well as detected.

Let us see how a single-bit error in an *n*-bit word can be corrected. The key to error correction is that it must be possible to identify the bit in error. In order to achieve this, we must append *k* check bits, $c_k, \ldots, c_2, c_1$, to the *n*-bit word such that we can uniquely determine which of $(n + k)$ bits are in error as well as the error-free condition. The *k* check bits required for single-error correction are determined from the relationship (known as the **Hamming relationship**)

$$2^k \geq m + k + 1$$

where
m = number of data bits
k = number of check bits

The resulting code is known as the **Hamming code.**

Let us consider the construction of the Hamming code for information bits of length 4 ($d_4\,d_3\,d_2\,d_1$), i.e., $m = 4$. It can be seen from the Hamming relationship that if $m = 4$, $k = 3$. Thus, 3 check bits have to be appended to the 4 data bits in order for the code to be single-error correcting. Let c_1, c_2, c_3 be the check bits. The bit positions of the code are labeled with numbers 1 through 7

Bit positions	7	6	5	4	3	2	1
Bit names	d_4	d_3	d_2	c_3	d_1	c_2	c_1

The bit positions corresponding to the powers of 2 are used as check bits c_1, c_2, and c_3. Thus, the $2^0 (=1)$ position is assigned to c_1, the $2^1 (=2)$ position is assigned to c_2, and the $2^2 (=4)$ position is assigned to c_3. The other bit positions correspond to the data bits d_1 to d_4. Thus, as shown in Figure 2.1, the check bit c_1 establishes even parity for the positions 1, 3, 5, 7; similarly, c_2 and c_3 provide even parity for the positions 2, 3, 6, 7 and 4, 5, 6, 7 respectively.

■ **EXAMPLE 2.3**

Let us construct the Hamming code for the information bits $d_4 d_3 d_2 d_1 = 1110$. The check bits are generated as follows:

	7	6	5	4	3	2	1	
	d_4	d_3	d_2	c_3	d_1	c_2	c_1	
	1	1	1		0			Information bits
	1		1		0		0	Generate c_1
	1	1			0	0		Generate c_2
	1	1	1	1				Generate c_3
	1	1	1	1	0	0	0	Code word

Let us now consider how a single bit error is detected and corrected in Hamming code. A binary number involving the parity check bits c_1, c_2, and c_3, and the appropriate bit positions they check, is then formed. If the check shows even (correct) parity, a 0 is entered in the corresponding position of the binary number; if the check shows odd (incorrect) parity, a 1 is entered. The position of the bit in error is indicated by the resulting binary number. ■

■ | **EXAMPLE 2.4**

Assume that the bit in position 5 in code word 1111000 is changed from 1 to 0. Thus, the code word changes to

7	6	5	4	3	2	1
1	1	0	1	0	0	0
		↑				

The binary number corresponding to the erroneous bit in the code word can be derived as follows:

	7	6	5	4	3	2	1	
	1	1	0	1	0	0	0	Erroneous code word
	1		0		0		0	Odd (incorrect) parity
	1	1			0	0		Even (correct) parity
	1	1	0	1				Odd (incorrect) parity

Thus, the position of the erroneous bit is

$$2^2 \quad 2^1 \quad 2^0$$
$$1 \quad \;\; 0 \quad \;\; 1 \quad \text{i.e., position 5}$$

This error can be corrected by changing the bit in this position, i.e., d_2 from 0 to 1. If the

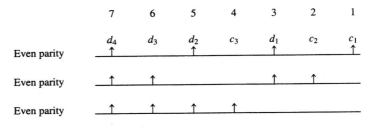

FIGURE 2.1
Check-bit generation for four information bits

resultant binary number derived by parity checking is zero (000), there is no bit error in the code word. ∎

EXAMPLE 2.5

Let us encode the information bits $d_8d_7d_6d_5d_4d_3d_2d_1 = 10010111$, as described above. There are 8 information bits, so 4 check bits ($c_1c_2c_3c_4$) are needed. The encoded data will have 12 bits. Check bit c_1 is assigned a value such that the parity for bits in positions 1,3,5,7,9 and 11 is even. Similarly, c_2, c_3, and c_4 are chosen to provide even parity for bits in positions (2,3,6,7,10,11), (4,5,6,7,12), and (8,9,10,11,12) respectively, as indicated below:

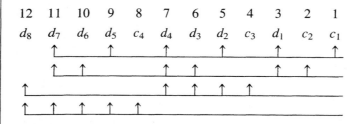

The check bits for $d_8d_7d_6d_5d_4d_3d_2d_1 = 10010111$ are derived as follows:

c_1 = even parity for bits $d_7d_5d_4d_2d_1$ (= 01011) = 1
c_2 = even parity for bits $d_7d_6d_4d_3d_1$ (= 00011) = 0
c_3 = even parity for bits $d_8d_4d_3d_2$ (= 1011) = 1
c_4 = even parity for bits $d_8d_7d_6d_5$ (=1001) = 0

Hence, the resulting codeword is 100100111101.

Let us now assume that c_1 (i.e., bit 1 in the code word) is changed from 1 to 0. The binary number identifying the position of the erroneous bit is derived below:

```
12   11   10   9    8    7    6    5    4    3    2    1
d8   d7   d6   d5   c4   d4   d3   d2   c3   d1   c2   c1
1    0    0    1    0    0    1    1    1    1    0    0
```

     ↑   ↑   ↑   ↑   ↑   ↑   odd (incorrect) parity
          ↑   ↑   ↑   ↑   ↑   ↑   even (correct) parity
     ↑   ↑   ↑   ↑   ↑   even (correct) parity
     ↑   ↑   ↑   ↑   ↑   even (correct) parity

Thus, the position of the erroneous bit is

$$2^3 \quad 2^2 \quad 2^1 \quad 2^0$$
$$0 \quad\ \ 0 \quad\ \ 0 \quad\ \ 1$$

In other words, the bit in position 1 (i.e., bit c_1) has to be inverted. ∎

It should be noted that any attempt to execute double-bit error correction with distance-3 Hamming code will result in false correction; i.e., the value of a correct bit will be changed.

■ **EXAMPLE 2.6**

Let us assume that bit 4 (c_3) and bit 7 (d_4) in the code word of Example 2.5 are erroneous. Thus, the code word changes to

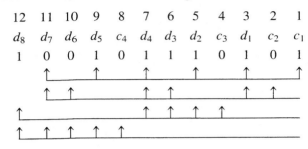

Thus, the locations of the erroneous bits are identified as

$$\begin{array}{cccc} 2^3 & 2^2 & 2^1 & 2^0 \\ 0 & 0 & 1 & 1 \end{array} \rightarrow \text{position 3, i.e., bit } d_1$$

Obviously this is wrong, since bits c_3 and d_4 were assumed to be erroneous. ■

EXERCISES

1. Encode each of the ten decimal digits using the weighted binary codes.
    a. 4, 4, 1, −2
    b. 7, 4, 2, −1
2. Given the following weighted codes
    a. (8, 4, −3, −2)
    b. (7, 5, 3, −6)
    c. (6, 2, 2, 1)
    determine whether any one of these is self-complementing.
3. Represent the decimal numbers 535 and 637 in
    a. BCD code
    b. Excess-3 code
    Add the resulting numbers so that the sum in each case is appropriately encoded.
4. Subtract 423 from 721, assuming the numbers are represented in excess-3 code.
5. Determine two forms for a cyclic code other than the one shown in Table 2.3.
6. Assign a binary code, using the minimum number of bits, to encode the 26 letters in the alphabet. Determine the odd parity bit and the even parity bit for each code word.
7. Determine the number of code words in the *m*-out-of-*n* coding scheme when
    a. $n = 5, m = 2$

b. $n = 6, m = 3$
   c. $n = 8, m = 1$
8. Encode the ten decimal digits using an m-out-of-n code where $n = 2m - 1$; use the minimum number of bits.
9. Calculate the residue of the following decimal numbers, assuming the modulus is 9.
   a. 1472943
   b. 236798
   c. 124011
10. Derive the residue codes for 4-bit information bits using a modulus of 3.
11. Consider an 8-bit residue code with modulus 3. For each of the following patterns, determine whether a single-bit error is present. Identify which bit may be erroneous. The rightmost 2 bits in each pattern are the check bits.
    a. 10111001
    b. 01010110
    c. 01110001
12. Write the code words of a 2-out-of-5 code, and derive the Berger code word for each of the resulting patterns.
13. Consider a 10-bit Berger code. For each of the following, determine whether or not there is any erroneous bit present (the 3 rightmost bits in each code word are the check bits).
    a. 0100100011
    b. 1111000100
    c. 0101111111
    d. 1000000010
14. Obtain a set of six code words in each of which a single-bit error can be corrected.
15. A 7-bit Hamming code (4 information bits and 3 check bits) is constructed as discussed in the text. For each of the following 7-bit code words, determine whether or not an erroneous bit is present. Identify the bit in error.
    a. 1110010
    b. 0111100
    c. 0101001
    d. 1000001
16. Show how the excess-3 codewords representing the decimal digits can be encoded as a 7-bit Hamming code as described in the text, but assuming the parity sense as odd for the check bits.

Further Reading

1. T. R. N. Rao and E. Fujiwara, *Error-Control Coding for Computer Systems*, Prentice Hall, Englewood Cliffs, 1989.

2. J. F. Wakerly, *Error Detecting Codes, Self-Checking Circuits, and Applications,* North Holland Publishing, Elsevier, Inc., New York, 1978.

3
Sets, Relations, Graphs, and Boolean Algebra

The objective of this chapter is to familiarize students with the basic concepts of discrete mathematics necessary to understand materials presented in later chapters. Boolean algebra, a mathematical system used for the analysis and design of electrical switching circuits (popularly known as logic circuits), is also discussed. We are mainly interested in the application of Boolean algebra in constructing circuits using elements called **gates.**

3.1 SET THEORY

A set is a collection of objects. The objects comprising a set must have at least one property in common—e.g., all male students in a class, all integers less than 15 but greater than 5, inputs to a logic circuit. A set that has a finite number of elements is described by listing the elements of the set between brackets. Thus, the set of all positive integers less than 15 but greater than 5 can be written as

$$\{6,7,8,9,10,11,12,13,14\}$$

The order in which the elements of a set are listed is not important. This set can also be represented as follows:

$$\{9,10,6,7,14,13,8,11,12\}$$

In general, uppercase letters are used to represent a set and lowercase letters are used to represent the numbers (or elements) of a set.

If a is an element of a set S, we represent the fact by writing

$$a \in S$$

On the other hand, if a is not an element of S we can indicate that by writing

$$a \notin S$$

For example, if $S = \{5,7,8,9\}$ then $5 \in S$, $7 \in S$, $8 \in S$, and $9 \in S$, but $6 \notin S$.

A set may have only one element; it is then called a **singleton.** For example, $\{6\}$ is the set with 6 as its only element. There is also a set that contains no element; it is known as **empty** or **null set,** and is denoted by ϕ.

A set may also be defined in terms of some property which all members of the set are required to have. Thus, the set $A = \{2,4,6,8,10\}$ may be defined as

$$A = \{x \mid x \text{ is an even positive integer not greater than } 10\}$$

In general, a set of objects that share common properties may be represented as

$$\{x \mid x \text{ possesses certain properties}\}$$

Sets with an infinite number of members are specified in this way, e.g.,

$$B = \{x \mid x \text{ is an even number}\}$$

A set P is a subset of a set Q if every element of P is also an element in Q. This may also be defined as P **is included in** Q and is represented with a new symbol,

$$P \subseteq Q$$

For example, the set $\{x,y\}$ is a subset of the set $\{a,b,x,y\}$ but is not a subset of the set $\{x,c,d,e\}$. Note that every set is a subset of itself and the null set is a subset of every set.

If P is a subset of Q and there is at least one element in Q that is not in P, then P is said to be a **proper subset** of Q. For example, the set $\{a,b\}$ is a proper subset of the set $\{x,y,a,b\}$. The notation $P \subset Q$ is used to denote that P is a proper subset of Q. When two sets P and Q contain exactly the same elements (in whatever order), they are equal. For example the two sets $P = \{w,x,y,z\}$ and $Q = \{x,y,w,z\}$ are equal. One may also say that two sets are equal if $P \subseteq Q$ and $Q \subseteq P$. The collection of all subsets of a set A is itself a set, called the **power set** of A, and is denoted by $P(A)$. For example, if $A = \{x,y,z\}$ then $P(A)$ consists of the following subsets of A:

$$\phi, \{x\}, \{y\}, \{z\}, \{x,y\}, \{x,z\}, \{y,z\}, \{x,y,z\}$$

Sets can be combined by various operations to form new sets. These operations are analogous to such familiar operations as addition on numbers. One such operation is **union**. The union of two sets P and Q, denoted by $P \cup Q$, is the set consisting of all elements that belong to either P or Q (or both). For example, if

$$P = \{d,c,e,f\} \qquad Q = \{b,c,a,f\}$$

then

$$P \cup Q = \{a,b,c,d,e,f\}$$

The **intersection** of two sets P and Q, denoted by $P \cap Q$, is the set that contains those elements that are common to both sets. For example, if

$$P = \{a,c,e,g\} \qquad Q = \{c,e,i,k\}$$

then

$$P \cap Q = \{c,e\}$$

Two sets are also **disjoint** if they have no element in common; i.e., their intersection is empty. For example, if

$$B = \{b,e,f,r,s\} \qquad C = \{a,t,u,v\}$$

then
$$B \cap C = \phi$$

In other words, there are no elements that belong to both P and Q.

If P and Q are two sets, then the **complement of Q with respect to P**, denoted by $P - Q$, is the set of elements that belong to P but not to Q. For example if
$$P = \{u,w,x\} \quad \text{and} \quad Q = \{w,x,y,z\}$$
then
$$P - Q = \{u\} \quad \text{and} \quad Q - P = \{y,z\}$$

Certain properties of the set operations follow easily from their definitions. For example, if P, Q, and R are sets, the following laws hold.

Commutative Law

1. $P \cup Q = Q \cup P$
2. $P \cap Q = Q \cap P$

Associative Law

3. $P \cup (Q \cup R) = (P \cup Q) \cup R$
4. $P \cap (Q \cap R) = (P \cap Q) \cap R$

Distributive Law

5. $P \cap (Q \cup R) = (P \cap Q) \cup (P \cap R)$
6. $P \cup (Q \cap R) = (P \cup Q) \cap (P \cup R)$

Absorption Law

7. $P \cap (P \cup Q) = P$
8. $P \cup (P \cap Q) = P$

Idempotent Law

9. $P \cup P = P$
10. $P \cap P = P$

DeMorgan's Law

11. $P - (Q \cup R) = (P - Q) \cap (P - R)$
12. $P - (Q \cap R) = (P - Q) \cup (P - R)$

3.2 RELATIONS

The word *relation* is used to express some association that may exist between certain objects. For example, among a group of students, two students may be considered to be related if they have common last names. This section studies the concept of binary relation and different properties such a relation may possess. To formally define a binary relation it is helpful to first define ordered pairs of elements. An **ordered pair** of elements is a pair of elements arranged in a prescribed order. The notation (a,b) is used to denote the ordered pair in which the first element is a and the second element is b. The order of elements in an ordered pair is important. Thus the ordered pair (a,b) is not the same as the ordered pair

FIGURE 3.1
Graphical form of the relation $\{(a,1),(a,4),(b,2),(b,3)\}$

(b,a). Besides, the two elements of an ordered pair need not be distinct. Thus, (a,a) is a valid ordered pair. Two ordered pairs (a,b) and (c,d) are equal only when $a = c$ and $b = d$.

The **Cartesian product** of two sets A and B, denoted by $A \times B$ is the set of all ordered pairs (a,b) with $a \in A$ and $b \in B$. For example, if $A = \{a,b\}$ and $B = \{1,2,3,4\}$ then $A \times B = \{(a,1),(a,2),(a,3),(a,4),(b,1),(b,2),(b,3),(b,4)\}$. A **binary relation** from A to B is a subset of $A \times B$. For example $\{(a,1),(a,4),(b,2),(b,3)\}$ is a binary relation on $A = \{a,b\}$ and $B = \{1,2,3,4\}$. A binary relation can be represented in a graphical form as shown in Figure 3.1, where the points in the left-hand side are the elements in A, the points in the right-hand side are the elements in B, and an arrow indicates that the corresponding element in A is related to the corresponding element in B.

The points used to represent the elements in sets A and B are known as **vertices** of the graph in Figure 3.1, and the directed lines $a \to 1$, $a \to 4$, $b \to 2$, and $b \to 3$ are called the **edges** of the graph. Thus, it is possible to represent any relation with a directed graph. For example, if $A = \{0,1,2\}$ and we consider the relation "not equal to" on set A and itself, then the directed graph representation of the relation can be derived directly from the Cartesian product of A with itself

$$A \times A = \{(0,0),(0,1),(0,2),(1,0),(1,1),(1,2),(2,0),(2,1),(2,2)\}$$

Figure 3.2 shows the resulting directed graph. Note the absence of the loop from a node to itself; this is because it is not possible for an element in A not to be equal to itself.

In general, in many applications we encounter only binary relations on a set and itself, rather than relations from one set to a different set. The remainder of this section deals with the properties these relations satisfy. A relation R on a set A (i.e., $R \subseteq A \times A$) is **reflexive** if (a,a) is in R for every a in A. In other words, in a reflexive relation every element in A is related to itself. For example let $A = \{1,2,3,4\}$ and let $R = \{(1,1), (2,4), (3,3), (4,1), (4,4)\}$. R is not a reflexive relation, since $(2,2)$ does not belong to R. Note that all ordered pairs (a,a) must belong to R in order for R to be reflexive. A relation that is not reflexive is called **irreflexive**. As another example, let A be a set of triangles and let the relation R on A be "a is similar to b". The relation is reflexive, since every triangle is similar to itself.

A relation R on a set is **symmetric** if $(a,b) \in R$ whenever $(b,a) \in R$; that is, if a is related to b, then b is also related to a. For example, let $A = \{1,2,3,4\}$ and let $R = \{(1,3),$

FIGURE 3.2
Directed graph representation of the relation "not equal to" on set $A = \{0,1,2\}$ and itself

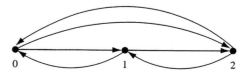

(4,2), (2,4), (2,3), (3,1)}. R is not a symmetric relation, since

$$(2,3) \in R \quad \text{but} \quad (3,2) \notin R.$$

On the other hand, if A is a set of people and the relation R on A is defined as

$$R = \{(x,y) \in A \times A \mid x \text{ is a cousin of } y\}$$

then R is symmetric, because if x is a cousin of y, y is a cousin of x.

A relation R on a set A is **antisymmetric** if

$$(a,b) \in R \quad \text{and} \quad (b,a) \in R \quad \text{implies } a = b$$

In other words, if $a \neq b$, then a may be related to b or b may be related to a, but not both. For example, if A is a set of people, then the relation

$$\{(a,b) \in A \times A \mid a \text{ is the father of } b\}$$

is antisymmetric because the reverse (i.e., b is the father of a) is not true. As a further example, let N be the set of natural numbers and let R be the relation on N defined by "x divides y." The relation R is antisymmetric, since x divides y and y divides x, which implies $x = y$.

A relation R on a set A is **transitive** if

$$(a,b) \in R \quad \text{and} \quad (b,c) \in R \quad \text{implies } (a,c) \in R$$

In other words, if a is related to b and b is related to c, then a is related to c. For example, let R be the relation "perpendicular to" on the set of lines $\{a,b,c\}$ in a plane. R is not transitive, since if a,b,c are lines such that "a is perpendicular to b," and "b is perpendicular to c," then a is not perpendicular to c; in fact, a is parallel to c. On the other hand, if R is the relation "less than or equal to" on the set of integers $A = \{4,6,8\}$ then $R = \{(4,6), (6,8), (4,8), (6,6)\}$ is transitive.

A relation R on a set A is called an **equivalence relation** if

i. R is reflexive, i.e., for every $a \in A$, $(a,a) \in R$.
ii. R is symmetric; i.e., $(a,b) \in R$ implies $(b,a) \in R$.
iii. R is transitive; i.e., $(a,b) \in R$ and $(b,c) \in R$ implies $(a,c) \in R$.

For example, the relation "working in the same office" on a set of employees in a given company is an equivalence relation (assuming that no employee works in more than one office).

1. It is reflexive, because each employee works in the same office as himself.
2. It is symmetric, because if a works in the same office as b, then certainly b works in the same office as a.
3. It is transitive, because if a works in the same office as b and b works in the same office as c, then a works in the same office as c.

3.3 PARTITIONS

A **partition,** denoted by Π, on any set A is a collection of nonempty subsets $A_1, A_2, \ldots, A_k$ of A such that their set union is A. In other words,

$$\Pi = (A_1, A_2, \ldots, A_k)$$

such that
i. $A_1 \cup A_2 \cup \cdots A_k = A$
ii. $A_i \cap A_j = \phi$ if $i \neq j$

The subsets A_i are called **blocks** of the partition.

For example, let $A = \{s,t,u,v,w,x,y,z\}$. Let us consider the following subsets of A:

$$A_1 = \{s,t,w,x\} \quad A_2 = \{u,v,y,z\} \quad A_3 = \{s,w,x\}$$

Then (A_1, A_3) is not a partition, since $A_1 \cap A_3 \neq \phi$. Also, $A_2 \cap A_3$ is not a partition, since $t \notin A_2$, and $t \notin A_3$. The collection (A_1, A_2) is a partition of A. The partition in which each block has a single member of a set A is known as the **null partition**, denoted as $\Pi(0)$. The partition in which all elements of a set are in a single block is known as **unity partition,** denoted as $\Pi(I)$.

For example, let $A = \{w,x,y,z\}$; then,

$$\Pi(0) = \{(w),(x),(y),(z)\} \quad \text{and} \quad \Pi(I) = \{(w,x,y,z)\}$$

It is evident that an equivalence relation on a set A induces a partition on A, since every two elements in a block are related. The blocks in the partition are called the **equivalence classes.** Conversely a partition of a nonempty set A induces an equivalence relation on A, two elements are related if they belong to the same block of the partition. For example, let $A = \{a,b,c,d,e\}$ and Π_1 be a partition on A. Then,

$$\Pi_1 = \{(a,c,e),(b,d)\}$$

The equivalence classes of the elements of A are the given blocks of the partition, i.e., (a,c,e) and (b,d). Let R be an equivalence relation that induces the preceding partition. From the equivalence class (a,c,e) and the fact that R is an equivalence relation, we find that

$$(a,a), (a,c), (a,e), (c,c), (c,a), (c,e), (e,e), (e,a), (e,c) \in R$$

Similarly, equivalence class $(b,d) \in R$. Hence,

$$R = \{(a,a),(a,c),(a,e),(c,e),(c,a),(c,e),(e,e),(e,a),(e,c),(b,b),(b,d),(d,d),(d,b)\}$$

A partition Π_1 on a set A **is equal to or less than** a partition Π_2 on A ($\Pi_1 \leq \Pi_2$) if and only if each block of Π_1 is contained in some block of Π_2. For example, consider the set A and the two partitions on A:

$$A = \{a,b,c,d,e,f,g,h,i\}$$
$$\Pi_1 = \{(a,b),(c,d),(e,f),(g,h,i)\}$$
$$\Pi_2 = \{(a,b,e,f),(c,d),(g,h,i)\}$$

In this case $\Pi_1 \leq \Pi_2$.

The **greatest lower bound** (glb) of a pair of partitions Π_1 and Π_2 is defined as $\Pi_1 \cdot \Pi_2 = \Pi_3$. The **product** ($\Pi_1 \cdot \Pi_2$) is obtained by intersecting their blocks. For example, let

$$\Pi_1 = \{(a,b,c),(e,f,g),(d,h)\}$$
$$\Pi_2 = \{(a,b,g),(e,f),(d,c,h)\}$$

Then

$$\Pi_3 = \Pi_1 \cdot \Pi_2 = \{(a,b),c,g,(e,f),(d,h)\}$$

The **lowest upper bound** (lub) of a pair of partitions Π_1 and Π_2 is defined as $\Pi_1 + \Pi_2 = \Pi_3$. The **sum** ($\Pi_1 + \Pi_2$) is obtained by including in every block those elements of Π_1 and Π_2 that are **chain connected**. Two blocks B_1 and B_n of a partition are chain connected if and only if there exists a sequence of blocks $B_1, B_2, ..., B_n$ such that

$$B_i \cap B_{i+1} \neq \phi \quad \text{where } i = 1, 2, ..., n - 1$$

For example, let

$$\Pi_1 = \{(a,b,c,d),(e,f,g),h\}$$
$$\Pi_2 = \{(a,b,c,h),d,(e,f),g\}$$

Then

$$\Pi_3 = \Pi_1 + \Pi_2 = \{(a,b,c,d,h),(e,f,g)\}$$

3.4 FUNCTIONS

A function is a special type of relation. The words **mapping** and **transformation** are sometimes used as synonyms for functions. A **function** f is defined to be a relation R from a set A to a set B such that for every $a \in A$ there is exactly one $b \in B$ such that $(a,b) \in R$. A is called the **domain** and B the **range** of the function. For example, let $A = \{2,4,6,8\}$ and $B = \{1,2,3,4\}$, and let R be the relation from A to B defined by

$$R = \{(x,y) \mid x \in A, y \in B, x + 2y = 10\}$$

Then $R = \{(2,4),(4,3),(6,2),(8,1)\}$ is a function. Thus, the domain of R is $\{2,4,6,8\}$ and the range of R is $\{4,3,2,1\}$.

A function f from A to B is said to be an **onto** function if every element $b \in B$ is the second element in some ordered pair (a,b) belonging to the function. For example, let $A = \{w,x,y\}$ and $B = \{u,v\}$. Then the function

$$f = \{(w,v),(x,v),(y,u)\}$$

is an onto function. An onto function is also known as a **surjection**.

A function f from A to B is said to be a **one-to-one** function if no two elements belonging to A have the same second element. For example, let $A = \{w,x,y\}$ and $B = \{a,b,c,d\}$. Then the function

$$f = \{(w,b),(x,a),(y,d)\}$$

is a one-to-one function. Note that this is not an onto function. A one-to-one function is also known as an **injection**.

A function f from A to B is said to be **one-to-one onto** if it is both one-to-one and onto. For example, let $A = \{w,x,y\}$ and $B = \{a,b,c\}$. Then the function

$$f = \{(w,b),(x,c),(y,a)\}$$

is a one-to-one onto function. A one-to-one onto function is called a **bijection** or a **one-to-one correspondence**.

FIGURE 3.3

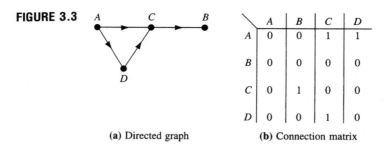

(a) Directed graph   (b) Connection matrix

3.5 GRAPHS

A graph consists of a set of points called **nodes** or **vertices** and a set of interconnecting line segments called **arcs** or **edges.** If the edges in a graph have directions (orientation), then the graph is **directed:** otherwise it is **non-directed.** An example of a directed graph, also known as a **digraph,** is shown in Figure 3.3a. Here the vertices are represented by points and there is a directed edge heading from A to C, A to D, D to C, and C to B. If there is a directed edge from vertex u to vertex v, u is **adjacent** to v. Thus, in Figure 3.3a A is adjacent to C, C is adjacent to B, and so on.

All connections in a directed graph can be described by the **connection matrix** of the directed graph. The connection matrix **T** is a square matrix of dimension n, where n is equal to the number of vertices in the directed graph. The entry t_{ij} at the intersection of row i and column j is 1 if there is an edge from node i to node j; otherwise, t_{ij} is 0. The connection matrix for the directed graph of Figure 3.3a is shown in Figure 3.3b.

In a directed graph, a **path** is a sequence of edges such that the terminal vertex of an edge coincides with the initial vertex of the following edge. For example, in Figure 3.4 ADCB is a path. A path with the same initial and final vertices is known as a **cycle.** In Figure 3.4 ACDA is a cycle. A directed graph without cycles is said to be **acyclic.**

The **in-degree** of a vertex in a directed graph is the number of edges terminating in the vertex. For example in Figure 3.4 the in-degree of vertex C is 3, the in-degree of vertex D is 0, and so on. The number of edges leaving a vertex is called its **out-degree.** For example, the out-degree of vertex D is 2. An acyclic graph in which one vertex has an in-degree of 0 and all other vertices have an in-degree of 1 is called a **tree.** The vertex with in-degree 0 is called the **root** of the tree. The vertices with out-degree 0 are called the **leaves** of the tree. The edges in a tree are known as **branches.** Figure 3.5 shows a tree in which vertex A has an in-degree of 0, and vertices B,C,F,G, and H have an out-degree of 0. Thus, A is the root of the tree, and B,C,F,G, and H are its leaves. Note that vertices D and E are not leaves of the tree because they have an out-degree of 1 and 3 respectively.

FIGURE 3.4
A directed graph

3.6 BOOLEAN ALGEBRA

Boolean algebra may be defined for a set A (which could be finite or infinite) in terms of two binary operations $+$ and $\cdot$. The symbols $+$ and $\cdot$ are called the inclusive OR and AND respectively; they should not be confused with the addition and multiplication operations in conventional algebra. The operations in Boolean algebra are based on the following axioms or postulates, known as Huntington's postulates:

1. If $x,y \in A$, then
$$x + y \in A; x \cdot y \in A$$

   This is called the **closure property**.
2. If $x,y \in A$, then
$$x + y = y + x; x \cdot y = y \cdot x$$

   that is, $+$ and $\cdot$ operations are commutative.
3. If $x,y,z \in A$, then
$$x + (y \cdot z) = (x + y) \cdot (x + z)$$
$$x \cdot (y + z) = (x \cdot y) + (x \cdot z)$$

   that is, $+$ and $\cdot$ operations are distributive.
4. Identity elements, denoted as 0 and 1, must exist such that $x + 0 = x$ and $x \cdot 1 = x$ for all elements of A.
5. For every element x in A there exists an element $\bar{x}$, called the **complement** of x, such that
$$x + \bar{x} = 1 \qquad x \cdot \bar{x} = 0$$

Note that the basic postulates are grouped in pairs. One postulate can be obtained from the other by simply interchanging all OR and AND operations, and the identity elements 0 and 1. This property is known as **duality**. For example,

$$x + (y \cdot z) = (x + y) \cdot (x + z)$$
$$\downarrow \quad \downarrow \qquad \downarrow \quad \downarrow \quad \downarrow$$
$$x \cdot (y + z) = (x \cdot y) + (x \cdot z)$$

FIGURE 3.5
A tree

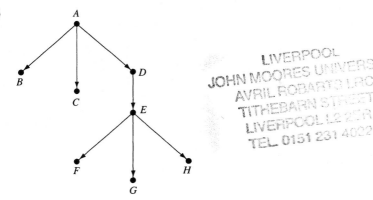

We shall now prove several theorems used for the manipulation of Boolean algebra. The proofs are based on Postulates 1–5 as well as principles of substitution and duality.

Theorem 1. The identity elements 0 and 1 are unique.

Proof: We first prove that the element 0 is unique. Let there be two elements 0_a and 0_b such that for every element x we have

$$x + 0_a = x \quad \text{and} \quad x + 0_b = x \quad \text{(by Postulate 4)}$$

Letting $x = 0_b$ in the first and $x = 0_a$ in the second of these equations, we have

$$0_b + 0_a = 0_b \quad \text{and} \quad 0_a + 0_b = 0_a$$

But

$$0_b + 0_a = 0_a + 0_b \quad \text{(by Postulate 2)}$$

Thus, $0_b = 0_a$; therefore, it can be proved by the principle of duality that the element 1 is unique.

Theorem 2. *The Idempotent Laws*

i. $x + x = x$
ii. $x \cdot x = x$

Proof: We first prove that $x + x = x$

$$\begin{aligned}
x + x &= (x + x)1 & \text{(by Postulate 4)} \\
&= (x + x)(x + \bar{x}) & \text{(by Postulate 5)} \\
&= x + x \cdot \bar{x} & \text{(by Postulate 3)} \\
&= x + 0 & \text{(by Postulate 5)} \\
&= x & \text{(by Postulate 4)}
\end{aligned}$$

The second part can be proved by duality.

Theorem 3.

i. $x + 1 = 1$
ii. $x \cdot 0 = 0$

Proof: Let us again prove part i:

$$\begin{aligned}
x + 1 &= (x + 1) \cdot 1 & \text{(by Postulate 4)} \\
&= (x + 1)(x + \bar{x}) & \text{(by Postulate 5)} \\
&= x + 1 \cdot \bar{x} & \text{(by Postulate 3)} \\
&= x + \bar{x} & \text{(by Postulate 4)} \\
&= 1 & \text{(by Postulate 5)}
\end{aligned}$$

Thus part i of this theorem is valid; therefore, part ii is valid by the principle of duality.

Theorem 4. *The Absorption Laws*

i. $x + xy = x$
ii. $x(x + y) = x$

Proof: We prove part i first

$$x + xy = x \cdot 1 + xy \quad \text{(by Postulate 4)}$$
$$= x(1 + y) \quad \text{(by Postulate 3)}$$
$$= x \cdot 1 \quad \text{(by Theorem 3)}$$
$$= x \quad \text{(by Postulate 4)}$$

Part ii can be proved by duality.

Theorem 5. Every element in set A has a unique complement.

Proof: Let $x \in A$ and let $\bar{x}_1$ and $\bar{x}_2$ both be complements of x such that $x + \bar{x}_1 = x + \bar{x}_2 = 1$ and $x \cdot \bar{x}_1 = x \cdot \bar{x}_2 = 0$. Then,

$$\bar{x}_2 = 1 \cdot \bar{x}_2$$
$$= (x + \bar{x}_1) \bar{x}_2 \quad \text{(by Postulate 5)}$$
$$= x \cdot \bar{x}_2 + \bar{x}_1 \cdot \bar{x}_2 \quad \text{(by Postulate 3)}$$
$$= 0 + \bar{x}_1 \cdot \bar{x}_2 \quad \text{(by Postulate 5)}$$
$$= \bar{x}_1 \cdot x + \bar{x}_1 \cdot \bar{x}_2 \quad \text{(by Postulate 5)}$$
$$= \bar{x}_1 (x + \bar{x}_2) \quad \text{(by Postulate 3)}$$
$$= \bar{x}_1 \cdot 1 \quad \text{(by Postulate 5)}$$
$$= \bar{x}_1 \quad \text{(by Postulate 4)}$$

Theorem 6. *Involution Law*

$$\overline{(\bar{x})} = x$$

Proof: Since we already know that

$$x + \bar{x} = 1 \quad \text{and} \quad x \cdot \bar{x} = 0 \quad \text{(by Postulate 5)}$$

then the complement of $\bar{x}$ is x. In other words

$$\overline{(\bar{x})} = x$$

Theorem 7.

i. $x + \bar{x}y = x + y$
ii. $x(\bar{x} + y) = xy$

For this theorem we shall prove part ii.

$$x(\bar{x} + y) = x\bar{x} + xy \quad \text{(by Postulate 3)}$$
$$= 0 + xy \quad \text{(by Postulate 5)}$$
$$= xy \quad \text{(by Postulate 4)}$$

Part i is valid by duality.

Theorem 8. *DeMorgan's Law*

i. $\overline{(x + y)} = \bar{x} \cdot \bar{y}$
ii. $\overline{xy} = \bar{x} + \bar{y}$

We shall prove part i of the theorem. By definition of the complement (Postulate 5) and its uniqueness (Theorem 5) it is obvious that

$$(x + y) + \bar{x}\bar{y} = 1 \quad \text{and} \quad (x + y)\bar{x}\bar{y} = 0$$

$$\begin{aligned}
(x + y) + \bar{x}\bar{y} &= [(x + y) + \bar{x}][(x + y) + \bar{y}] && \text{(by Postulate 3)} \\
&= [y + (x + \bar{x})][x + (y + \bar{y})] && \text{(by associativity)} \\
&= [y + 1][x + 1] && \text{(by Postulate 5)} \\
&= 1 \cdot 1 && \text{(by Theorem 3)} \\
&= 1
\end{aligned}$$

$$\begin{aligned}
(x + y)\bar{x}\bar{y} &= \bar{x}\bar{y}(x + y) && \text{(by commutativity)} \\
&= \bar{x}\bar{y} \cdot x + \bar{x}\bar{y} \cdot y && \text{(by distributivity)} \\
&= \bar{y}(\bar{x}x) + \bar{x}(\bar{y}y) && \text{(by associativity)} \\
&= \bar{y}(x\bar{x}) + \bar{x}(y\bar{y}) && \text{(by commutativity)} \\
&= \bar{y} \cdot 0 + \bar{x} \cdot 0 && \text{(by Postulate 5)} \\
&= 0
\end{aligned}$$

Since $(x + y) + \bar{x}\bar{y} = 1$ and $(x + y) \cdot \bar{x}\bar{y} = 0$, by Postulate 5 $\overline{(x + y)} = \bar{x}\bar{y}$.

The theorem may be generalized to include more than two elements:

(i) $\quad \overline{a + b + \cdots + z} = \bar{a}\bar{b} \cdots \bar{z}$

(ii) $\quad \overline{ab \cdots z} = \bar{a} + \bar{b} + \cdots + \bar{z}$

■ EXAMPLE 3.1

Let us complement the following expression using Theorem 8

$$\overline{a + b(\bar{c} + u\bar{v})}$$

$$\begin{aligned}
\overline{a + b(\bar{c} + u\bar{v})} &= \bar{a} \cdot \overline{b(\bar{c} + u\bar{v})} \\
&= \bar{a}(\bar{b} + \overline{(\bar{c} + u\bar{v})}) \\
&= \bar{a}(\bar{b} + \overline{\bar{c}}\,\overline{(u\bar{v})}) \\
&= \bar{a}(\bar{b} + c(\bar{u} + v))
\end{aligned}$$ ■

It can be seen from the example that the complement of an expression can be obtained by replacing + (OR) with · (AND) and vice versa, and replacing each element by its complement.

Theorem 9. *Consensus*

i. $xy + \bar{x}z + yz = xy + \bar{x}z$
ii. $(x + y)(\bar{x} + z)(y + z) = (x + y)(\bar{x} + z)$

Proof:

$$\begin{aligned}
xy + \bar{x}z + yz &= xy + \bar{x}z + 1 \cdot yz && \text{(by Postulate 4)} \\
&= xy + \bar{x}z + (x + \bar{x})yz && \text{(by Postulate 5)} \\
&= xy + \bar{x}z + xyz + \bar{x}yz && \text{(by Postulate 3)}
\end{aligned}$$

$$= (xy + xyz) + (\bar{x}z + \bar{x}zy)$$
$$= xy + \bar{x}z \quad \text{(by Theorem 4)}$$

EXAMPLE 3.2

Let us simplify the following expression using Theorem 9

$$(\bar{x} + y)\,wz + x\bar{y}v + vwz$$

Assume $x\bar{y} = a$ and $wz = b$. Then,

$$(\bar{x} + y)\,wz + x\bar{y}v + vwz = \bar{a}b + av + bv$$
$$= \bar{a}b + av \quad \text{(by Theorem 9i)}$$
$$= (\bar{x} + y)wz + x\bar{y}v \quad \text{(by replacing the values of } a \text{ and } b)$$

Theorem 10.

i. $xy + x\bar{y}z = xy + xz$
ii. $(x + y)(x + \bar{y} + z) = (x + y)(x + z)$

Proof:

$$xy + x\bar{y}z = x(y + \bar{y}z) \quad \text{(by Postulate 3)}$$
$$= x(y + z) \quad \text{(by Theorem 7i)}$$
$$= xy + xz \quad \text{(by Postulate 3)}$$

EXAMPLE 3.3

The following expression can be represented in a simplified form using Theorem 10.

$$(a + \bar{b})(a + b + c) = (a + \bar{b})(a + c)$$
$$= a + \bar{b}c \quad \text{(by Postulate 3)}$$

Theorem 11.

i. $xy + \bar{x}z = (x + z)(\bar{x} + y)$
ii. $(x + y)(\bar{x} + z) = xz + xy$

Proof:

$$xy + \bar{x}z = (xy + \bar{x})(xy + z) \quad \text{(by Postulate 3)}$$
$$= (x + \bar{x})(y + \bar{x})(x + z)(y + z) \quad \text{(by Postulate 3)}$$
$$= 1 \cdot (\bar{x} + y)(x + z)(y + z) \quad \text{(by Postulate 5)}$$
$$= (\bar{x} + y)(x + z)(y + z) \quad \text{(by Postulate 4)}$$
$$= (\bar{x} + y)(x + z) \quad \text{(by Theorem 9ii)}$$

EXAMPLE 3.4

Let us show the application of Theorem 11 in changing the form of the following Boolean expression.

$$(\bar{a}b + ac)(a + \bar{b})(\bar{a} + \bar{c})$$
$$= (\bar{a} + c)(a + b)(a + \bar{b})(\bar{a} + \bar{c}) \quad \text{(by Theorem 11i)}$$
$$= (\bar{a} + c)(\bar{a} + \bar{c})(a + b)(a + \bar{b}) \quad \text{(by Postulate 2)}$$
$$= (\bar{a} + c \cdot \bar{c})(a + b \cdot \bar{b}) \quad \text{(by Postulate 3)}$$
$$= (\bar{a} + 0)(a + 0) \quad \text{(by Postulate 5)}$$
$$= \bar{a} \cdot a \quad \text{(by Postulate 5)}$$
$$= 0$$

∎

The postulates and theorems of Boolean algebra presented here relate to elements of a finite set A. If the set A is restricted to contain just two elements 0 and 1, then Boolean algebra is useful in the analysis and design of digital circuits. The two elements 0 and 1 are not binary numbers, they are just two symbols that are used to represent data in digital circuits. Two-valued Boolean algebra is often referred to as **switching algebra**. Henceforth, we shall use Boolean algebra for the set $\{0,1\}$ unless otherwise specified.

3.7 BOOLEAN FUNCTIONS

In Boolean algebra, symbols are used to represent statements or propositions that may be true or false. These statements or propositions are connected together by operations AND, OR, and NOT. If 0 is used to denote a false statement and 1 is used to denote a true statement, then the AND ($\cdot$) combination of two statements can be written as follows:

$$0 \cdot 0 = 0$$
$$0 \cdot 1 = 0$$
$$1 \cdot 0 = 0$$
$$1 \cdot 1 = 1$$

In tabular form it can be represented as

$\cdot$	0	1
0	0	0
1	0	1

The AND combination of two statements is known as the **product** of the statements.
The OR ($+$) combination of two statements can be written as follows:

$$0 + 0 = 0$$
$$0 + 1 = 1$$
$$1 + 0 = 1$$
$$1 + 1 = 1$$

In tabular form it can be represented as

+	0	1
0	0	1
1	1	1

The OR combination of two statements is known as the **sum** of the statements.

The NOT operation of a statement is true if and only if the statement is false. The NOT operation on a statement can be stated as follows:

$$\overline{0} = 1$$
$$\overline{1} = 0$$

The NOT operation is also known as **complementation**.

A Boolean function $f(x_1, x_2, x_3, \ldots, x_n)$ is a function of n individual statements x_1, $x_2, x_3, \ldots, x_n$ combined by AND, OR, and NOT operations. The statements $x_1, x_2, \ldots, x_n$ are also known as **Boolean variables** and can be either **true** or **false**. In other words, each statement represents either the element 0 or 1 of the Boolean algebra. For example,

$$f(x,y,z) = xy + \overline{x}z + \overline{y}\overline{z}$$

is a function of three Boolean variables, x, y, and z. In this function if $x = 0$, $y = 0$, and $z = 1$, then $f = 1$ as verified below

$$f(0,0,1) = 0 \cdot 0 + \overline{0} \cdot 1 + \overline{0} \cdot \overline{1}$$
$$= 0 \cdot 0 + 1 \cdot 1 + 1 \cdot 0$$
$$= 0 + 1$$
$$= 1$$

A Boolean function of n variables may also be described by a **truth table**. Since each of the Boolean variables can independently assume either a true (1) or a false (0) value, there are 2^n combinations of values for n variables. For each combination of values, a function can have a value of either 0 or 1. A truth table displays the value of a function for all possible 2^n combinations of its variables. The truth table for the function $f(x, y, z) = xy + \overline{x}z + \overline{y}\overline{z}$ is shown in Table 3.1.

TABLE 3.1
Truth table for $f(x,y,z) = xy + \overline{x}z + \overline{y}\overline{z}$

x	y	z	$f(x,y,z)$
0	0	0	1
0	0	1	1
0	1	0	0
0	1	1	1
1	0	0	1
1	0	1	0
1	1	0	1
1	1	1	1

It should be noted that a truth table describes only one Boolean function, although this function may be expressed in a number of ways.

The complement function $\bar{f}(x_1, x_2, \ldots, x_n)$ of a Boolean function $f(x_1, x_2, \ldots, x_n)$ has a value 1 whenever the value of f is 0 and a value 0 when $f = 1$. The truth table of a function can be used to derive the complement of the function by complementing each entry in column f. For example, by replacing each entry in the column f of Table 3.1 by its complement, we get the truth table of Table 3.2. It is the truth table for the function $f(x,y,z) = \bar{x}y\bar{z} + x\bar{y}z$, which is the complement of the function $f(x,y,z) = xy + \bar{x}z + \bar{y}\bar{z}$. This can be proved by deriving the complement of the function $f(x,y,z) = xy + \bar{x}z + \bar{y}\bar{z}$ using DeMorgan's theorem:

$$\overline{f(x,y,z)} = \overline{xy + \bar{x}z + \bar{y}\bar{z}}$$
$$= \overline{(xy)} \cdot \overline{(\bar{x}z)} \cdot \overline{(\bar{y}\bar{z})}$$
$$= (\bar{x} + \bar{y})(x + \bar{z})(y + z)$$
$$= (\bar{x}x + \bar{x}\bar{z} + \bar{y}x + \bar{y}\bar{z})(y + z)$$
$$= (0 + \bar{x}\bar{z} + \bar{y}x + \bar{y}\bar{z})(y + z)$$
$$= (\bar{x}\bar{z} + \bar{y}x + \bar{y}\bar{z})(y + z)$$
$$= \bar{x}\bar{z}y + \bar{y}xy + \bar{y}\bar{z}y + \bar{x}\bar{z}z + \bar{y}xz + \bar{y}\bar{z}z$$
$$= \bar{x}y\bar{z} + x\bar{y}y + y\bar{y}\bar{z} + \bar{x}\bar{z}z + x\bar{y}z + \bar{y}\bar{z}z$$
$$= \bar{x}y\bar{z} + x \cdot 0 + 0 \cdot \bar{z} + \bar{x} \cdot 0 + x\bar{y}z + \bar{y} \cdot 0$$
$$= \bar{x}y\bar{z} + 0 + 0 + 0 + x\bar{y}z + 0$$
$$= \bar{x}y\bar{z} + x\bar{y}z$$

3.8 DERIVATION AND CLASSIFICATION OF BOOLEAN FUNCTIONS

The behavior of a digital circuit is usually specified in plain English. Therefore, this specification must be formulated into a truth table format before the circuit can actually be designed. As an example, let us consider the behavior of a circuit.

A circuit for controlling the lighting of a room is to be designed. The lights may be switched on or off from any of three switch points. Let the three on/off switches be X, Y,

TABLE 3.2

x	y	z	f(x,y,z)
0	0	0	0
0	0	1	0
0	1	0	1
0	1	1	0
1	0	0	0
1	0	1	1
1	1	0	0
1	1	1	0

FIGURE 3.6
Lighting circuit to be designed

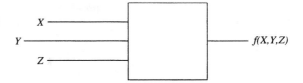

and Z and the light on or off condition be represented $f = 1$ or 0 respectively. The desired circuit is shown as a black box in Figure 3.6. From the word description it is possible to tabulate the output condition (i.e., light on or off) for each combination of switch inputs. This is shown in Table 3.3.

Note that if the light is already on, turning on a switch will turn the light off. The truth table may be used to derive the Boolean function by extracting from the table those combinations of XYZ that make $Z = 1$ (i.e., turn the light on). Thus,

$$f(X,Y,Z) = \overline{X}\overline{Y}Z + \overline{X}Y\overline{Z} + X\overline{Y}\overline{Z} + XYZ$$

A Boolean function in this form is known as a **sum of products**. A product term is a Boolean product (i.e., AND) of complemented or uncomplemented variables. As can be seen, the above function is the sum (OR) of the products of the variables X, Y, and Z. The following expression is another example of the sum-of-product form of a Boolean function

$$f(A,B,C,D) = AB\overline{C}\overline{D} + A\overline{B}CD + \overline{A}BCD + \overline{A}BC\overline{D}$$

This function has four variables and four product terms. The **product-of-sums** form of a Boolean function can be derived from the truth table by extracting those combinations of input variables that produce an output of logic 0. For example, in the truth table of Figure 3.3, the product terms $\overline{X}\overline{Y}\overline{Z}$, $\overline{X}YZ$, $X\overline{Y}Z$, and $XY\overline{Z}$ make the output $f = 0$ (i.e., do not turn the light on). Thus,

$$\overline{f(X,Y,Z)} = \overline{X}\overline{Y}\overline{Z} + \overline{X}YZ + X\overline{Y}Z + XY\overline{Z}$$

By applying DeMorgan's theorem we obtain

TABLE 3.3
Truth table for the lighting circuit (switch on = 1; switch off = 0)

X	Y	Z	f(X,Y,Z)
0	0	0	0
0	0	1	1
0	1	0	1
0	1	1	0
1	0	0	1
1	0	1	0
1	1	0	0
1	1	1	1

$$\overline{f(X,Y,Z)} = \overline{\overline{X}\overline{Y}\overline{Z} + \overline{X}YZ + X\overline{Y}Z + XY\overline{Z}}$$
$$f(X,Y,Z) = \overline{\overline{X}\overline{Y}\overline{Z}} \cdot \overline{\overline{X}YZ} \cdot \overline{X\overline{Y}Z} \cdot \overline{XY\overline{Z}}$$
$$= (X + Y + Z)(X + \overline{Y} + \overline{Z})(\overline{X} + Y + \overline{Z})(\overline{X} + \overline{Y} + Z)$$

the product-of-sums form of the Boolean function for the lighting circuit.

Either the sum-of-products or the product-of-sums form may be used to represent a Boolean function. A product-of-sums form of a Boolean function can be obtained from its sum-of-products form by using the following steps:

i. Derive the dual of the sum-of-products expression.
ii. Convert the resulting product-of-sums expression into a sum-of-products expression.
iii. Derive the dual of the sum-of-products expression to obtain the product-of-sums expression.

■ **EXAMPLE 3.5**

Let us convert the following sum-of-products expression into the product-of-sums form using the procedure just given.

$$f(X,Y,Z) = XY + X\overline{Z} + YZ$$

The dual of the given expression is

$$(X + Y)(X + \overline{Z})(Y + Z)$$

The corresponding sum-of-products expression is

$$XY + XZ + Y\overline{Z}$$

The dual of this expression is the desired product of sums

$$(X + Y)(X + Z)(Y + \overline{Z})$$ ■

3.9 CANONICAL FORMS OF BOOLEAN FUNCTIONS

For any Boolean function, either sum-of-products or product-of-sums, there exists a standard or canonical form. In these two alternative forms, every variable appears, in either complemented or uncomplemented form, in each product term or sum term. A product that has this property is known as a **minterm**, whereas a sum term possessing this property is known as a **maxterm**.

A Boolean function composed completely of minterm is said to be in **canonical sum-of-products** form. For example,

$$f(X,Y,Z) = \overline{X}\overline{Y}Z + \overline{X}Y\overline{Z} + X\overline{Y}Z + XYZ \tag{3.1}$$

is a canonical function of three variables. Every product term in this expression contains all the variables in the function. A variable v and its complement $\overline{v}$ in an expression are known as **literals**. Note that although v and $\overline{v}$ are not two different variables, they are considered to be different literals.

If a Boolean function is composed completely of maxterms, then it is said to be in **canonical product-of-sums** form. For example,

$$f(X,Y,Z) = (X + Y + Z)(X + \overline{Y} + \overline{Z})(\overline{X} + Y + \overline{Z})(\overline{X} + \overline{Y} + Z) \tag{3.2}$$

is a canonical function of three variables with four maxterms.

3.9 CANONICAL FORMS OF BOOLEAN FUNCTIONS

For a Boolean function of n variables, there are 2^n minterms and 2^n maxterms. For example minterms and maxterms for the three-variable Boolean function $f(X,Y,Z)$ are

Minterms	Maxterms
$\overline{X}\overline{Y}\overline{Z}$	$X + Y + Z$
$\overline{X}\overline{Y}Z$	$X + Y + \overline{Z}$
$\overline{X}Y\overline{Z}$	$X + \overline{Y} + Z$
$\overline{X}YZ$	$X + \overline{Y} + \overline{Z}$
$X\overline{Y}\overline{Z}$	$\overline{X} + Y + Z$
$X\overline{Y}Z$	$\overline{X} + Y + \overline{Z}$
$XY\overline{Z}$	$\overline{X} + \overline{Y} + Z$
XYZ	$\overline{X} + \overline{Y} + \overline{Z}$

It can be noticed from this minterm/maxterm list that the complement of any minterm is a maxterm and vice versa.

In order to simplify the notation for minterms, they are usually coded in decimal numbers. This is done by assigning 0 to a complemented variable and 1 to an uncomplemented variable, which results in the binary representation of a minterm. The corresponding decimal number d is derived, and the minterm is represented by m_d. For example the minterm $X\overline{Y}\overline{Z}$ may be written as 1 0 0 (=4); hence, the minterm can be denoted by m_4.

Thus the Boolean function represented by Equation (3.1) may be written as

$$f(X,Y,Z) = m_1 + m_2 + m_5 + m_7$$

This equation can be written in the **minterm list form** as

$$f(X,Y,Z) = \Sigma m\ (1,2,5,7)$$

The notation for maxterms is also simplified by coding them in decimal numbers. However in this case, 0 is assigned to an uncomplemented variable and 1 to a complemented variable. The maxterms are denoted by M_d, where d is the decimal equivalent of the binary number. Thus, the Boolean function represented by Equation (3.2) may be written as

$$f(X,Y,Z) = M_0 \cdot M_3 \cdot M_5 \cdot M_6$$

The above equation can be written in the **maxterm list form** as

$$f(X,Y,Z) = \Pi M(0,3,5,6)$$

■ EXAMPLE 3.6

Let us derive the minterm list and the maxterm list for the Boolean function specified by the following truth table:

X	Y	Z	$f(X,Y,Z)$
0	0	0	0
0	0	1	1
0	1	0	1
0	1	1	0
1	0	0	1
1	0	1	0
1	1	0	1
1	1	1	1

The sum-of-products form of the function is

$$f(X,Y,Z) = XYZ + XY\overline{Z} + X\overline{Y}\,\overline{Z} + \overline{X}Y\overline{Z} + \overline{X}\,\overline{Y}Z$$

$$ \underbrace{111}\ \ \underbrace{110}\ \ \underbrace{100}\ \ \underbrace{010}\ \ \underbrace{001}$$

$$= m_7 + m_6 + m_4 + m_2 + m_1$$
$$= \Sigma m(1,2,4,6,7)$$

The product-of-sums form of the function is

$$f(X,Y,Z) = \overline{\overline{X}\,\overline{Y}\,\overline{Z} + \overline{X}YZ + X\overline{Y}Z}$$
$$= \overline{\overline{X}\,\overline{Y}\,\overline{Z}} \cdot \overline{\overline{X}YZ} \cdot \overline{X\overline{Y}Z}$$
$$= (X + Y + Z)(X + \overline{Y} + \overline{Z})(\overline{X} + Y + \overline{Z})$$
$$\ \ 0\ \ \ 0\ \ \ 0\ \ \ \ \ \ 0\ \ \ 1\ \ \ 1\ \ \ \ \ \ 1\ \ \ 0\ \ \ 1 \quad \text{Maxterm code}$$
$$= M_0 \cdot M_3 \cdot M_5$$
$$= \Pi M(0,3,5) \qquad\blacksquare$$

It can be seen from the example that the minterm and maxterm list of a Boolean function can be written directly from the truth table by inspection. The minterm list is the summation of all minterms for which the function has a value 1, whereas the maxterm list is the product of all decimal integers that are missing from the minterm list. Thus, the conversion from one canonical form to the other is straightforward. The following function is expressed in sum-of-products form

$$f(X,Y,Z) = \Sigma m(1,3,6,7)$$

Its conversion to product-of-sums form results in

$$f(X,Y,Z) = \Pi M(0,2,4,5)$$

A non-canonical Boolean function can be expanded to canonical form through repeated use of Postulate 5 (Sec. 3.6).

■ | **EXAMPLE 3.7**

Let us expand the following non-canonical sum-of-products form of the Boolean function

$$f(X,Y,Z) = XY + X\overline{Z} + YZ$$
$$= XY(Z + \overline{Z}) + X(Y + \overline{Y})\overline{Z} + (X + \overline{X})YZ$$
$$= XYZ + XY\overline{Z} + \cancel{XY\overline{Z}} + X\overline{Y}\overline{Z} + \cancel{XYZ} + \overline{X}YZ$$

Duplicate terms have been deleted, since by Theorem 2, $x + x = x$. Hence,

$$f(X,Y,Z) = XYZ + XY\overline{Z} + X\overline{Y}\,\overline{Z} + \overline{X}YZ$$
$$\ 111\ \ \ \ 110\ \ \ \ 100\ \ \ \ 011$$
$$= m_7 + m_6 + m_4 + m_3$$
$$= \Sigma m(3,4,6,7) \qquad\blacksquare$$

Similarly, the following non-canonical product-of-sums form of the Boolean function may be expanded into the canonical form.

$$f(X,Y,Z) = (X + Y)(\overline{X} + Z)$$

In the first sum term Z is missing, and in the second Y is missing. Since by Theorem 2 $x \cdot x = 0$, we introduce $Z\overline{Z} = 0$ and $Y\overline{Y} = 0$ in the first and second sum terms respectively. Hence,

$$\begin{aligned} f(X,Y,Z) &= (X + Y + Z\overline{Z})(\overline{X} + Y\overline{Y} + Z) \\ &= (X + Y + Z)(X + Y + \overline{Z})(\overline{X} + Y + Z)(\overline{X} + \overline{Y} + Z) \\ &\ \ 0\ \ \ \ 0\ \ \ 0\ \ 0\ \ \ \ \ \ 0\ \ \ 1\ 1\ \ \ \ \ \ 0\ \ \ 0\ 1\ \ \ \ \ \ 1\ \ \ 0 \\ &= M_0 \cdot M_1 \cdot M_4 \cdot M_6 \\ &= \Pi M(0,1,4,6) \end{aligned}$$

The canonical sum-of-products and product-of-sums forms of a Boolean function are unique. This property can be used to determine whether two non-canonical forms of a Boolean function are equal or not.

■ **EXAMPLE 3.8**

Let us determine whether or not the following Boolean expressions are equal

$$f(X,Y,Z) = XY + YZ + \overline{X}Z + \overline{X}Y \quad (3.3)$$

$$f(X,Y,Z) = XY + \overline{X}Y + \overline{X}YZ \quad (3.4)$$

We can expand Equation (3.3) to its canonical form:

$$\begin{aligned} f(X,Y,Z) &= XY(Z + \overline{Z}) + (X + \overline{X})YZ + \overline{X}(Y + \overline{Y})Z + \overline{X}Y(Z + \overline{Z}) \\ &= XYZ + XY\overline{Z} + \cancel{XYZ} + \overline{X}YZ + \cancel{\overline{X}YZ} + \overline{X}\overline{Y}Z + \cancel{\overline{X}YZ} + \overline{X}YZ \\ &= XYZ + XY\overline{Z} + \overline{X}YZ + \overline{X}\overline{Y}Z + \overline{X}Y\overline{Z} \end{aligned}$$

Next we expand Equation (3.4) to its canonical form

$$\begin{aligned} f(X,Y,Z) &= XY + \overline{X}Y + \overline{X}YZ \\ &= XY(Z + \overline{Z}) + \overline{X}Y(Z + \overline{Z}) + \overline{X}YZ \\ &= XYZ + XY\overline{Z} + \overline{X}YZ + \overline{X}Y\overline{Z} + \overline{X}YZ \end{aligned}$$

Since the canonical forms of both Equations (3.3) and (3.4) are identical, they represent the same Boolean function. ■

3.10 LOGIC GATES

The circuit elements used to realize Boolean functions are known as logic gates. There are AND, OR, and NOT (inverter) gates corresponding to AND, OR, and NOT operations respectively.

AND. The AND gate produces a 1 output if and only if *all* outputs are 1's. For example, the circuit in Figure 3.7 requires that both switches X and Y, which are normally open, must be closed before the light (L) comes on. In terms of Boolean algebra a closed switch

FIGURE 3.7
AND circuit

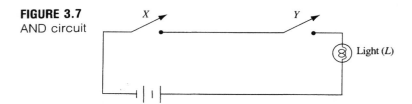

may correspond to a 1 and an open switch to a 0. Similarly the light off and on condition may be represented by 0 and 1 respectively. Thus, the truth table for the AND gate is shown in Table 3.4.

The AND circuit is symbolized as

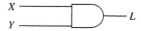

The AND gate shown above has a *fan-in* of two, i.e., two inputs. However, it is possible for an AND gate to have more than two inputs; all inputs must be 1 for the output to be 1. Under any other condition the output will be 0.

OR. The OR gate produces a 1 output if at least one of the inputs is 1. This is represented by switches in parallel, as shown in Figure 3.8. The truth table corresponding to the OR circuit is shown in Table 3.5.

The gate circuit for OR is

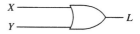

TABLE 3.4
Truth table for AND circuit

X	Y	L
0	0	0
0	1	0
1	0	0
1	1	1

FIGURE 3.8
OR circuit

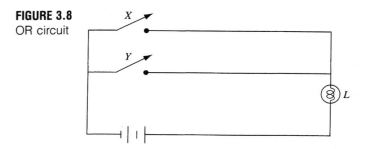

TABLE 3.5
Truth table for OR circuit

X	Y	L
0	0	0
0	1	1
1	0	1
1	1	1

Again, the OR gate shown above has a fan-in of two; theoretically, an OR gate can have any number of inputs. In such a gate the output will be 0 only if all the inputs are 0; otherwise the output will be 1.

NOT. The NOT gate produces an output of 1 when the input is 0, and an output of 0 when the input is 1. The circuit representation of the NOT gate is shown in Figure 3.9. The switch X is normally closed. If it is not operated, the light remains on. In terms of Boolean algebra this means if $X = 0$, $L = 1$. When the switch is operated, i.e., $X = 1$, the circuit is broken, causing the light to be off, i.e., $L = 0$. The truth table of the NOT circuit is shown in Table 3.6. The NOT gate is also known as an **inverter** and is represented by the following symbol.

The three basic gates give rise to two further compound gates: the *NOR* gate and the *NAND* gate. A NOR gate is formed by combining a NOT gate with an OR gate (Figure 3.10). The truth table for the NOR gate is shown in Table 3.7. It can be seen from the truth table that if a 1 appears at any input, the output will be 0. The output is 1 if and

FIGURE 3.9
NOT circuit

TABLE 3.6
Truth table for NOT circuit

X	L
0	1
1	0

FIGURE 3.10
NOR gate

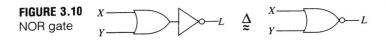

TABLE 3.7
Truth table for NOR gate

X	Y	L
0	0	1
0	1	0
1	0	0
1	1	0

FIGURE 3.11
Alternative form of NOR gate

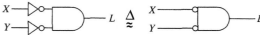

only if both inputs are 0 (i.e., $L = \overline{X}\,\overline{Y}$). Thus, a NOR gate may also be formed by combining two NOT gates with an AND gate as shown in Figure 3.11.

A NAND gate is formed by combining a NOT gate with an AND gate (Fig. 3.12). The truth table for the NAND gate is shown in Table 3.8. It can be seen from the truth table that the output of the NAND gate will be 1 if at least one of the inputs is 0. The output is 0 if and only if the inputs are 1 (i.e., $L = \overline{X \cdot Y}$). By DeMorgan's law $L = \overline{X \cdot Y} = \overline{X} + \overline{Y}$. Thus, a NAND gate may also be formed by combining two NOT gates with an OR gate, as shown in Figure 3.13.

We have so far discussed five logic gates: OR, AND, NOT, NOR, and NAND. These gates can be used to design any digital circuit. Two additional types of gates are also frequently used in digital circuit design; they are exclusive-OR (EX-OR) and exclusive-NOR (EX-NOR) gates. The EX-OR gate produces a 1 output when *either* of the inputs is 1, but *not both*. This is different from the traditional OR gate, which produces a 1 output when either one or both of the inputs are 1. The truth table for an EX-OR gate is shown in Table 3.9.

FIGURE 3.12
NAND gate

TABLE 3.8
Truth table for NAND gate

X	Y	L
0	0	1
0	1	1
1	0	1
1	1	0

FIGURE 3.13
Alternative form of NAND gate

3.10 LOGIC GATES

TABLE 3.9
Truth table for EX-OR gate

X	Y	L
0	0	0
0	1	1
1	0	1
1	1	0

FIGURE 3.14
EX-OR gate

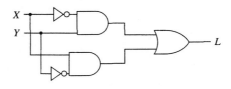

The EX-OR gate is symbolized as

In order to distinguish an EX-OR from the conventional or *inclusive* OR, a different symbol ($\oplus$) is used for an EX-OR operation.

The Boolean function corresponding to the truth table for the EX-OR gate is

$$f(X,Y) = \overline{X}Y + X\overline{Y} = X \oplus Y$$

Thus, an EX-OR gate can be formed from a combination of AND, OR, and NOT gates as shown in Figure 3.14.

An EX-NOR gate can be formed by combining a NOT gate with an EX-OR gate (Fig. 3.15). The truth table for an EX-NOR gate is shown in Table 3.10. It can be seen from the truth table that the output of the EX-NOR gate is 1 only when both the inputs are either 0 or 1. For this reason an EX-NOR gate is also known as an **equivalence** or **coincidence** gate. It is represented by $\odot$.

FIGURE 3.15
EX-NOR gate

TABLE 3.10
Truth table for EX-NOR gate

X	Y	L
0	0	1
0	1	0
1	0	0
1	1	1

64 □ CHAPTER 3 / SETS, RELATIONS, GRAPHS, AND BOOLEAN ALGEBRA

EXERCISES

1. Let $S = \{a,b,c\}$. What are the subsets of S?
2. Consider the following sets: $W = \{a,b,c\}$, $Y = \{a,b\}$, and $Z = \{c\}$.
   a. Which of these sets are subsets of others or of themselves?
   b. How many proper subsets does each set have?
3. Write a specification by properties for each of the following sets:
   a. $\{4,8,12,16, \ldots\}$ b. $\{3,4,7,8,11,12,15,16,19,20,\ldots\}$ c. $\{3,13,23,33,\ldots\}$
4. Let $W = \{w,x,y\}$, $X = \{y,d\}$, and $Y = \{y,e,f\}$. Determine
   a. $W \cup (X \cap Y)$
   b. $W \cap X \cap Y$
   c. $W - X$
5. Describe in words the following sets:
   a. $\{x \mid x \text{ is oddly divisible by 3 and } 25 < x\}$
   b. $\{x \mid x^2 - x - 6 = 0\}$
   c. $\{5,6,(2,3)\}$
6. If $A \subseteq B$, what is $B \cup A$? What is $A \cap B$?
7. If C and D are disjoint, what is $C \cap D$? What is $C - D$?
8. Determine which of the properties reflexive, transitive, and symmetric apply to the following relations between integers x and y.
   a. $x \leq y$ b. $x < y$ c. $x = y$
9. Let $S = \{1,2,3,4,\ldots,14,15\}$, and let $a\,R\,b$ mean $a = b$ mod n where n is a positive integer.
   a. Prove R is an equivalent relation on S.
   b. List the equivalence classes into which R partitions S.
10. Let $S = \{1,2,3,4,5,6,7,8,9,10\}$ and let
    $\Pi_1 = \{1,2,3,4\}$ $\Pi_2 = \{5,6,7\}$
    $\Pi_3 = \{4,5,7,9\}$ $\Pi_4 = \{4,8,10\}$
    $\Pi_5 = \{8,9,10\}$ $\Pi_6 = \{1,2,3,6,8,10\}$
    Which of the following are partitions of S?
    a. $\{\Pi_1,\Pi_2,\Pi_5\}$ b. $\{\Pi_1,\Pi_3,\Pi_5\}$ c. $\{\Pi_3,\Pi_6\}$ d. $\{\Pi_4,\Pi_3,\Pi_2\}$
11. Let $\Pi_1 = (1,2)(4,5)(6,7)(3,8,9)$ and $\Pi_2 = (4,5,7)(1,2)(3,9)(8)$
    Find
    a. $\Pi_1 + \Pi_2$ b. $\Pi_1 \cdot \Pi_2$
12. Let $A =$ the set of people belonging to a sports club. Let $a\,R\,b$ if and only if a and b play tennis; let $a\,S\,b$ if and only if a and b play golf. Determine $R \cap S$. (R and S are relations)
13. In the directed graph shown below, identify
    a. The set of vertices   b. The set of arcs

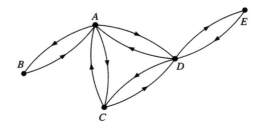

14. Find the number of distinct cycles of length 3 in the following graph? Are there any cycles of length 4 and length 5?

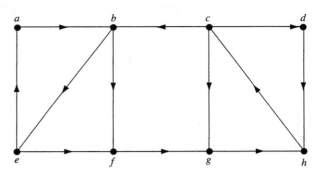

15. The set of vertices (V) and the set of edges (E) for three separate graphs are given below. In each case determine if the graph is a tree, and if it is, find the root.

$\quad$ G1: $\quad V = \{a,b,c,d,e,f\}$
$\quad\quad\quad\quad E = \{a - d, b - c, c - a, d - e\}$
$\quad$ G2: $\quad V = \{a,b,c,d,e,f\}$
$\quad\quad\quad\quad E = \{a - b, c - e, f - a, f - c, f - d\}$
$\quad$ G3: $\quad V = \{a,b,c,d,e,f\}$
$\quad\quad\quad\quad E = \{b - a, b - c, c - d, d - e, d - f\}$

16. Simplify the following expressions using the postulates and theorems of Boolean algebra.
    a. $ab + \bar{a}c + \bar{a}b\bar{c}$
    b. $(a + \bar{c}) + abc + ac\bar{d} + cd$
    c. $\overline{(\bar{a}(b + \bar{c})}(a + \bar{b} + \bar{c})\overline{(abc)}$
    d. $(a + b)(a + c)b$
    e. $\bar{a}\bar{b}(ac + \bar{b}) + (a + b)(a\bar{b}\bar{c} + \bar{a}bc)$

17. Derive the dual of the following Boolean functions
    a. $f(a,b,c) = \bar{a}\bar{b}c + b\bar{c} + ac$
    b. $f(a,b,c,d) = (\bar{a}\bar{c} + d)(ab + \bar{c})(\bar{b} + d)$
    c. $f(a,b,c,d,e) = (a\bar{c} + bd + e)(\bar{a} + de) + (a + \bar{d}\bar{e})(b + \bar{c})$

18. Prove that the complement of the EX-OR function is equal to its dual.

19. Find the complements of the following Boolean functions
    a. $f(a,b,c) = ab\bar{c} + a\bar{b}c + \bar{a}c$
    b. $f(a,b,c) = ab \oplus c \oplus \bar{b}\bar{c}$
    c. $f(a,b,c,d) = (b\bar{c} + \bar{a}d)(\bar{b}\bar{d} + ac)(ab + cd)$

20. Represent each of the following switching circuits using AND, OR, and NOT gates. Derive the truth table for each circuit.

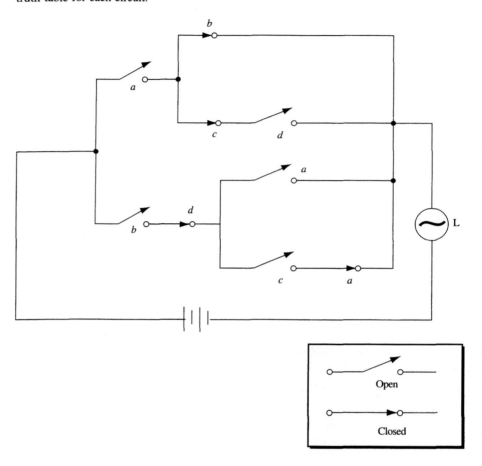

21. Derive the truth tables for the following gates:
    a. 3-input NAND   b. 3-input NOR
22. A 3-input EX-OR gate can be formed as shown:

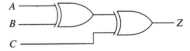

Derive the truth table for the gate.

23. A 4-input AOI (AND-OR-INVERT) gate is shown:

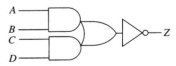

Derive the truth table for the gate.

24. Prove that the NAND function can be implemented as shown:

Further Reading

1. Z. Kohavi, *Switching and Finite Automata Theory*, McGraw-Hill, New York, 1978.

2. B. Kolman and R. C. Busby, *Discrete Mathematical Structures for Computer Science*, Prentice Hall, Englewood Cliffs, 1984.

3. C. L. Sheng, *Introduction to Switching Logic*, International Text Book Co., Toronto, 1972.

4
Combinational Logic Design

4.1 INTRODUCTION

Logic circuits are classified into two categories:

i. Combinational
ii. Sequential

In a combinational logic circuit the output is a function of the present input only. It does not depend upon the past values of the inputs. If the output is a function of past inputs (memory) as well as the present inputs, then the circuit is known as a sequential logic circuit. This chapter is concerned with combinational circuit design.

The main objective of combinational circuit design is to construct a circuit utilizing the minimum number of gates and inputs from the behavioral specification of the circuit. The first step in the design process is to construct a truth table of the circuit from its specification. The sum-of-products or product-of-sums form of the Boolean expression is then derived from the truth table and simplified where possible. The simplified expression is then implemented into the actual circuit by using appropriate gates.

Let us consider the design of a combinational circuit to meet the following specification: "The circuit has four inputs and one output. The output will be 1 if any two or more inputs are 1, otherwise the output will be 0." We begin by constructing the truth table that shows all the possible input combinations and the resulting output (Table 4.1). It is assumed that A, B, C, and D are the four inputs to the circuit and Z is the output. Writing down those input combinations that produce an output of 1, we obtain the canonical sum-of-products Boolean expression for the circuit.

$$Z = \overline{A}\overline{B}CD + \overline{A}B\overline{C}D + \overline{A}BC\overline{D} + \overline{A}BCD + A\overline{B}\overline{C}D + A\overline{B}C\overline{D}$$
$$+ A\overline{B}CD + AB\overline{C}\overline{D} + AB\overline{C}D + ABC\overline{D} + ABCD$$

The next step in the design process is to simplify the sum-of-products expression (if possible). The expression can be rewritten in the following form, introducing certain redundant terms (underlined):

TABLE 4.1
Truth table for the specified circuit

Input				Output
A	B	C	D	Z
0	0	0	0	0
0	0	0	1	0
0	0	1	0	0
0	0	1	1	1
0	1	0	0	0
0	1	0	1	1
0	1	1	0	1
0	1	1	1	1
1	0	0	0	0
1	0	0	1	1
1	0	1	0	1
1	0	1	1	1
1	1	0	0	1
1	1	0	1	1
1	1	1	0	1
1	1	1	1	1

$$Z = \overline{A}\overline{B}CD + \overline{A}BCD + \overline{A}B\overline{C}D + ABCD + AB\overline{C}\overline{D} + AB\overline{C}D + ABC\overline{D} + ABC\overline{D}$$
$$+ \overline{A}BCD + AB\overline{C}D + \overline{A}BCD + ABCD + A\overline{B}\overline{C}D + A\overline{B}CD + A\overline{B}C\overline{D} + A\overline{B}CD$$
$$+ \overline{A}BC\overline{D} + \overline{A}BCD + AB\overline{C}\overline{D} + ABCD + A\overline{B}C\overline{D} + \overline{A}BCD + ABC\overline{D} + ABCD$$

$$Z = (\overline{A}\overline{B} + \overline{A}B + A\overline{B} + AB)CD + AB(\overline{C}\overline{D} + \overline{C}D + CD + C\overline{D}) + BD(\overline{A}\overline{C}$$
$$+ A\overline{C} + \overline{A}C + AC) + AD(\overline{B}\overline{C} + \overline{B}C + B\overline{C} + BC) + BC(\overline{A}\overline{D} + \overline{A}D + A\overline{D} + AD)$$
$$+ AC(\overline{B}\overline{D} + \overline{B}D + B\overline{D} + BD) = CD + AB + BD + AD + BC + AC$$

Note that the inclusion of redundant terms does not affect the expression, since by Theorem 2 (Chap. 3) of Boolean algebra, $x + x = x$.

As can be seen from the final expression, these additional terms helped considerably in simplifying the original Boolean expression. It can be implemented in AND and OR gates to give the required combinational logic circuit as shown in Figure 4.1.

4.2 MINIMIZATION OF BOOLEAN EXPRESSIONS

The formal specification of combinational logic circuits leads to canonical Boolean expressions. In most cases, these expressions must be simplified in order to reduce the number of gates required to implement the corresponding circuits. There are two steps that may be used to simplify a Boolean expression:

i. Reduce the number of terms in the expression.
ii. Reduce the number of literals in the expression.

The first step corresponds to the reduction of the number of gates; the second step corre-

4.2 MINIMIZATION OF BOOLEAN EXPRESSIONS □ 71

FIGURE 4.1
Implementation of the Boolean expression $Z = CD + AB + BD + AD + BC + AC$

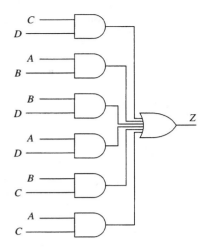

sponds to the reduction of inputs to the gates in the resulting combinational logic circuit.

As an illustration, let us consider the minimization of the following Boolean expression:

$$Z = ABD + AB\overline{D} + \overline{A}C + \overline{A}BC + ABC$$

The expression has 5 product terms and 14 literals. Direct implementation of the expression would require 5 AND gates and 1 OR gate, assuming that the complemented variables are already available. Figure 4.2a shows the direct implementation of the expression.

The expression can be simplified in the following way:

$$\begin{aligned}
Z &= ABD + AB\overline{D} + \overline{A}C + \overline{A}BC + ABC \\
&= AB(D + \overline{D}) + \overline{A}C(1 + B) + ABC \\
&= AB + \overline{A}C + ABC \quad \text{(by Postulate 5 and Theorem 3, Chap. 3)} \\
&= AB(1 + C) + \overline{A}C \\
&= AB + \overline{A}C
\end{aligned}$$

The minimized expression has two product terms and four literals. The variable D has been found to be redundant and has been eliminated. The minimized circuit is shown in Figure 4.2b.

So far we have considered the minimization of the sum-of-products form of the Boolean expression. The product-of-sums expressions can be minimized in a similar manner. For example, let us minimize the following expression:

$$Z = (A + B + C)(A + \overline{B} + C)(A + \overline{B} + \overline{C})(\overline{A} + \overline{B} + \overline{C})$$

We first take the complement of the product-of-sums expression,

$$\begin{aligned}
\overline{Z} &= \overline{(A + B + C)(A + \overline{B} + C)(A + \overline{B} + \overline{C})(\overline{A} + \overline{B} + \overline{C})} \\
&= \overline{A}\,\overline{B}\,\overline{C} + \overline{A}B\overline{C} + \overline{A}BC + ABC \\
&= \overline{A}\,\overline{C}(\overline{B} + B) + (\overline{A} + A)BC \\
&= \overline{A}\,\overline{C} + BC
\end{aligned}$$

FIGURE 4.2
Circuit implementation of sum-of-products expression

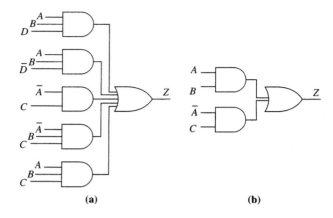

The next step is to take the complement of the resulting sum-of-products expression,

$$\overline{Z} = \overline{AC} + BC$$
$$Z = \overline{\overline{AC}\ \overline{BC}}$$
$$= (A + C)(\overline{B} + \overline{C})$$

Thus,

$$Z = (A + B + C)(A + \overline{B} + C)(A + \overline{B} + \overline{C})(\overline{A} + \overline{B} + \overline{C})$$
$$= (A + C)(\overline{B} + \overline{C})$$

The original product-of-sums expression had 4 sum terms and 12 literals, whereas the minimized expression has 2 sum terms and 4 literals. Figures 4.3a and b show the implementation of the original and minimized expression respectively.

Sometimes the dual of an expression provides an easier way of minimizing a product-of-sums expression. The expression to be minimized is first converted to its dual. The dual expression is minimized and then converted to *its* dual. For example, the Boolean expression

$$Z = (\overline{A} + B)(\overline{A} + C)(\overline{B} + \overline{C})$$

can be minimized by simplifying its dual expression,

FIGURE 4.3
Circuit implementation of product-of-sums expression

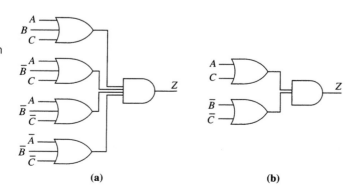

$$Z_d = \overline{A}B + \overline{A}C + \overline{B}\,\overline{C}$$
$$= \overline{A}(B + C) + \overline{B}\,\overline{C}$$

The dual of Z_d is

$$Z = (\overline{A} + BC)(\overline{B} + \overline{C})$$
$$= \overline{A}\,\overline{B} + \overline{A}\,\overline{C} + BC \cdot \overline{B} + BC \cdot \overline{C}$$
$$= \overline{A}\,\overline{B} + \overline{A}\,\overline{C}$$
$$= \overline{A}(\overline{B} + \overline{C})$$

4.3 KARNAUGH MAPS

Boolean expressions can be graphically depicted and simplified with the use of Karnaugh maps. In a Karnaugh map 2^n possible minterms of an n-variable Boolean function are represented by means of separate squares or cells on the map. For example, the Karnaugh map for two variables A and B will consist of 2^2 squares—one for each possible combination of A and B as shown in Figure 4.4. Each square of the Karnaugh map is designated by a decimal number written on the right-hand upper corner of the square. The decimal number corresponds to the minterm number of the Boolean function.

For a Boolean function of n variables, the Karnaugh map is a $2^{n/2} \times 2^{n/2}$ square array if n is even. Thus, if $n = 2$ the Karnaugh map is a 2×2 array as shown in Figure 4.4. If n is odd, the Karnaugh map is a $2^{(n-1)/2} \times 2^{(n+1)/2}$ rectangular array. Thus, for a three-variable function the Karnaugh map is a 2×4 array as shown in Figure 4.5. Figure 4.6 shows the Karnaugh map for a four-variable Boolean function, which contains 16 squares.

FIGURE 4.4
Karnaugh map for a two-variable Boolean function

	$\overline{B}$	B
$\overline{A}$	0 $\overline{A}\,\overline{B}$	1 $\overline{A}B$
A	2 $A\overline{B}$	3 AB

FIGURE 4.5
Karnaugh map for a three-variable Boolean function

	$\overline{B}\,\overline{C}$	$\overline{B}C$	BC	$B\overline{C}$
$\overline{A}$	0	1	3	2
A	4	5	7	6

FIGURE 4-6
Karnaugh map for a four-variable Boolean function

	$\overline{C}\,\overline{D}$	$\overline{C}D$	CD	$C\overline{D}$
$\overline{A}\,\overline{B}$	0	1	3	2
$\overline{A}B$	4	5	7	6
AB	12	13	15	14
$A\overline{B}$	8	9	11	10

FIGURE 4.7
Karnaugh map for $Z = A\bar{B}\bar{C} + \bar{A}BC + A\bar{B}C + \bar{A}B\bar{C}$

	$\bar{B}\bar{C}$	$\bar{B}C$	BC	$B\bar{C}$
$\bar{A}$	0 / 0	1 / 0	3 / 1	2 / 1
A	4 / 1	5 / 1	7 / 0	6 / 0

Boolean expressions may be plotted on Karnaugh maps if they are expressed in canonical form. For example, the following Boolean expression may be represented by the Karnaugh map shown in Figure 4.7:

$$Z(A,B,C) = A\bar{B}\bar{C} + \bar{A}BC + A\bar{B}C + \bar{A}B\bar{C}$$

Note that 1's are entered in cells 4, 3, 5, and 2, which correspond to the minterms $A\bar{B}\bar{C}$, $\bar{A}BC$, $A\bar{B}C$, and $\bar{A}B\bar{C}$ respectively; 0's are entered in all other cells.

A further simplification is frequently made on Karnaugh maps by representing zeros by blank squares. Thus, a blank square means that the corresponding minterm is not included in the Boolean function. For example, the function

$$Z(A,B,C,D) = \bar{A}\bar{B}\bar{C}\bar{D} + \bar{A}\bar{B}C\bar{D} + \bar{A}\bar{B}CD + \bar{A}BC\bar{D} + A\bar{B}\bar{C}D + A\bar{B}CD$$
$$= \Sigma m(0,2,3,5,9,11)$$

can be plotted on a Karnaugh map as shown in Figure 4.8.

The main feature of a Karnaugh map is that each square on the map is logically adjacent to the square that is physically adjacent to it. In other words, minterms corresponding to physically adjacent squares differ by a single variable. For example, in Figure 4.8 squares 9 and 11 are physically adjacent; square 9 represents minterm $A\bar{B}\bar{C}D$ and square 11 represents $A\bar{B}CD$, which are the same except in variable C. It should be noted that the first and last rows and the first and last columns in a Karnaugh map are also logically adjacent. For example square 3 (minterm $\bar{A}\bar{B}CD$) and square 11 (minterm $A\bar{B}CD$) are logically adjacent; similarly, square 0 (minterm $\bar{A}\bar{B}\bar{C}\bar{D}$) and square 2 (minterm $\bar{A}\bar{B}C\bar{D}$) are also adjacent.

Boolean functions on the Karnaugh maps can be simplified by using the property of adjacency. Thus, two minterms that are similar in all but one of their variables can be replaced by their common factor.

EXAMPLE 4.1

The Boolean function represented by the Karnaugh map of Figure 4.8 can be reduced to

FIGURE 4.8
Karnaugh map for $Z = \Sigma m(0,2,3,5,9,11)$

	$\bar{C}\bar{D}$	$\bar{C}D$	CD	$C\bar{D}$
$\bar{A}\bar{B}$	0 / 1	1	3 / 1	2 / 1
$\bar{A}B$	4	5 / 1	7	6
AB	12	13	15	14
$A\bar{B}$	8	9 / 1	11 / 1	10

FIGURE 4.9
Looping of adjacent squares

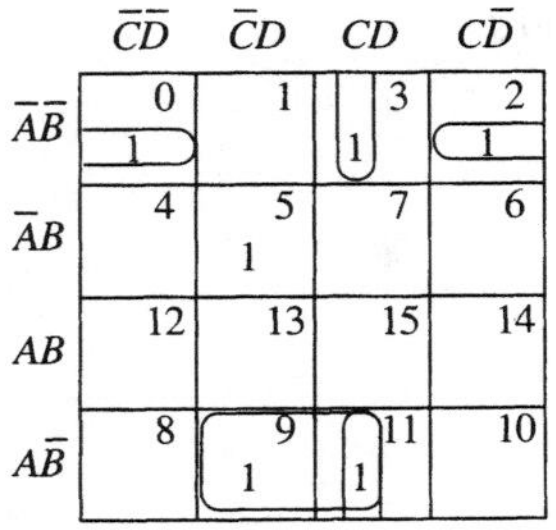

	$\overline{C}\overline{D}$	$\overline{C}D$	CD	$C\overline{D}$
$\overline{A}\overline{B}$	0 1	1 1	3 1	2 1
$\overline{A}B$	4	5 1	7	6
AB	12	13	15	14
$A\overline{B}$	8	9 1	11 1	10

$$Z(A,B,C,D) = \overline{A}\,\overline{B}\,\overline{D} + \overline{B}CD + A\overline{B}D + \overline{A}B\overline{C}D$$

Five of the six minterms that make up Z combine into three pairs:

$$m_0 + m_2 = \overline{A}\,\overline{B}\,\overline{C}\,\overline{D} + \overline{A}\,\overline{B}C\overline{D} = \overline{A}\,\overline{B}\,\overline{D}$$
$$m_3 + m_{11} = \overline{A}\,\overline{B}CD + A\overline{B}CD = \overline{B}CD$$
$$m_9 + m_{11} = A\overline{B}\,\overline{C}D + A\overline{B}CD = A\overline{B}D$$

Minterm 5 cannot be combined with any other minterm. Note that minterm 11 has been combined with two separate minterms, 3 and 9; this is possible because of Theorem 2 (Chap. 3). Thus, a cell may be used in as many pairings as desired. The pairings of the minterms in the Karnaugh map for Z are shown by the loops in Figure 4.9. These loops indicate which two minterms have been combined to produce a simpler term. As can be seen in Figure 4.9, when combining two minterms within a loop, the variable that changes from 0 to 1, or vice versa, is eliminated from the minterms. Thus, if we combine minterms 3 and 11, the variable A is eliminated because it changes from 0 to 1. Similarly, when we combine minterms 9 and 11 the variable C is eliminated, and combining minterms 0 and 2 eliminates the variable C. In other words, the variables that are constant for a loop define the product term corresponding to the loop. ∎

So far, we have considered only the grouping of two cells which are adjacent, either vertically or horizontally. Large numbers of cells can also be grouped, provided the number of cells in the group is a power of 2, i.e., 4 cells, 8 cells, etc. In fact, the larger the group of cells, the fewer will be the number of literals in the resulting product term. To illustrate, let us plot the four-variable Boolean function $f(A,B,C,D) = \Sigma m(0,2,5,7,8,10,13,15)$ on the Karnaugh map (Fig. 4.10).

There are two groups, each containing four cells on the map, as shown by the loops in Figure 4.11. We have enclosed the terms $\overline{A}\,\overline{B}\,\overline{C}\,\overline{D}$, $\overline{A}BCD$, $AB\overline{C}D$, and $ABCD$; these com-

FIGURE 4.10
Karnaugh map for $f = \Sigma m(0,2,5,7,8,10,13,15)$

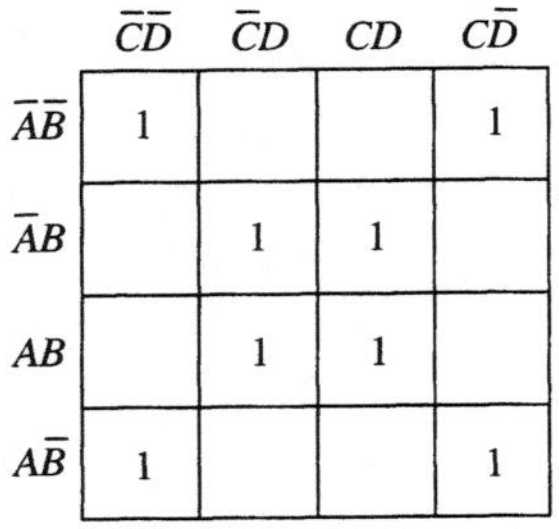

	$\overline{C}\overline{D}$	$\overline{C}D$	CD	$C\overline{D}$
$\overline{A}\overline{B}$	1			1
$\overline{A}B$		1	1	
AB		1	1	
$A\overline{B}$	1			1

FIGURE 4.11
Looping of squares

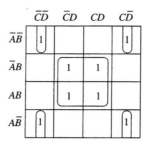

bine to form $\overline{A}BD(C + \overline{C})$ and $ABD(\overline{C} + C)$, the results of which may then be combined to give $BD(\overline{A} + A) = BD$. Now, grouping terms 0 and 8 together and terms 2 and 10 together eliminates the variable A from each pair. As can be seen in the map (Fig. 4.11) these four terms reduce to the two terms—$\overline{B}\,\overline{C}\overline{D}$ and $\overline{B}C\overline{D}$. These two terms, which represent combined minterms 0 and 8, and 2 and 10, respectively, can be grouped to eliminate the variable C, thus reducing the four terms to one term, $\overline{B}\,\overline{D}$. Note that four terms have been combined to eliminate two literals. Thus, the reduced form of the above Boolean function is

$$f(A,B,C,D) = BD + \overline{B}\,\overline{D}$$

EXAMPLE 4.2

Simplify the following four-variable function

$$f(A,B,C,D) = \Sigma m(0,1,4,5,7,8,9,12,13,15)$$

The Karnaugh map for the function is shown in Figure 4.12. The reduced form of the function can be derived directly from the Karnaugh map

$$f(A,B,C,D) = \overline{C} + BD$$

FIGURE 4.12

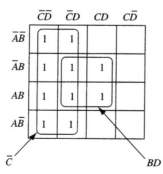

In four-variable Karnaugh maps, the top and bottom rows are logically adjacent and so are the left and right columns. We saw one example of grouping four cells that were not physically adjacent (Fig. 4.11). Figure 4.13 shows a few more left-right column, top-bottom row adjacencies.

FIGURE 4.13

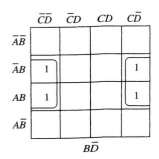
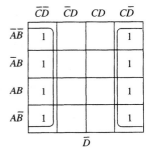
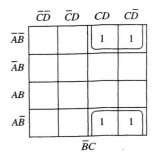

Don't Care Conditions

In certain Boolean functions it is not possible to specify the output for some input combinations. It means that these particular input combinations have no relevant effect on the output. These input combinations or conditions are called **don't care conditions**, and the minterms corresponding to these input combinations are called don't care terms. Functions that include don't care terms are said to be **incompletely specified** functions. The don't care minterms are labeled d instead of m.

■ **EXAMPLE 4.3**

Let us consider the following function

$$f(A,B,C) = \Sigma m(0,4,7) + d(1,2,6)$$

where 1, 2, and 6 are the don't care terms. Since the don't care combinations cannot occur, the output corresponding to these input combinations can be assigned either 0 or 1 at will. It is often possible to utilize the don't care terms to aid in the simplification of Boolean functions. For example, the Karnaugh map resulting from the above Boolean function is as shown in Figure 4.14a. Note that the simplified function

$$f(A,B,C) = \overline{C} + AB$$

is obtained by grouping minterms 0 and 4 with don't cares 2 and 6, and grouping minterm 7 with don't care 6. Don't care 2 is not used.

If the don't care terms are not included in the minimization, only minterms 0 and

FIGURE 4.14

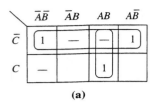

4 can be grouped, as shown in Figure 4.14b. In that case, the function simplifies to

$$f(A,B,C) = \overline{B}\,\overline{C} + ABC$$

which contains more literals than the simplified function obtained by considering the don't care terms. ∎

The best way to utilize don't cares in minimizing Boolean functions is to assume don't cares to be 1 if that results in grouping a larger number of cells on the map than would be possible otherwise. In other words, only those don't cares that aid in the simplification of a function are taken into consideration.

■ **EXAMPLE 4.4**

Let us minimize the following Boolean function using a Karnaugh map

$$f(A,B,C,D) = \Sigma m(0,1,5,7,8,9,12,14,15) + d(3,11,13)$$

The Karnaugh map is shown in Figure 4.15. From this the minimized function is given by

$$f(A,B,C,D) = D + AB + \overline{B}\,\overline{C}$$

FIGURE 4.15

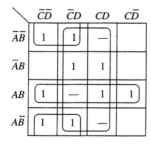

∎

Again, it is emphasized that while considering don't care terms on Karnaugh maps, it is not necessary to use all the terms (or even one term), unless their inclusion in a group would assist the minimization process.

The Complementary Approach

Sometimes it is more convenient to group the 0's on a Karnaugh map rather than the 1's. The resultant sum of products is the complement of the desired expression. This sum of products is then complemented by using DeMorgan's theorem, which results in a minimum product-of-sums expression. This is known as the complementary approach.

EXAMPLE 4.5

Let us consider the three-variable Karnaugh map shown in Figure 4.16. The grouping of 0's yields the function

$$\overline{f(A,B,C)} = A\overline{C} + \overline{B}C$$

Hence, $f = \overline{\overline{f}} = \overline{A\overline{C} + \overline{B}C} = (\overline{A} + C)(B + \overline{C})$

FIGURE 4.16

	$\overline{B}\overline{C}$	$\overline{B}C$	BC	$B\overline{C}$
$\overline{A}$	1	0	1	1
A	0	0	1	0

Occasionally the complement function gives a better minimization.

EXAMPLE 4.6

The minimized form of the Boolean function

$$f(A,B,C,D) = \Sigma m(0,1,2,4,5,6,8,9,10)$$

is $f = \overline{A}\overline{C} + \overline{B}\overline{C} + \overline{A}\overline{D} + \overline{B}\overline{D}$; this is derived from the Karnaugh map of Figure 4.17a.

FIGURE 4.17

(a)

	$\overline{C}\overline{D}$	$\overline{C}D$	CD	$C\overline{D}$
$\overline{A}\overline{B}$	1	1	0	1
$\overline{A}B$	1	1	0	1
AB	0	0	0	0
$A\overline{B}$	1	1	0	1

(b)

	$\overline{C}\overline{D}$	$\overline{C}D$	CD	$C\overline{D}$
$\overline{A}\overline{B}$	1	1	0	1
$\overline{A}B$	1	1	0	1
AB	0	0	0	0
$A\overline{B}$	1	1	0	1

The complement function is derived by grouping the 0's in the Karnaugh map (Fig. 4.17b). Thus, the complement function is

$$\overline{f} = AB + CD$$

Inverting the complement function,

$$\overline{\overline{f}} = f = \overline{AB + CD} = (\overline{A} + \overline{B})(\overline{C} + \overline{D})$$

The implementation of the minimized sum-of-products and the product-of-sums functions are shown in Figures 4.18a and b respectively. The product-of-sums form gives a simpler solution in this case. Thus, both the simplified sum-of-products and the product-of-sums forms for a given function must be examined before a decision can be made as to which form will be cheaper to implement.

FIGURE 4.18

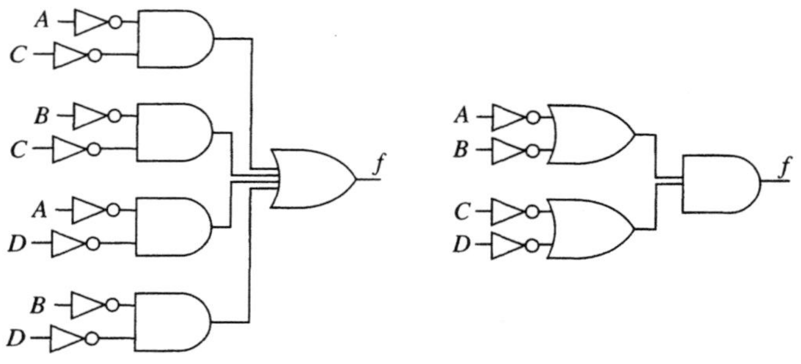

(a) (b)

The complementary approach may also be utilized to expand a general product-of-sums function to canonical form. Let us illustrate this by obtaining the canonical product-of-sums form for the function

$$f(A,B,C) = (A + C)(A + B + \overline{C})(\overline{A} + B)$$

The complement of the function is

$$\overline{f} = \overline{A}\,\overline{C} + \overline{A}\,\overline{B}C + A\overline{B}$$

which leads to the Karnaugh map shown in Figure 4.19. The canonical form of $\overline{f}$ is

$$\overline{f} = \overline{A}\,\overline{B}\,\overline{C} + \overline{A}\,B\,\overline{C} + \overline{A}\,\overline{B}C + A\overline{B}\,\overline{C} + A\overline{B}C$$

Thus, the canonical product-of-sums form for f is

$$\begin{aligned}f = \overline{\overline{f}} &= \overline{\overline{A}\,\overline{B}\,\overline{C} + A\overline{B}\,\overline{C} + \overline{A}\,\overline{B}C + A\overline{B}C + \overline{A}B\overline{C}}\\ &= (A + B + C)(\overline{A} + B + C)(A + B + \overline{C})(\overline{A} + B + \overline{C})(A + \overline{B} + C)\end{aligned}$$

A Karnaugh map with four input variables is quite straightforward. With five or more variables, however, the map becomes complicated and the adjacencies are difficult to recognize. However, it is possible to minimize a five-variable Boolean function with a four-variable Karnaugh map.

■ **EXAMPLE 4.7**

Let us minimize the Boolean function

$$f(A,B,C,D,E) = \Sigma m(4,5,10,11,15,18,20,24,26,30,31) + d(9,12,14,16,19,21,25)$$

The Karnaugh map for the function is shown in Figure 4.20. The five variables are divided

FIGURE 4.19

	$\overline{B}\,\overline{C}$	$\overline{B}C$	BC	$B\overline{C}$
$\overline{A}$	0	0	1	0
A	0	0	1	1

between two four-variable maps. Note that the difference between the $A = 0$ map and $A = 1$ map is that for $A = 0$, the entry in cell $B\overline{C}DE$ is 1, whereas the corresponding entry is 0 in the $A = 1$ map. In addition, the entries in cells $\overline{B}\,\overline{C}\,\overline{D}\,\overline{E}$, $\overline{B}\,\overline{C}DE$, and $BC\overline{D}\,\overline{E}$ in the $A = 1$ map are 1, and 1 respectively, whereas the corresponding entries in the $A = 0$ map are 0's. Thus, the four-variable Karnaugh map with $\overline{A}$ in cell $B\overline{C}DE$ and A in cells $\overline{B}\,\overline{C}DE$ and $BC\overline{D}\,\overline{E}$ shown in Figure 4.21 is equivalent to the maps of Figure 4.20.

FIGURE 4.20

(a) $A=0$

	$\overline{D}\,\overline{E}$	$\overline{D}E$	DE	$D\overline{E}$
$\overline{B}\,\overline{C}$				
$\overline{B}C$	1	1		
BC	—		1	—
$B\overline{C}$		—	1	1

(b) $A=1$

	$\overline{D}\,\overline{E}$	$\overline{D}E$	DE	$D\overline{E}$
$\overline{B}\,\overline{C}$	—		—	1
$\overline{B}C$	1	—		
BC			1	1
$B\overline{C}$	1	—		1

FIGURE 4.21
Equivalent Karnaugh map for Figures 4.20a and b

	$\overline{D}\,\overline{E}$	$\overline{D}E$	DE	$D\overline{E}$
$\overline{B}\,\overline{C}$	—		—	A
$\overline{B}C$	1	—		
BC	—		1	—
$B\overline{C}$	A	—	$\overline{A}$	1

The map is then reduced in two steps [1]:

Step 1: Group all terms employing 1's and —s. The letter variable terms are ignored at this step. Figure 4.22 shows the relevant groupings on the map of Figure 4.21.

FIGURE 4.22
$f(A,B,C,D,E) = \overline{B}C\overline{D} + BCD$

	$\overline{D}\,\overline{E}$	$\overline{D}E$	DE	$D\overline{E}$
$\overline{B}\,\overline{C}$	—		—	A
$\overline{B}C$	1	—		
BC	—		1	—
$B\overline{C}$	A	—	$\overline{A}$	1

Step 2: Group the letter variable(s) with the adjacent 1's and —s. The resulting terms are then ORed with the terms derived in step 1 to obtain the minimized function (shown in Figure 4.23).

FIGURE 4.23
$f(A,B,C,D,E) = A\overline{C}\overline{E} + \overline{B}C\overline{D} + BCD + \overline{A}BD$

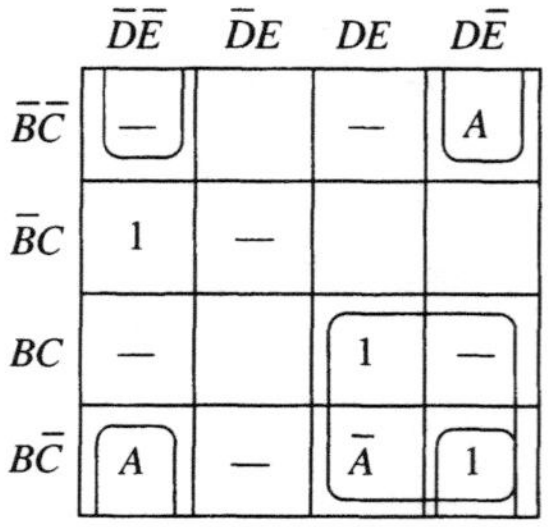

	$\overline{D}\overline{E}$	$\overline{D}E$	DE	$D\overline{E}$
$\overline{B}\overline{C}$	—		—	A
$\overline{B}C$	1	—		
BC	—		1	—
$B\overline{C}$	A	—	$\overline{A}$	1

4.4 THE QUINE-McCLUSKEY METHOD

The Karnaugh map approach is not suitable for minimizing Boolean functions having more than six variables. For functions with a large number of variables, a tabular method known as Quine-McCluskey method is much more effective. The method consists of two steps:

1. Generation of all prime implicants
2. Selection of a minimum subset of prime implicants, which will represent the original function

A **prime implicant** is a product term that cannot be combined with any other product term to generate a term with fewer literals than the original term.

As an example consider a Boolean function

$$f(A,B,C) = ABC + AB\overline{C} + A\overline{B}C + \overline{A}BC + \overline{A}\,\overline{B}\,\overline{C}$$

which after minimization becomes

$$f(A,B,C) = AB + BC + AC + \overline{A}\,\overline{B}\,\overline{C}$$

The product terms AB, BC, AC, and $\overline{A}\,\overline{B}\,\overline{C}$ are all prime implicants because none of them can be combined with any other term in the function to yield a term with fewer literals. A prime implicant is called an **essential prime implicant** if it covers at least one minterm that is not covered by any other prime implicant of the function.

■ **EXAMPLE 4.8**

Let us minimize the following Boolean function

$$f(A,B,C,D) = \Sigma m(1,4,5,10,12,13,14,15)$$

The Karnaugh map for the function is shown in Figure 4.24.

FIGURE 4.24

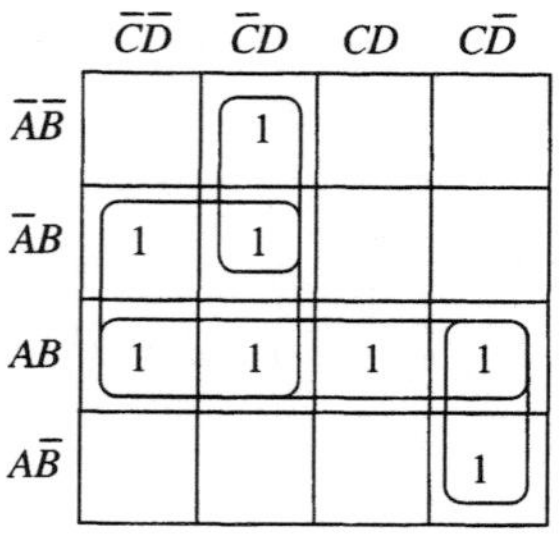

	$\overline{C}\overline{D}$	$\overline{C}D$	CD	$C\overline{D}$
$\overline{A}\overline{B}$		1		
$\overline{A}B$	1	1		
AB	1	1	1	1
$A\overline{B}$				1

The prime implicants for the function are $B\overline{C}$, AB, $\overline{A}\,\overline{C}D$, and $AC\overline{D}$. The minimized function is

$$f(A,B,C,D) = \overline{A}\,\overline{C}D + B\overline{C} + AB + AC\overline{D}$$

The prime implicant $\overline{A}\,\overline{C}D$ is an essential prime implicant because it covers minterm $\overline{A}\,\overline{B}\,\overline{C}D$, which is not covered by any other prime implicant. Similarly, only $AC\overline{D}$ covers minterm $A\overline{B}C\overline{D}$, $B\overline{C}$ covers $\overline{A}B\overline{C}\,\overline{D}$, and AB covers $ABCD$; in other words, $AC\overline{D}$, $B\overline{C}$, and AB are also essential prime implicants. ∎

The Quine-McCluskey method for minimization can be formulated as follows:

Step 1: Tabulate all the minterms of the function by their binary representations.

Step 2: Arrange the minterms into groups according to the number of 1's in their binary representation. For example, if the first group consists of minterms with n 1's, the second group will consist of minterms with $(n + 1)$ 1's etc. Lines are drawn between different groups to simplify identification.

Step 3: Compare each minterm in a group with each of the minterms in the group below it. If the compared pair is adjacent (i.e., if they differ by one variable only), they are combined to form a new term. The new term has a dash in the position of the eliminated variable. Both combining terms are checked off in the original list indicating that they are not prime implicants.

Step 4: Repeat the above step for all groups of minterms in the list. This results in a new list of terms with dashes in place of eliminated variables.

Step 5: Compare terms in the new list in search for further combinations. This is done by following step 3. In this case a pair of terms can be combined only if they have dashes in the same positions. As before, a term is checked off if it is combined with another. This step is repeated until no new list can be formed. All terms that remain unchecked are prime implicants.

Step 6: Select a minimal subset of prime implicants that cover all the terms of the original Boolean function.

EXAMPLE 4.9

Let us minimize the following Boolean function using the Quine-McCluskey procedure

$$f(A,B,C,D,E) = \Sigma m(0,1,2,9,11,12,13,27,28,29)$$

The minterms are first tabulated according to step 1.

Minterm	A	B	C	D	E
0	0	0	0	0	0
1	0	0	0	0	1
2	0	0	0	1	0
9	0	1	0	0	1
11	0	1	0	1	1
12	0	1	1	0	0
13	0	1	1	0	1

27	1	1	0	1	1
28	1	1	1	0	0
29	1	1	1	0	1

The minterms are then grouped according to the number of 1's contained in each term, as specified in step 2. This results in list 1 of Figure 4.25. In list 1, terms of group 1 are combined with those of group 2, terms of group 2 are combined with those of group 3, etc., using step 3. For example, 0(00000) is adjacent to 1(00001), so they are combined to form 0000– which is the first term in list 2. Both combined terms are checked off in list 1. Since 0(00000) is also adjacent to 2(00010) they are combined to form the term 000–0, which is also entered in list 2. A line is then drawn under the two terms in list 2 in order to identify them as a distinct group.

The next step is to compare the two terms in group 2 of list 1 with the two terms in group 3. Only terms 1(00001) and 9(01001) combine to give 0–001; all other terms differ in more than one variable and therefore do not combine. As a result, the second group of list 2 contains only one combination. The two terms in group 3 are now compared with the three terms in group 4. Terms 9(01001) and 11(01011) combine to give 010–1, terms 9(01001) and 13(01101) combine to give 01–01, terms 12(01100) and 13(01101) combine to give 0110–, and terms 12(01100) and 28(11100) combine to give –1100. Thus, the third group of list 2 contains four terms. Finally, the three terms in group 4 of list 1 are compared with the two terms in group 5. Terms 13(01101) and 29(11101) combine to give –1101, terms 11(01011) and 27(11011) combine to give –1011, and terms 28(11100) and 29(11101) combine to give 1110–. Therefore, the fourth group of list 2 contains three terms.

The process of combining terms in adjacent groups is continued for list 2. This results in list 3. It can be seen in Figure 4.25 that certain terms cannot be combined further in list 2. These correspond to the prime implicants of the Boolean function and are labeled PI_1, ..., PI_7.

The final step of the Quine-McCluskey procedure is to find a minimal subset of the

		List 1			List 2			List 3		
	Minterm	ABCDE		Minterms	ABCDE			Minterms	ABCDE	
Group 1	0	00000	✓	0,1	0000—	PI_2		12,13,28,29	—110—	PI_1
	1	00001	✓	0,2	000—0	PI_3				
Group 2	2	00010	✓	1,9	0—001	PI_4				
	9	01001	✓	9,13	01—01	PI_5				
Group 3	12	01100	✓	9,11	010—1	PI_6				
	13	01101	✓	12,13	0110—	✓				
Group 4	11	01011	✓	12,28	—1100	✓				
	28	11100	✓	13,29	—1101	✓				
	29	11101	✓	11,27	—1011	PI_7				
Group 5	27	11011	✓	28,29	1110—	✓				

FIGURE 4.25
Determination of prime implicants

prime implicants which can be used to realize the original function. The complete set of prime implicants for the given function can be derived from Figure 4.25, these are

$$(BC\overline{D}, \overline{A}\overline{B}\overline{C}D, \overline{A}\overline{B}\overline{C}E, \overline{A}\overline{C}\overline{D}E, \overline{A}B\overline{D}E, \overline{A}B\overline{C}E, B\overline{C}DE)$$

In order to select the smallest number of prime implicants that account for all the original minterms, a **prime implicant chart** is formed as shown in Figure 4.26. A prime implicant chart has a column for each of the original minterms and a row for each prime implicant. For each prime implicant row, an X is placed in the columns of those minterms that are accounted for by the prime implicant. For example, in Figure 4.26 prime implicant PI_1, comprising minterms 12, 13, 28, and 29, has X's in columns 12, 13, 28, and 29. To choose a minimum subset of prime implicants, it is first necessary to identify the essential prime implicants. A column with a single X indicates that the prime implicant row is the only one covering the minterm corresponding to the column; therefore, the prime implicant is essential and must be included in the minimized function. Figure 4.26 has three essential prime implicants, and they are identified by asterisks. The minterms covered by the essential prime implicants are marked with asterisks.

The next step is to select additional prime implicants that can cover the remaining column terms. This is usually done by forming a reduced prime implicant chart that contains only the minterms that have not been covered by the essential prime implicants. Figure 4.27 shows the reduced prime implicant chart derived from Figure 4.26.

FIGURE 4.26
Prime implicant chart

	0	1	2	9	11	12	13	27	28	29
PI_1*						X	X		X	X
PI_2	X	X								
PI_3*	X		X							
PI_4		X		X						
PI_5				X			X			
PI_6				X	X					
PI_7*					X			X		

FIGURE 4.27

	1	9
PI_2	X	
PI_4	X	X
PI_5		X
PI_6		X

Prime implicant PI_4 covers the minterms 1 and 9. Therefore, the minimum sum-of-products equivalent to the original function is

$$f(A,B,C,D,E) = PI_1 + PI_3 + PI_4 + PI_7$$
$$= -110- + 000-0 + 0-001 + -1011$$
$$= BC\overline{D} + \overline{A}\overline{B}\overline{C}E + \overline{A}\overline{C}\overline{D}E + B\overline{C}DE$$ ■

For some functions, the prime implicant chart may not contain any essential prime implicants. In other words, in every column of a prime implicant chart there are two or more X's. Such a chart is said to be **cyclic**.

EXAMPLE 4.10

The following Boolean function has a cyclic prime implicant chart

$$f(A,B,C) = \Sigma m(1,2,3,4,5,6)$$

The prime implicants of the function can be derived as shown in Figure 4.28. The resulting prime implicant chart as shown in Figure 4.29 is cyclic; all columns have two X's. As can be seen, there is no simple way to select the minimum number of prime implicants from the cyclic chart. We can proceed by selecting prime implicant PI_1, which covers minterms 1 and 3. After crossing out row PI_1 and columns 1 and 3, we see that PI_4 and PI_5 cover the remaining columns (Fig. 4.30). Thus, the minimum sum-of-products form of the given Boolean function is

$$f(A,B,C) = PI_1 + PI_4 + PI_5$$
$$= \overline{A}C + B\overline{C} + A\overline{B}$$

This is not a unique minimum sum of products for the function. For example,

$$f(A,B,C) = PI_6 + PI_2 + PI_3$$
$$= A\overline{C} + \overline{B}C + \overline{A}B$$

is also a minimal form of the original function. It can be verified from the Karnaugh map

FIGURE 4.28

Minterm	ABC		Minterm	ABC	
1	001	✓	1,3	0—1	PI_1
2	010	✓	1,5	—01	PI_2
4	100	✓	2,3	01—	PI_3
3	011	✓	2,6	—10	PI_4
5	101	✓	4,5	10—	PI_5
6	110	✓	4,6	1—0	PI_6

FIGURE 4.29

	1	2	3	4	5	6
PI_1	X		X			
PI_2	X				X	
PI_3		X	X			
PI_4		X				X
PI_5				X	X	
PI_6				X		X

FIGURE 4.30

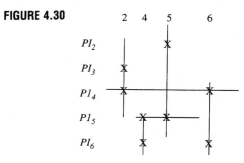

FIGURE 4.31

	$\bar{B}\bar{C}$	$\bar{B}C$	BC	$B\bar{C}$
$\bar{A}$		1	1	1
A	1	1		1

	$\bar{B}\bar{C}$	$\bar{B}C$	BC	$B\bar{C}$
$\bar{A}$		1	1	1
A	1	1		1

of the function (Fig. 4.31) that these are the minimum sum-of-products forms. Note that each minterm in the Karnaugh map can be grouped within two different loops, which indicates that two different prime implicants can cover the same minterm. ■

Simplification of Boolean Functions with Don't Cares

The Quine-McCluskey procedure for minimizing Boolean functions containing don't care minterms is similar to the conventional procedure in that all the terms, including don't cares, are used to produce the complete set of prime implicants. However, don't care terms are not listed as column headings in the prime implicant chart because they need not be included in the final expression.

EXAMPLE 4.11

Let us minimize the following Boolean function

$$f(A,B,C,D) = \Sigma m(3,7,9,14) + d(1,4,6,11)$$

Both the minterms and don't cares are listed in the minimizing table and combined in the manner discussed previously.

List 1			List 2					
Minterm	ABCD		Minterm	ABCD		Minterm	ABCD	
1	0001 ✓		1,3	00—1 ✓		1,3,9,11	—0—1	PI_1
4	0100 ✓		1,9	—001 ✓				
3	0011 ✓		4,6	01—0	PI_2			
6	0110 ✓		3,7	0—11	PI_3			
9	1001 ✓		3,11	—011 ✓				
7	0111 ✓		6,7	011—	PI_4			
14	1110 ✓		6,14	—110	PI_5			
11	1011 ✓		9,11	10—1 ✓				

A prime implicant chart is then obtained that contains only the minterms.

	3	7	9	14
PI_1			X	
PI_3	X	X		
PI_4		X		
PI_5				X

It can be seen from the chart that PI_1 and PI_5 are essential prime implicants. Since only minterm 7 is not covered by the essential prime implicants, a reduced prime implicant chart is not required. Thus, a minimal form of the Boolean function is

$$f(A,B,C,D) = PI_1 + PI_3 + PI_5 = \bar{B}D + \bar{A}CD + BC\bar{D}$$

■

*4.5 ESPRESSO

Although the Quine-McCluskey method guarantees optimal minimization in terms of two-level logic, it is practical only for functions with a small number of input variables. The search for techniques that can be used to minimize functions with large numbers of variables has resulted in a technique known as ESPRESSO [2], which provides very good results without necessarily guaranteeing minimal two-level logic. It is a **heuristic** technique[1] and requires less computation time than any other two-level minimization techniques currently available. In order to understand the basic concept of ESPRESSO, it is necessary to be familiar with the cube notation for representing Boolean functions. A **cube** is a product of literals; a literal is a variable or its complement.

■ **EXAMPLE 4.12**

For the Boolean function

$$F(w,x,y,z) = wx\bar{y}z + \bar{w}\bar{x} + \bar{w}x + w\bar{x}\bar{z}$$

we have the cubes

$$wx\bar{y}z = (1101) \quad \text{4 literals}$$
$$\bar{w}\bar{x} = (00\text{-}\text{-}) \quad \text{2 literals}$$
$$\bar{w}x = (01\text{-}\text{-}) \quad \text{2 literals}$$
$$w\bar{x}\bar{z} = (10\text{-}0) \quad \text{3 literals}$$

■

Note that a cube can represent one or more minterms. The number of minterms that a cube can represent is 2^d, where d is the number of don't cares (–) in a cube. A cube c_1 **covers** cube c_2 if every minterm represented by c_2 is also represented by c_1. In that case, c_2 is considered to be a subcube of c_1. Two cubes c_1 and c_2 are intersecting if both have at least one common minterm.

A Boolean function f in general can be considered as the set of minterms that makes $f = 1$; this set is identified as the **on-set** of f. Similarly, the set of minterms that makes $f = 0$ is known as the **off-set** of f. There may also be set of minterms for which the function f is not specified; this set of minterms is known as the don't care set of f. A **prime cube**, usually called a prime implicant, of a Boolean function is a cube that cannot be expanded (by eliminating some literal) without including a minterm belonging to off-set of the Boolean function. For example, in the following function

$$f(a,b,c) = \Sigma m(0,1,5) + d(2,7)$$

the cube (–01) is a prime cube.

ESPRESSO performs two-level logic minimization using the following steps in a loop:

i. **Expand**: expand a cube so that it is prime.
ii. **Irredundant_cover**: derive an irredundant set of cubes by removing a maximal set of nonessential prime cubes.
iii. **Reduce**: replace each cube by the smallest cube while still covering the on-set of the function.

If a logic circuit consists of two-level AND and OR gates, then expanding a cube will correspond to deleting some literal (i.e., disconnecting some inputs from an AND gate).

---
[1] A heuristic technique is an informal, trial-and-error approach to solve a problem.

Replacement of a nonessential prime cube corresponds to disconnecting an input of an OR gate. On the other hand, reduction of each cube by the smallest cube will add literals to the cubes. However, if the **Reduce** operation is followed by the **Expand** operation, the number of gates or connections will not be increased.

It should be emphasized that ESPRESSO has been designed for rapid minimization of Boolean functions, and hence its potential can be fully utilized only in a computer-aided-design environment. The SIS synthesis tool developed at UC Berkeley uses ESPRESSO for minimizing logic functions of two-level (i.e., AND-OR) implementation.

EXAMPLE 4.13

Let us minimize the following six-variable function using ESPRESSO

$$f(u,v,w,x,y,z) = \Sigma m(0,2,4,7,8,10,11,16,24,32,34,45,52) + d(5,19,23,29,51,58)$$

In order for ESPRESSO to minimize logic expressions, a source file corresponding to these expressions must be created. Figure 4.32 shows the source file for the above expression; this file represents the expression in PLA (truth table) format. The first 2 rows on the file represent the number of inputs and outputs in the Boolean expression to be minimized, which for the expression under consideration are 6 and 1 respectively. The next 13 rows represent the on-set of the expression, and the final 6 rows correspond to the don't care set.

```
.i 6
.o 1
.type fd
000000    1
000010    1
000100    1
000111    1
001000    1
001010    1
001011    1
010000    1
011000    1
100000    1
100010    1
101101    1
110100    1
000101    -
010011    -
010111    -
011101    -
110101    -
111010    -
.e
```

FIGURE 4.32
Source file for ESPRESSO

The total number of literals in the sum-of-products representation of the expression before minimization is 78. After the minimization of the expression using ESPRESSO, the number of literals is reduced to 34. The minimized expression is written in the following form:

$$A = \bar{u}\bar{v}\bar{w}\bar{y}\bar{z} \quad B = \bar{u}\bar{w}xyz \quad C = u\bar{v}wx\bar{y}z \quad D = uv\bar{w}x\bar{y} \quad E = \bar{u}\bar{v}w\bar{x}y$$
$$F = \bar{u}\bar{x}\bar{y}\bar{z} \quad G = \bar{v}\bar{w}\bar{x}\bar{z}$$
$$f(u,v,w,x,y,z) = A + B + C + D + E + F + G$$

*4.6 MINIMIZATION OF MULTIPLE-OUTPUT FUNCTIONS

Multi-input/multi-output combinational logic blocks are frequently used in complex logic systems, e.g., VLSI (very large scale integrated) chips. The area occupied by such combinational logic blocks has a significant impact on the VLSI design objective of putting the maximum amount of logic in the minimum possible area.

In minimizing multiple-output functions, instead of considering individual single-output functions, the emphasis is on deriving product terms that can be shared among the functions. This results in a circuit having fewer gates than if each function is minimized independently. For example, if the following two functions are individually minimized, the resulting circuit will be as shown in Figure 4.33a.

$$f_1 = ab\bar{c}\bar{d} + ab\bar{c}d + abcd + a\bar{b}cd$$
$$f_2 = ab\bar{c}\bar{d} + ab\bar{c}d + abcd + a\bar{b}cd + a\bar{b}\bar{c}\bar{d} + a\bar{b}\bar{c}d$$

However, if the functions are minimized as shown in Figure 4.33b, one product term (i.e., acd) can be shared among the functions, resulting in a circuit with one fewer gate.

As is clear from this example, the determination of **shared** product terms form among many Boolean functions is an extremely complicated task. This can only be efficiently done by using a computer-aided minimization technique. For example, the two-level minimizer ESPRESSO, in general, identifies the shared terms reasonably well.

FIGURE 4.33a

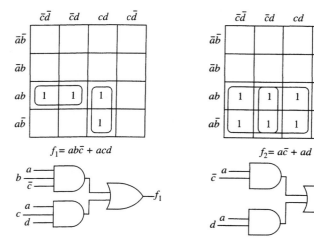

FIGURE 4.33b

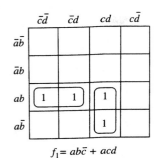

$f_1 = ab\bar{c} + acd$

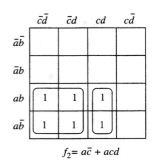

$f_2 = a\bar{c} + acd$

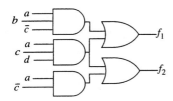

■ **EXAMPLE 4.14**

Let us consider the following functions:

$$f_1 = \Sigma m(0,2,4,5,9,10,11,13,15) \quad (4.1)$$
$$f_2 = \Sigma m(2,5,10,11,12,13,14,15) \quad (4.2)$$
$$f_3 = \Sigma m(0,2,3,4,9,11,13,14,15) \quad (4.3)$$

The Karnaugh maps for the functions are shown in Figure 4.34. Individual minimization of the functions results in the following shared terms:

Term	Functions
$\bar{a}\bar{c}\bar{d}$	f_1, f_3
$b\bar{c}d$	f_1, f_2
ac	f_2, f_3
$\bar{b}c\bar{d}$	f_1, f_2

There are 12 product terms in the original expressions, out of which 4 can be shared. The expression can be rewritten as follows:

$$f_1 = W + X + ad + Z$$
$$f_2 = ab + X + Y + Z$$
$$f_3 = W + \bar{a}\bar{b}c + Y + abd$$

where $W = \bar{a}\bar{c}\bar{d}$, $X = b\bar{c}d$, $Y = ac$, and $Z = \bar{b}c\bar{d}$.

The SIS system includes a minimizer for multi-output functions [3]. This is a multi-level minimizer activated by the command **full-simplify**. The expressions obtained by full-simplifying the functions under consideration are

FIGURE 4.34
Karnaugh maps for functions f_1, f_2, and f_3

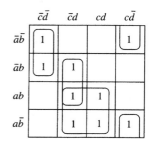

$$f_1 = \bar{a}\bar{c}\bar{d} + b\bar{c}d + ad + \bar{b}c\bar{d}$$

$$f_2 = ab + b\bar{c}d + ac + \bar{b}c\bar{d}$$

$$f_3 = \bar{a}\bar{c}\bar{d} + \bar{a}\bar{b}c + abd + ac$$

$$f_1 = \bar{c}f_3 + df_2 + \bar{b}f_2$$
$$f_2 = b\bar{c}d + \bar{b}c\bar{d} + ac + ab$$
$$f_3 = \bar{a}\bar{c}\bar{d} + \bar{a}\bar{b}c + f_2bc + ad$$

There are 27 literals in these equations, which can be further reduced to 23 by using factoring. The resulting circuit is shown in Figure 4.35. ∎

4.7 NAND-NAND AND NOR-NOR LOGIC

So far, we have discussed various ways of simplifying Boolean functions. A simplified sum-of-products function can be implemented by AND-OR logic, and a simplified product-of-sums function can be implemented by OR-AND logic. This section shows that any logic function can be implemented using only one type of gate—either the NAND or the NOR gate. This has the advantage of standardizing the components required to realize a circuit.

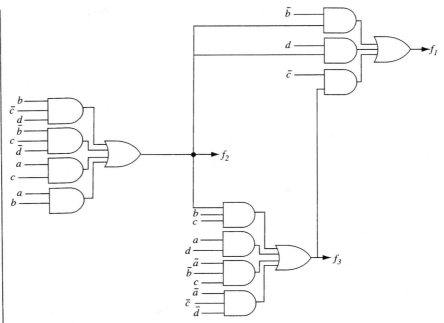

FIGURE 4.35
Multilevel implementation of the simplified expressions

NAND-NAND Logic

All Boolean functions in sum-of-products form can be realized by a two-level logic involving only NAND gates. Consider the following Boolean function in sum-of-products form:

$$f(A,B,C) = \overline{A}\,\overline{B}C + \overline{A}B\overline{C} + \overline{A}BC + A\overline{B}C + AB\overline{C}$$

The first step in the design is to minimize the function. This can be done by algebraic manipulation:

$$\begin{aligned}f(A,B,C) &= \overline{A}\,\overline{B}C + \overline{A}B\overline{C} + \overline{A}BC + A\overline{B}C + AB\overline{C} \\ &= \overline{A}C(\overline{B} + B) + (\overline{A} + A)B\overline{C} + (\overline{A} + A)\overline{B}C \\ &= \overline{A}C + B\overline{C} + \overline{B}C\end{aligned}$$

The minimized function can be directly realized in AND-OR logic as shown in Figure 4.36. The NAND-NAND form of the function can be directly derived from the AND-OR form as illustrated in Figure 4.37. First, two inverter gates in series are inserted at each input of the OR gate (Fig. 4.37a). The output of the circuit remains unchanged in spite of the incorporation of the inverter gates because the output of each AND gate is inverted twice. Then the first level of inverter gates are combined with the AND gates to for NAND gates. The second level of inverter gates are combined with the OR gate to form a NAND gate (since by DeMorgan's theorem $i_1 + i_2 + i_3 = \overline{i_1} \cdot \overline{i_2} \cdot \overline{i_3}$) as shown in Figure 4.37b.

Thus, any Boolean function in sum-of-products form can be implemented by two levels of NAND gates. However, it should be noted that such an implementation is based on the

FIGURE 4.36

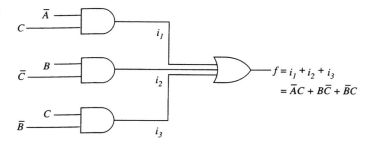

FIGURE 4.37

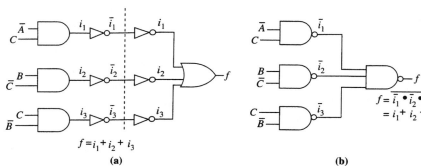

FIGURE 4.38
Three-level NAND gate representation of the circuit of Fig. 4.36

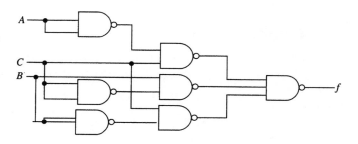

assumption that double-rail inputs are available. **Double-rail inputs** to a circuit indicate that each variable and its complement can be used as inputs to the circuit. If the complements of the variables are not available, the circuit inputs are **single-rail**. Any Boolean function can be realized with three-level NAND gates using only single-rail inputs. For example, the circuit of Figure 4.36 can be implemented in three-level NAND gates as shown in Figure 4.38.

The two-level NAND implementation of a sum-of-products Boolean function can be obtained by using the following steps in sequence:

i. Take the complement of the minimized expression.
ii. Take the complement of the complemented expression. Eliminate the OR operator from the resulting expression by applying DeMorgan's theorem.

■ **EXAMPLE 4.15**

Let us implement the following function using two-level NAND logic

$$f(A,B,C,D) = \Sigma m(2,3,4,6,9,11,12,13)$$

The minimal form of the given function can be derived from its Karnaugh map:

	$\bar{C}\bar{D}$	$\bar{C}D$	CD	$C\bar{D}$
$\bar{A}\bar{B}$			1	1
$\bar{A}B$	1			1
AB	1	1		
$A\bar{B}$		1	1	

$$f = B\bar{C}\bar{D} + A\bar{C}D + \bar{B}CD + \bar{A}C\bar{D}$$

Hence

$$\bar{f} = \overline{B\bar{C}\bar{D} + A\bar{C}D + \bar{B}CD + \bar{A}C\bar{D}}$$

The complement of the expression for $\bar{f}$ is then derived:

$$\bar{\bar{f}} = \overline{\overline{B\bar{C}\bar{D} + A\bar{C}D + \bar{B}CD + \bar{A}C\bar{D}}}$$

$$\therefore \quad f = \overline{\overline{B\bar{C}\bar{D}} \cdot \overline{A\bar{C}D} \cdot \overline{\bar{B}CD} \cdot \overline{\bar{A}C\bar{D}}}$$

The NAND-NAND realization of the above expression is shown in Figure 4.39.

FIGURE 4.39
NAND-NAND realization of
$f = \Sigma m(2,3,4,6,9,11,12,13)$

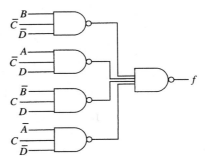

NOR-NOR Logic

All Boolean functions expressed in product-of-sums form can be implemented using two-level NOR logic. Let us assume we have the OR-AND implementation of the Boolean function $f(A,B,C) = (A + C)(A + \bar{B})(\bar{A} + C)$. Figure 4.40 shows the OR-AND implementation. The OR-AND circuit can be converted to the circuit of Figure 4.41a by inserting a cascade of two inverters at each input of the AND gate. Then a NOR-NOR realization of the circuit can be obtained as shown in Figure 4.41b.

The NOR-NOR implementation of a Boolean function expressed in sum-of-products form can be obtained by the following two steps:

i. Derive the complementary sum-of-products version of the original expression.

FIGURE 4.40
OR-AND logic implementation

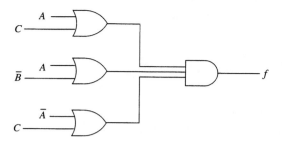

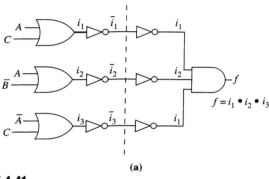

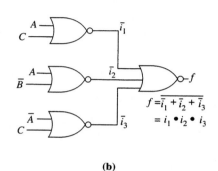

FIGURE 4.41

ii. Take the complement of this complemented sum-of-product expression. Eliminate the AND operators from the resulting expression by using DeMorgan's theorem.

■ **EXAMPLE 4.16**

Let us implement the following Boolean function in NOR-NOR logic
$$f(A,B,C,D) = \overline{A}B + \overline{A}C + A\overline{C}D + A\overline{B}D$$

The complementary sum-of-products expression, $\bar{f}$, can be obtained from the Karnaugh map of the function by grouping the 0's.

	$\overline{C}\overline{D}$	$\overline{C}D$	CD	$C\overline{D}$
$\overline{A}\overline{B}$	0	0	1	1
$\overline{A}B$	1	1	1	1
AB	0	1	0	0
$A\overline{B}$	0	1	1	0

$$\bar{f} = A\overline{D} + \overline{A}\overline{B}\overline{C} + ABC$$

Hence

$$f = \overline{AD + \overline{A}\,\overline{B}\,\overline{C} + ABC}$$
$$= \overline{(\overline{A} + D) + (A + B + C) + (\overline{A} + \overline{B} + \overline{C})}$$

The resulting NOR-NOR logic circuit is shown in Figure 4.42.

FIGURE 4.42
NOR-NOR implementation of $f(A,B,C,D) = \overline{A}B + \overline{A}C + A\overline{C}D + A\overline{B}D$

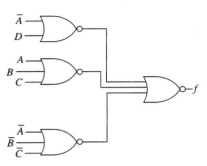

■

*4.8 MULTILEVEL LOGIC DESIGN

Multilevel logic, as the name implies, uses more than two levels of logic to implement a function. Two-level implementation of a function is often difficult to implement at the gate level because of the fan-in restrictions. For example, two-level implementation of the following minimized function,

$$f(u,v,w,x,y,z) = \overline{u}\,\overline{v}\,\overline{w}\,\overline{y}\,\overline{z} + \overline{u}\,\overline{w}xyz + u\overline{v}wx\overline{y}z + uv\overline{w}x\overline{y} + \overline{u}\,\overline{v}w\overline{x}y + \overline{u}\,\overline{x}\,\overline{y}\,\overline{z} + \overline{v}\,\overline{w}\,\overline{x}\,\overline{z} \quad (4.4)$$

will require seven AND gates (one with fan-in of 6, four with fan-in of 5, and two with fan-in of 4), and one OR gate with fan-in of 7. However, by increasing the number of levels in the circuit, the fan-in of the gates can be reduced.

The starting point of the multilevel implementation of a function is the minimized two-level representation of the function. Several operations are used to manipulate this two-level representation; these include **decomposition, extraction, substitution, collapsing,** and **factoring** [4].

Decomposition is the process of representing a single expression as a collection of several subfunctions. For example, the decomposition of the function

$$f = ad + bd + \overline{a}\,\overline{b}c + bc\overline{d} + bce$$

will result in the following subfunctions

$$f = dY + c\overline{Y} + X(\overline{d} + e)$$
$$X = bc$$
$$Y = a + b$$

In general, the decomposition increases the number of levels of a logic circuit while decreasing the fan-in of the gates used to implement the circuit.

The extraction operation creates some intermediate node variables for a given set of functions. These node variables together with the original variables are then used to re-express the given functions. The extraction operation applied to the following functions will yield

$$z_1 = ae + be + c$$
$$z_2 = af + bf + d$$

$$z_1 = Xe + c$$
$$z_2 = Xf + d$$
$$X = a + b$$

where X is a fan-out node.

The substitution process is used to determine whether a given function can be expressed as a function of its original inputs and other functions. As an example, let us consider the functions

$$X = ab + bd + ac + cd$$
$$Y = b + c$$

Substituting Y in X produces

$$X = Y(a + d)$$

This is in fact an example of **algebraic substitution** since $(b + c)$ is an algebraic divisor of X. If the expression is multiplied out, the resulting expression will be identical to the original form.

Another type of substitution is the **Boolean substitution,** which creates logic functions by using Boolean division. Thus, the original and the substituted expression may not have the same form but they are logically equivalent. For example, algebraic substitution does not simplify the following functions

$$X = b + ac + \bar{a}\bar{b}$$
$$Y = b + c$$

However, by using Boolean substitution X can be rewritten as

$$X = (b + c)(a + b) + \bar{a}\bar{b}$$

The inverse operation of substitution is known as **collapsing** or **flattening.** For example, if

$$X = Y(c + d) + e$$
$$Y = a + b$$

then collapsing Y into X results in

$$X = ac + bc + ad + bd + e$$
$$Y = a + b$$

Thus, if Y is an internal node in the circuit, it can be removed.

Factoring is the conversion of a function in the sum-of-products form to a form with parentheses and having a minimum number of literals [4]. A straightforward approach for deriving the factored form of a function from its given two-level representation is to select a literal that is common to the maximum number of product terms. This results in partitioning of the product terms into two sets—one set contains the product terms having the literal, the other containing the rest of the product terms. If the literal is factored out from the first set, a new two-level expression results, which is then ANDed with the literal. Similarly, the second set of product terms is evaluated for a common literal. This process is repeated for the newly generated two-level expressions until they cannot be factored any further. The resulting expression is a multilevel representation of the original two-level form. By using this approach the factored version of the two-level expression (4.4) is derived

$$f(u,v,w,x,y,z) = \bar{y}[\bar{u}\bar{z}(\bar{v}\bar{w} + \bar{x}) + ux(\bar{v}wz + v\bar{w})] + \bar{u}y(\bar{w}xz + \bar{v}w\bar{x}) + \bar{v}\bar{w}\bar{x}\bar{z}$$

Note that the original two-level expression has 34 literals, whereas the factored form has 25 literals. As can be seen from the factored expression, such a representation of a Boolean function automatically leads to the multilevel realization of the function.

In general, a factored-form representation of a two-level function is not unique. For example, the preceding six-variable function can also be represented in the factored form as

$$f(u,v,w,x,y,z) = \bar{z}[\bar{u}\bar{y}(\bar{w}\bar{v} + \bar{x}) + \bar{v}\bar{w}\bar{x}] + xz(\bar{u}\bar{w}y + u\bar{v}wy) + uv\bar{w}x\bar{y} + \bar{u}\bar{v}wxy$$

which has 28 literals. Obviously, only the factored form with the fewest number of literals has to be selected in order to guarantee a minimal multilevel implementation.

Algebraic and Boolean Division

Let us assume two Boolean expressions f and g. If there is an operation which generates expressions h and r such that $f = gh + r$, where gh is an **algebraic product** (i.e., g and h have no common variable), then this operation is called an **algebraic division**. For example, if $f = wy + xy + yz$ and $g = w + x$, a polynomial division will yield

$$f = gh + r = y(w + x) + yz$$

Note that this factored-form representation is algebraically equivalent to the original sum-of-products expression. In other words, if the algebraic factor is expanded, exactly the same set of terms as in f will be obtained.

Another form of division used in factoring Boolean expressions uses the identities of Boolean algebra (e.g., $x\bar{x} = 0$, $xx = x$, and $x + \bar{x} = 1$ for variable x). Thus, if in the expression $f = gh + r$, gh is a **Boolean product** [i.e., g and h have one or more common variable(s)], then the division of f by g is called a **Boolean division**. For example, if $f = abd + bcd + \bar{a}c + \bar{b}d$ and $g = a + c$, the use of Boolean division will yield

$$f = gh + r = (bd + \bar{a})(a + c) + \bar{b}d$$

whereas algebraic division will produce

$$f = gh + r = bd(a + c) + \bar{a}c + \bar{b}d$$

Kernels

The quotient resulting from an algebraic division of an expression f by a cube c (i.e., f/c) is the kernel k of f, if there are at least two cubes in the quotient and the cubes do not have any common literal. The cube divisor c used to obtain the kernel is called its **co-**

kernel. Different cokernels may produce the same kernel; therefore, the cokernel of a kernel is not unique. If a kernel has no kernels except itself, it is said to be a **level-0 kernel.** A kernel is of level n if it has at least one level-$(n-1)$ kernel but no kernel, except itself, of level n or greater.

■ **EXAMPLE 4.17**

Let us consider the Boolean expression

$$f(a,b,c,d) = \bar{a}\bar{c}d + \bar{a}bc + abd + ab\bar{c} + bcd$$

The quotient of f and the cube a is

$$f/a = bd + b\bar{c}$$

but since literal b is common to both cubes, f/a is not a kernel of f.

The quotient of f and the cube $\bar{c}$ is

$$f/\bar{c} = \bar{a}d + ab$$

It has two cubes and no common literal, hence it is a kernel. The cokernel of $f/\bar{c}$ is $\bar{c}$. The kernels and the corresponding cokernels of the function are represented by the tree shown below, where the leaves of the tree are kernels and the branches are the corresponding cokernels. Note that all of these kernels are of level 0.

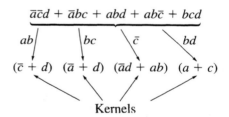

If the original expression is rewritten as

$$f = b[a(\bar{c} + d) + c(\bar{a} + d)] + \bar{c}(\bar{a}d + ab)$$

then $b(a + c) + \bar{a}\bar{c}$ is a kernel of level 1 corresponding to the cokernel d because it contains a level-0 kernel $(a + c)$. ■

As mentioned previously, the cokernel of a kernel is not unique. For example,

$$f = ad + bd + ac + bc$$

has a kernel $(a + b)$ obtained by using cokernels c and d.

Kernels can be used to derive common subexpressions in two Boolean expressions. (The intersection of two kernels k_1 and k_2 is defined as the set of cubes present in both k_1 and k_2.) If there is a kernel intersection of more than one cube, then two Boolean expressions will have common subexpressions of more than one cube.

■ **EXAMPLE 4.18**

Let us consider two Boolean expressions

$$f_1 = abc + a\bar{c}g + \bar{b}df + cde$$
$$f_2 = \bar{a}b\bar{d} + bc\bar{e} + \bar{b}de + \bar{b}\bar{g} + \bar{c}\bar{e}g$$

The kernels of f_1 and f_2 are shown in Table 4.2.

TABLE 4.2

Expression	Cokernel	Kernel
f_1	a	$bc + \bar{c}g$
f_1	c	$ab + de$
f_1	d	$\bar{b}f + ce$
f_2	b	$\bar{a}\bar{d} + c\bar{e}$
f_2	$\bar{b}$	$de + \bar{g}$
f_2	$\bar{e}$	$bc + \bar{c}g$

The kernels of expression f_1 are intersected with those of expression f_2 to find terms that are identical between pairs of kernels. For example, kernel $ab + de$ in f_1 intersects with $de + \bar{g}$ in f_2 to create common terms de. Note that f_1 and f_2 have a common kernel $bc + \bar{c}g$ corresponding to different cokernels a and $\bar{e}$. ■

The selection of the kernel intersection (i.e., a common subexpression) that, once substituted in the given Boolean expression, will result in the minimum number of literals can be considered as a **rectangular covering problem** [5]. Let us explain the rectangular covering formulation through the above Boolean expressions. The expressions are rewritten below, with each cube being uniquely identified by an integer.

$$f_1 = \underset{1}{abc} + \underset{2}{a\bar{c}g} + \underset{3}{\bar{b}df} + \underset{4}{cde}$$
$$f_2 = \underset{5}{\bar{a}b\bar{d}} + \underset{6}{bc\bar{e}} + \underset{7}{\bar{b}de} + \underset{8}{\bar{b}\bar{g}} + \underset{9}{\bar{c}\bar{e}g}$$

First we form the **cokernel cube matrix** for the set of expressions. Such a matrix shows all the kernels simultaneously and allows the detection of kernel intersections. A row in the matrix corresponds to a kernel, whose cokernel is the label for that row. Each column corresponds to a cube, which is the label for that column. The integer identifier of the cube, resulting from the product of the cokernel for row i and the cube for column j, is entered in position (i,j) of the matrix. As can be seen from Table 4.2, the unique cubes from all the kernels of the given equations are bc, $\bar{c}g$, ab, de, $\bar{b}f$, ce, $\bar{a}\bar{d}$, $c\bar{e}$, and $\bar{g}$; these are the labels of the columns of the matrix. There are six kernels; the corresponding cokernels are the labels of the rows of the matrix. Thus, the cokernel matrix for the given expressions is as shown in Table 4.3.

A **rectangle** (R,C), where R and C are sets of rows and columns respectively, is a submatrix of the cokernel cube matrix such that for each row $r_i \in R$ and each column $c_j \in C$, the entry (r_i,c_j) of the cokernel matrix is nonzero. A rectangle that has more than one row indicates a kernel intersection between the kernels corresponding to the rows in the rectangle. The columns in the rectangle identify the cubes of the kernel intersection. For example, in Table 4.2, the rectangle $\{R(1,6),C(1,2)\}$ indicates intersection between the

		1	2	3	4	5	6	7	8	9
		bc	$\bar{c}g$	ab	de	$\bar{b}f$	ce	$\bar{a}\bar{d}$	$c\bar{e}$	$\bar{g}$
1	a	1	2	0	0	0	0	0	0	0
2	c	0	0	1	4	0	0	0	0	0
3	d	0	0	0	0	3	4	0	0	0
4	b	0	0	0	0	0	0	5	6	0
5	$\bar{b}$	0	0	0	7	0	0	0	0	8
6	$\bar{e}$	6	9	0	0	0	0	0	0	0

TABLE 4.3

kernels corresponding to rows 1 and 6. The intersection between these two kernels generates a common subexpression ($bc + \bar{c}g$). A rectangle having more than one column will identify a kernel intersection of more than cube.

A set of rectangles form a **rectangular cover** of a matrix B, if an integer (i.e., a nonzero entry) in B is covered by at least one rectangle from the set. Once an integer is covered, it is not necessary to cover the same integer by any other rectangle; all other appearances of the integer in the matrix can be considered as don't cares. A covering for the above cokernel cube matrix is

$$\{R(1,6), C(1,2)\}, \{R(3), C(5,6)\}, \{R(4), C(7)\}, \{R(5), C(4,9)\}$$

The implementation resulting from this covering is shown in Figure 4.43; the corresponding Boolean expressions are

$$f_1 = aX + dY$$
$$f_2 = \bar{e}X + \bar{b}Z + \bar{a}b\bar{d}$$
$$X = bc + \bar{c}g$$
$$Y = \bar{b}f + ce$$
$$Z = de + \bar{g}$$

This implementation has 22 literals, compared to 26 literals in the original expressions. With larger subexpressions, the reduction in the number of literals is significantly higher.

*4.9 COMBINATIONAL LOGIC IMPLEMENTATION USING EX-OR GATES

Conventional gate-level design normally requires fairly complex interconnections. This is because gates are not very powerful in logic terms. For example a three-input NAND gate can have 8 ($= 2^3$) possible input combinations. However, the gate output changes for only one particular input combination (111). Similarly, any n-input gate can identify only one of the 2^n possible input combinations. This limited logic discrimination capability often leads to the use of a large number of gates to implement a circuit [6].

The EX-OR gate is logically more powerful than conventional gates and often requires fewer interconnections for realizing a logic circuit. A two-input EX-OR gate can

FIGURE 4.43
Multilevel implementation

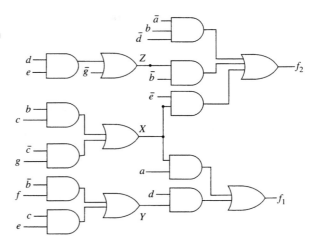

be made from AND, OR, and inverter gates as shown in Figure 3.14. However, the Boolean expression for the EX-OR function can be manipulated algebraically into another form, which can be easily implemented using four two-input NAND gates (e.g., a single SN7401 chip):

$$f(A,B) = A \oplus B = \overline{A}B + A\overline{B}$$
$$= \overline{A}B + B\overline{B} + A\overline{B} + A\overline{A}$$
$$= B(\overline{A} + \overline{B}) + A(\overline{A} + \overline{B})$$
$$= \overline{\overline{(B(\overline{A} + \overline{B}) + A(\overline{A} + \overline{B}))}}$$
$$= \overline{\overline{B(\overline{A} + \overline{B})} \cdot \overline{A(\overline{A} + \overline{B})}}$$
$$= \overline{B \cdot \overline{AB} \cdot A \cdot \overline{AB}}$$

The implementation of this expression using two-input NAND gates is shown in Figure 4.44. SSI chips containing four independent EX-OR gates are also available (e.g., SN7486). There are several rules associated with the EX-OR operations; Table 4.4 lists some of these.

TABLE 4.4
Rules for EX-OR operation

$$X \oplus X = 0$$
$$X \oplus \overline{X} = 1$$
$$1 \oplus X = \overline{X}$$
$$X + Y = X \oplus Y \oplus XY = X \oplus \overline{X}Y$$
$$X(Y \oplus Z) = XY \oplus XZ$$

FIGURE 4.44
NAND gate implementation of the EX-OR function

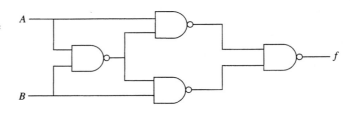

FIGURE 4.45
EX-OR circuit of four variables

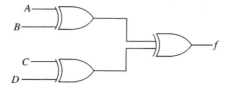

FIGURE 4.46
The EX-OR gate as a programmable inverter

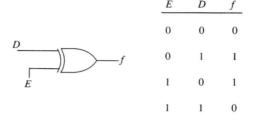

E	D	f
0	0	0
0	1	1
1	0	1
1	1	0

The EX-OR function can be expanded for any number of input variables by cascading EX-OR gates as shown in Figure 4.45. The output expression for the circuit is

$$f(A,B,C,D) = A \oplus B \oplus C \oplus D$$

The output of the circuit is 0 if all inputs are 0 or if the number of inputs at 1 is even; the output is 1 if the number of inputs at 1 is odd. For example, if $A = 0$, $B = 1$, $C = 0$, and $D = 1$, the output is 0 because there is an even number ($= 2$) of high inputs. If $A = 1$, $B = 1$, $C = 0$, and $D = 1$, the output will be 1 because there is an odd number ($= 3$) of high inputs.

An EX-OR gate can also be used as a **programmable inverter** as shown in Figure 4.46. If the control input E is 0, data on the D line is transferred to the output. However, if $E = 1$, the output of the gate is the inverse of the data input. This particular arrangement is widely used in field-programmable devices [7]. The implementation of combinational logic functions using EX-OR gates in certain cases requires fewer gates than are required by more usual AND/OR and NAND/NOR configurations.

■ **EXAMPLE 4.19**

Let us consider the following four-variable function:

$$f(A,B,C,D) = \Sigma m(1,2,4,7,13,14)$$

The Karnaugh map corresponding to this function is shown in Figure 4.47a. The implementation of the function using AND and OR gates and inverters is shown in Figure 4.47b. An alternative implementation is shown in Figure 4.47c.

Conventional Karnaugh maps are not suitable for designing combinational circuits using EX-OR gates. A variation of Karnaugh maps, known as **decomposition charts**, can directly lead to EX-OR implementation [8]. For a function of n variables, the decomposition chart consists of 2^{n-1} subcharts. A subchart is arranged as an array of 2^{n-s} rows and 2^s columns, where $s = \{(n-2),(n-1),n\}$.

*4.9 COMBINATIONAL LOGIC IMPLEMENTATION USING EX-OR GATES   105

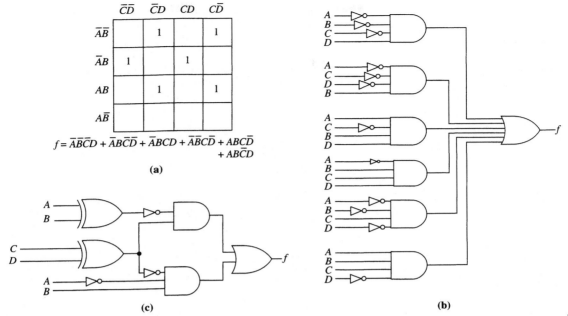

$$f = \overline{AB}\overline{CD} + \overline{AB}\overline{C}\overline{D} + \overline{A}BCD + \overline{A}B\overline{CD} + AB\overline{CD} + AB\overline{C}D$$

FIGURE 4.47

■ | **EXAMPLE 4.20**

Let us consider the four-variable function in Example 4.19. The decomposition subcharts for the function are shown in Figure 4.48(a)-(h). A 1 entry in a subchart corresponds to a minterm in the function. Each subchart is checked vertically as well as horizontally to determine whether it is possible to decompose the function. If the number of distinct columns in a subchart is 2 or fewer, the function is decomposable. The best choice for decomposition is the subchart with the highest number of identical columns.

We need to consider only the two-variable by two-variable subcharts; one-variable by three-variable subcharts can be ignored because three variables cannot be represented by an EX-OR gate. The AB-CD subchart (Fig. 4.48f) has two distinct columns: $\overline{C}D$ is the same as CD, and $\overline{CD}$ is the same as $C\overline{D}$. Hence, the function is decomposable.

Let us designate rows 0 and 2 of the AB-CD subchart as ϕ; consequently row 1 is $\overline{\phi}$ (because it is the complement of row 0), while row 3 is 0. Thus,

$$f(\phi,A,B) = \overline{A}\,\overline{B}\,\phi + \overline{A}B\overline{\phi} + AB\phi$$

and

$$\phi(C,D) = \overline{C}D + C\overline{D}$$

Next, all possible subcharts are formed using ϕ, A, and B as the variables (Fig. 4.49). Since all the subcharts of Figure 4.49 have more than two distinct columns, they cannot be further decomposed. Hence,

ABCD	0000	0001	0010	0011	0100	0101	0110	0111	1000	1001
	0	1	1	0	1	0	0	1	0	0

ABCD	1010	1011	1100	1101	1110	1111
	0	0	0	1	1	0

(a)

ABC	000	001	011	010	110	111	101	100
$\bar{D}$	0	1	0	1	0	1	0	0
D	1	0	1	0	1	0	0	0

(b)

ABD	000	001	011	010	110	111	101	100
$\bar{C}$	0	1	0	1	0	1	0	0
C	1	0	1	0	1	0	0	0

(c)

ACD	000	001	011	010	110	111	101	100
$\bar{B}$	0	1	0	1	0	0	0	0
B	1	0	1	0	1	0	1	0

(d)

BCD	000	001	011	010	110	111	101	100
$\bar{A}$	0	1	0	1	0	1	0	1
A	0	0	0	0	1	0	1	0

(e)

FIGURE 4.48
Decomposition subcharts

*4.9 COMBINATIONAL LOGIC IMPLEMENTATION USING EX-OR GATES   □   107

FIGURE 4.48
(Continued)

	$\overline{CD}$	$\overline{C}D$	CD	$C\overline{D}$
$\overline{A}\overline{B}$	0	1	0	1
$\overline{A}B$	1	0	1	0
AB	0	1	0	1
$A\overline{B}$	0	0	0	0

(f)

	$\overline{B}\overline{D}$	$\overline{B}D$	BD	$B\overline{D}$
$\overline{A}\overline{C}$	0	1	0	1
$\overline{A}C$	1	0	1	0
AC	0	0	0	1
$A\overline{C}$	0	0	1	0

(g)

	$\overline{B}\overline{C}$	$\overline{B}C$	BC	$B\overline{C}$
$\overline{A}\overline{D}$	0	1	0	1
$\overline{A}D$	1	0	1	0
AD	0	0	0	1
$A\overline{D}$	0	0	1	0

(h)

$$f(A,B,C,D) = (\overline{A}\,\overline{B} + AB)\phi + \overline{AB}\overline{\phi}$$
$$= (\overline{A}\,\overline{B} + AB)(\overline{C}D + C\overline{D}) + \overline{AB}(\overline{\overline{C}D + C\overline{D}})$$
$$= \overline{(A \oplus B)}(C \oplus D) + \overline{AB}\overline{(C \oplus D)}$$

This decomposition function has been implemented in Figure 4.47c.

	$\overline{A}\overline{B}$	$\overline{A}B$	AB	$A\overline{B}$
$\overline{\phi}$	0	1	0	0
ϕ	1	0	1	0

(a)

	$\overline{A}\overline{\phi}$	$\overline{A}\phi$	$A\phi$	$A\overline{\phi}$
$\overline{B}$	0	1	0	0
B	1	0	1	0

(b)

	$\overline{B}\overline{\phi}$	$\overline{B}\phi$	$B\phi$	$B\overline{\phi}$
$\overline{A}$	0	1	0	1
A	0	0	1	0

(c)

FIGURE 4.49
Subcharts with variables A, B, and ϕ ■

It is also possible to realize arbitrary combinational functions using AND and EX-OR gates. However, in order to do that, it is first necessary to express the function in a canonical form using AND and EX-OR. For example a canonical sum-of-products expression of two variables

$$f(A,B) = a_0\overline{A}\,\overline{B} + a_1\overline{A}B + a_2A\overline{B} + a_3AB \quad \text{where } a_i = 0 \text{ or } 1$$

can also be represented as

$$f(A,B) = a_0\overline{A}\,\overline{B} \oplus a_1\overline{A}B \oplus a_2A\overline{B} \oplus a_3AB$$

The OR operators can be replaced by EX-OR because at any time only one minterm in a sum-of-products expression can take the value of 1. The third line of Table 4.4 shows that the complemented variables can be replaced: $\bar{A} = 1 \oplus A$, $\bar{B} = 1 \oplus B$. Hence,

$$f(A,B) = a_0(1 \oplus A)(1 \oplus B) \oplus a_1(1 \oplus A)B \oplus a_2A(1 \oplus B) \oplus a_3AB$$
$$= a_0(1 \oplus A \oplus B \oplus AB) \oplus a_1(B \oplus AB) \oplus a_2(A \oplus AB) \oplus a_3AB$$
$$= a_0 \oplus (a_0 \oplus a_2)A \oplus (a_0 \oplus a_1)B \oplus (a_0 \oplus a_1 \oplus a_2 \oplus a_3)\,AB$$

A combinational function expressed as the EX-OR of the products of uncomplemented variables is said to be in **Reed-Muller canonical form**. Thus, the above expression represents the Reed-Muller form of two-variable functions. It can be rewritten as

$$f(A,B) = c_0 \oplus c_1A \oplus c_2B \oplus c_3AB$$

where $c_0 = a_0$, $c_1 = a_0 \oplus a_2$, $c_2 = a_0 \oplus a_1$, and $c_3 = a_0 \oplus a_1 \oplus a_2 \oplus a_3$. In general, any combinational function of n-variables can be expressed in Reed-Muller canonical form

$$f(x_1, x_2, \ldots, x_n) = c_0 \oplus c_1x_1 \oplus c_2x_2 + \cdots + c_nx_n \oplus c_{n+1}\,x_1x_2$$
$$\oplus c_{n+2}\,x_1x_3 \oplus \cdots \oplus c_{2^n-1}\,x_1x_2 \cdots x_n$$

As an example, let us derive the Reed-Muller form of the combinational function

$$f(A,B,C) = A\bar{B} + B(\bar{A} + C)$$
$$= A\bar{B} \oplus B(\bar{A} + C) \oplus A\bar{B} \cdot B(\bar{A} + C) \qquad \text{(by line 4, Table 4.4)}$$
$$= A\bar{B} \oplus B(\bar{A} + C) \oplus 0$$
$$= A\bar{B} \oplus B(\bar{A} \oplus C \oplus \bar{A}C) \qquad \text{(by line 4)}$$
$$= A(1 \oplus B) \oplus B[1 \oplus A \oplus C \oplus (1 \oplus A)C] \qquad \text{(by line 3)}$$
$$= A \oplus AB \oplus B \oplus AB \oplus BC \oplus BC \oplus ABC \qquad \text{(by line 5)}$$
$$= A \oplus B \oplus ABC$$

A direct implementation of the function is shown in Figure 4.50.

It has been shown that circuits realized using AND and EX-OR gates alone are easy to test. Further discussions on the easily testable combinational circuit design can be found in [9].

4.10 LOGIC DESIGN USING MSI CHIPS

So far, we have considered the implementation of combinational logic circuits by using basic gates. In practice, gate-by-gate implementation is rarely used. Instead, integrated circuits (often called **chips**) containing a number of gates are used as building blocks. Chips can be divided into four distinct groups based on the number of gates they contain:

FIGURE 4.50
Implementation of $f = A\bar{B} + B(\bar{A} + C)$

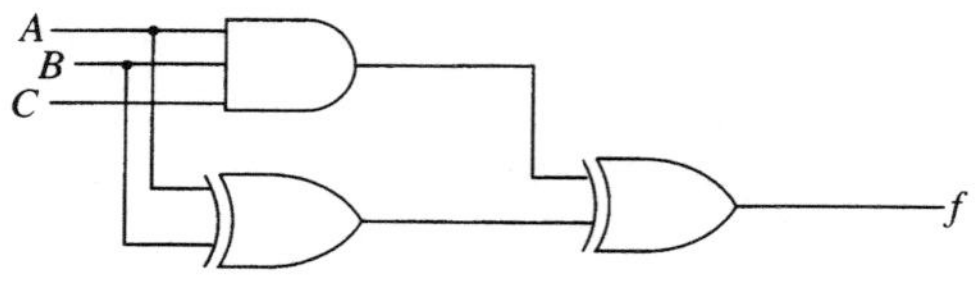

SSI (small scale integrated)
MSI (medium scale integrated)
LSI (large scale integrated)
VLSI (very large scale integrated)

Depending on the fabrication technologies, chips can be classified as transistor-transistor logic (TTL), *n*-channel metal oxide semiconductor (NMOS), complementary metal oxide semiconductor (CMOS), or emitter coupled logic (ECL) devices. We discuss some of the fabrication technologies in Chapter 10. Typical SSI chips contain two to six independent gates. The gates are individually used to implement combinational logic functions. Figure 4.51 shows some common SSI chips. These chips belong to the 7400-series TTL family chips originally developed by Texas Instruments and now marketed by several other semiconductor chip manufacturers.

MSI chips enable logic design at a higher level, where complex functions rather than basic gates are interconnected to implement the required system function. Many standard logic functions (such as encoding, decoding, multiplexing, etc.) can be directly implemented using a single MSI chip. Thus, it is possible to replace a number of SSI chips by a single MSI chip. In general, it is better to utilize standard MSI chips, even if this introduces redundant gates, rather than design optimized logic at the gate level. Fewer chips result in fewer interconnections among them, and there is a consequent decrease in the printed circuit board area occupied by the logic system.

Multiplexers

Multiplexers are typically described as data selectors and are frequently used in digital systems. In a standard multiplexing application, digital signals are connected to the multiplexer's input lines ($I_0, I_1, I_2, \ldots$) and binary control signals are fed to the select lines ($S_0, S_1, \ldots$). Figure 4.52a shows the block diagram of a 4-to-1 multiplexer (i.e., a multiplexer having four input lines—I_0, I_1, I_2, and I_3—and one output line Z). It also has two select lines (S_0 and S_1) and an enable line $\overline{E}$. The signals on the select lines specify which of the four input lines will be gated to the output. For example, if the select lines are $_0S_1 = 00$, the output is I_0; similarly, the select inputs 01, 10, and 11 give outputs I_1, I_2, and I_3 respectively. In order for the multiplexers to operate at all, the $\overline{E}$ line must be set to logic 0; otherwise, the multiplexer output will be 0 regardless of all other input combinations. The operation of the multiplexer can be described by the function table of Figure 4.52b. The implementation of the multiplexer circuit using AND and OR gates and inverters is shown in Figure 4.52c.

Multiplexers are available as catalog devices in quad 2-input (four 2-to-1 multiplexers in one chip), dual 4-input (two 4-to-1 multiplexers in one chip), or single 8- and 16-input devices. However, a multiplexer of any size can be formed by combining several multiplexers in a tree form. Figure 4.53 shows the implementation of a 32-to-1 multiplexer by interconnecting eight 4-to-1 and a single 8-to-1 multiplexers. It is assumed that the enable lines of all the multiplexers are connected to ground (i.e., they are at logic 0). By replacing all 4-to-1 multiplexers by their 8-to-1 counterparts, the circuit of Figure 4.53 can be converted to a 64-input multiplexer circuit without adding any additional delay.

Multiplexers can be used to implement combinational logic functions. For example, a 4-to-1 multiplexer can be used to generate any of the possible functions of three variables.

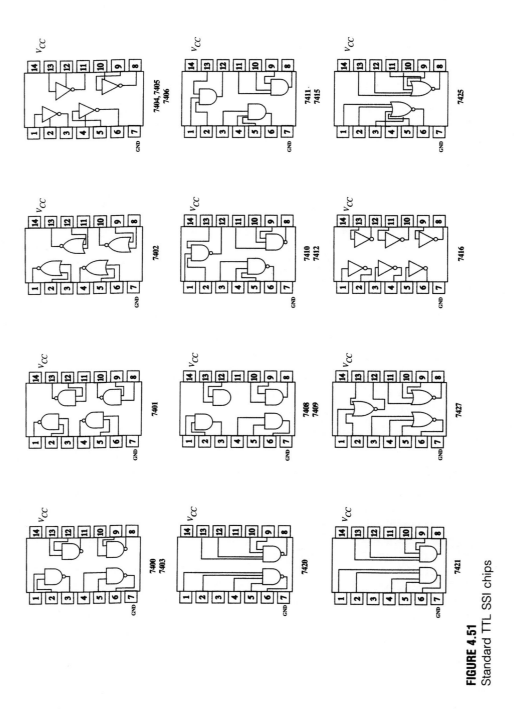

FIGURE 4.51
Standard TTL SSI chips

FIGURE 4.52

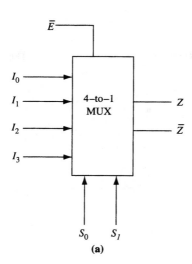

(a)

Enable	Select	Input	Inputs				Outputs	
$\bar{E}$	S_0	S_1	I_0	I_1	I_2	I_3	Z	$\bar{Z}$
1	–	–	–	–	–	–	0	1
0	0	0	0	–	–	–	0	1
0	0	0	1	–	–	–	1	0
0	0	1	–	0	–	–	0	1
0	0	1	–	1	–	–	1	0
0	1	0	–	–	0	–	0	1
0	1	0	–	–	1	–	1	0
0	1	1	–	–	–	0	0	1
0	1	1	–	–	–	1	1	0

(b)

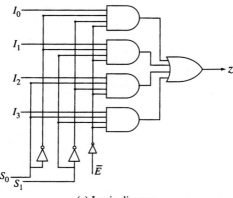

(c) Logic diagram

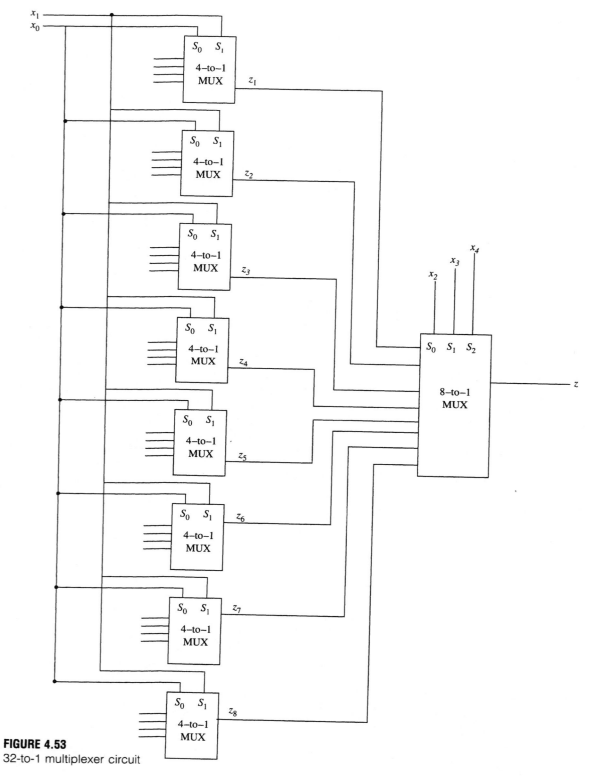

FIGURE 4.53
32-to-1 multiplexer circuit

EXAMPLE 4.21

Let us consider the following function of three variables, which is to be implemented with a 4-to-1 multiplexer

$$f(A,B,C) = BC + \overline{A}B + A\overline{B}\overline{C}$$

The truth table for the function is first derived as shown in Figure 4.54a. Two of three variables (e.g., B and C) are arbitrarily chosen to feed the two select lines of the 4-to-1 multiplexer; the remaining variable A or its complement $\overline{A}$ is to be connected to the input lines I_0, I_1, I_2, and I_3. The truth table shows that when B and C are 0, the output f has the same value as variable A. Hence, A must be connected to I_0. When $B = 0$ and $C = 1$, the truth table indicates that the output f will be 0 irrespective of the variable A. It means that I_1 must be connected to ground (i.e., to 0). For $B = 1$ and $C = 0$, the output f is the complement of A (i.e., $f = 1$ when $A = 0$ and $f = 0$ when $A = 1$). This means that $\overline{A}$ must be connected to I_2. Finally, when $B = 1$ and $C = 1$ the output f will be 1 irrespective of the variable A. Hence, input line I_3 must be connected to 1 (i.e., supply line V_{CC}). Implementing the same function using only NAND gates requires two 3-input NANDs, two 2-input NANDs, and inverters to derive the complement of the input variables, for a total of two SSI chips as shown in Figure 4.55. The multiplexer implementation of the same circuit requires half an MSI chip (SN74153). In general, any n-variable function can be implemented with a 2^{n-1}- to -1 multiplexer by using the technique just described. ∎

The action of the multiplexer circuit may be expressed more formally by noting that any logic function $f(x_1, x_2, \ldots, x_n)$ of n variables can be expanded with respect to any $(n - 1)$ of the n variables:

$$f(x_1, x_2, \ldots, x_{n-2}, x_{n-1}, x_n) = \overline{x}_1\overline{x}_2 \cdots \overline{x}_{n-2}\overline{x}_{n-1} f(0, 0, \ldots, 0, 0, x_n)$$
$$+ \overline{x}_1\overline{x}_2 \cdots \overline{x}_{n-2} x_{n-1} f(0, 0, \ldots 0, 1, x_n) + \overline{x}_1\overline{x}_2 \cdots x_{n-2}\overline{x}_{n-1} f(0, 0, \ldots 1, 0, x_n)$$
$$+ \overline{x}_1\overline{x}_2 \cdots x_{n-2} x_{n-1} f(0, 0, \ldots 1, 1, x_n)$$
$$\vdots$$
$$+ x_1 x_2 \cdots x_{n-2} x_{n-1} f(1, 1, \ldots 1, 1, x_n)$$

FIGURE 4.54

A	B	C	f
0	0	0	0
0	0	1	0
0	1	0	1
0	1	1	1
1	0	0	1
1	0	1	0
1	1	0	0
1	1	1	1

(a) Truth table

(b) Multiplexer implementation of $f(A,B,C) = BC + \overline{A}B + A\overline{B}\overline{C}$

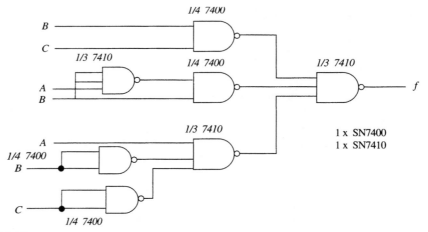

FIGURE 4.55
SSI implementation of $f(A,B,C) = BC + \overline{A}B + AB\overline{C}$

In this expression each of the $f(i_1, i_2, ..., i_{n-1}, x_n)$, where $i_i = 0$ or 1, is a **residue** function of a single variable x_n and assumes one of the four values 0, 1, x_n, or $\bar{x}_n$. The equation describes a 2^{n-1}-to-1 multiplexer, where the variable x_i is connected to the select line S_i and the quantity that represents the function $f(i_1, i_2, ..., i_{n-1}, x_n)$ is connected to the input line I_j (where j is the decimal equivalent of $i_1 i_2 \cdots i_{n-1}$).

■ **EXAMPLE 4.22**

Let us implement the following function of five variables

$$f(A,B,C,D,E) = \overline{A}\,\overline{B}\,\overline{C} + A\overline{B}\,\overline{C}D + \overline{A}CD + AB\overline{C}D + AE + \overline{B}C\overline{E} + \overline{C}E$$

Since $n = 5$, the function can be realized with a 16-to-1 multiplexer. We expand the function with respect to A, B, C, and D:

$$f(A,B,C,D,E) = \overline{A}\,\overline{B}\,\overline{C}\,\overline{D}f_0 + \overline{A}\,\overline{B}\,\overline{C}Df_1 + \overline{A}\,\overline{B}C\overline{D}f_2 + \overline{A}\,\overline{B}CDf_3 + \cdots + ABCDf_{15}$$

where $f_0 = f(0,0,0,0,E) = 1, f_1 = f(0,0,0,1,E) = 1, f_2 = f(0,0,1,0,E) = \overline{E}, f_3 = \overline{E}, f_4 = E, f_5 = 1, f_6 = 0, f_7 = 0, f_8 = 1, f_9 = E, f_{10} = 1, f_{11} = 1, f_{12} = E, f_{13} = 1, f_{14} = E$, and $f_{15} = E$. Figure 4.56 shows the implementation of the above expression where A, B, C, and D are control inputs, and E is the multiplexed variable. The information obtained by expanding the function about A, B, C, and D can also be derived by representing it in a tabular form as shown in Table 4.5.

If both the entries in a column have the same value, then this is the value of the corresponding residue function $f(i_1 = A, i_2 = B, i_3 = C, i_4 = D, E)$. If the entries in a column are different but are the same as the row numbers, then the corresponding residue function has the value E. However, if the entries in the column are opposite to the row numbers, then the residue function has the value $\overline{E}$. These values are also shown in Table 4.5.

E	$\bar{A}\bar{B}\bar{C}\bar{D}$	$\bar{A}\bar{B}\bar{C}D$	$\bar{A}\bar{B}C\bar{D}$	$\bar{A}\bar{B}CD$	$\bar{A}B\bar{C}\bar{D}$	$\bar{A}B\bar{C}D$	$\bar{A}BC\bar{D}$	$\bar{A}BCD$	$A\bar{B}\bar{C}\bar{D}$	$A\bar{B}\bar{C}D$	$A\bar{B}C\bar{D}$	$A\bar{B}CD$	$AB\bar{C}\bar{D}$	$AB\bar{C}D$	$ABC\bar{D}$	$ABCD$
0	1	1	1	1	0	1	0	0	1	0	1	1	0	1	0	0
1	1	0	0	1	1	0	0	0	1	1	1	1	1	1	1	1
$f(i_1=A,$ $i_2=B,$ $i_3=C,$ $i_4=D,$ $E)$	1	$\bar{E}$	$\bar{E}$	1	E	E	0	0	1	E	1	1	E	1	E	E

TABLE 4.5
Selection of control inputs and the multiplexed variable

FIGURE 4.56

Multiplexer implementation of the function
$f(A,B,C,D,E) = \overline{A}\overline{B}\overline{C} + AB\overline{C}\overline{D} + \overline{A}\overline{C}D + AB\overline{C}D + AE + \overline{B}C\overline{E} + \overline{C}E$

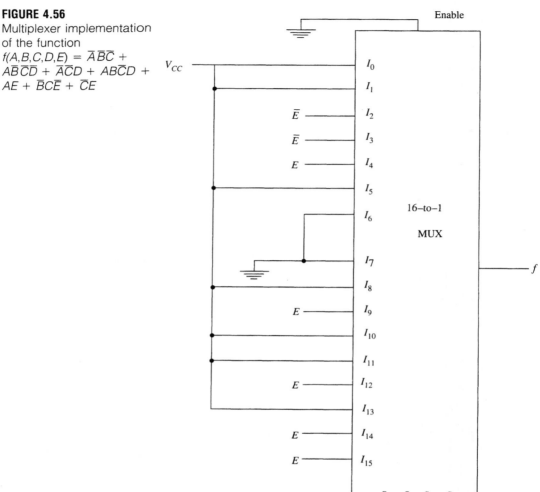

In general, a multiplexer with S select lines can realize a function of $S + 1$ variables. Many functions of $S + 1$ variables can be realized with a smaller multiplexer (i.e., a multiplexer with fewer than S select lines). It has been shown in [10] that an n-variable function may be implemented with a multiplexer having S select lines if the number of minterms corresponding to the function is a multiple of 2^{n-S-1}. For example, the function $f(A,B,C,D) = \overline{A}\overline{B}\overline{C} + \overline{A}D + BC\overline{D}$ cannot be realized with a 4-to-1 multiplexer. This is because for a 4-to-1 multiplexer, $S = 2$ and the number of minterms (=7) in the function ($n = 4$) is not a multiple of 2^{n-S-1} (=2).

If this condition is satisfied (i.e., an n-variable function can be realized with a multiplexer of S select lines), then the total number of 1's(0's) in the **minterm table** for a column corresponding to a select variable should be a multiple of 2^{n-S-1}. Thus, a function is realizable with a multiplexer having S select lines, provided there are at least S columns in the minterm table in which the number of 1's(0's) is a multiple of 2^{n-S-1}.

EXAMPLE 4.23

Let us consider the function

$$f(A,B,C,D) = A\overline{C}\overline{D} + \overline{B}D$$

It satisfies the necessary condition for realization with a 4-to-1 multiplexer because the number of minterms (=6) is a multiple of 2^{4-2-1} (=2). The minterm table for the function is shown in Table 4.6. It can be seen from the table that the number of 1's in columns A, C, and D is a multiple of 2.

TABLE 4.6
Minterm table for $f(A,B,C,D) = A\overline{C}\overline{D} + \overline{B}D$

A	B	C	D	f
1	0	0	0	1
1	1	0	0	1
0	0	0	1	1
0	0	1	1	1
1	0	0	1	1
1	0	1	1	1

If a function satisfies the required conditions for realizibility with a multiplexer having S select lines, it is expanded around the variables corresponding to the S columns. Table 4.7 shows the minterm tables of the reduced functions obtained by expanding $f(A,B,C,D) = A\overline{C}\overline{D} + \overline{B}D$ around the variables (A,C), (A,D), and (C,D).

If each of the reduced function obtained by expanding around the S variables satisfies one of the following two conditions, then the given function is realizable with 2^S-to-1 multiplexers:

i. The value of reduced function f_i is dependent only on the expanding variables (i.e., $f_i = 0$ or 1).
ii. The value of reduced function f_i is dependent on a single nonexpanding variable.

As can be seen from Table 4.7, each of the reduced functions obtained by expanding the given function around variables (A, D) satisfies one of the two conditions. By connecting the reduced function values and the expanding variables to the input lines and the select lines respectively of the 4-to-1 multiplexer, the desired function can be directly implemented as shown in Figure 4.57a. Similarly, the expanded function around the variables (C, D) can also be implemented with a 4-to-1 multiplexer as shown in Figure 4.57b.

Demultiplexers and Decoders

A **demultiplexer** performs the function opposite to that of a multiplexer. It is used to route data on a single input to one of several outputs which is determined by the choice of signals on the address lines. Figure 4.58a shows the block diagram of a demultiplexer with four output lines (D_0, D_1, D_2, and D_3), one input line (I), and two address lines (A_0, A_1). The operation of the demultiplexer can be described by the function table shown in

i. Expanding variables A, C

B D f_0	B D f_1	B D f_2	B D f_3
0 1 1	0 1 1	0 0 1	0 1 1
		0 1 1	
		1 0 1	
(a) $f_0(0,B,0,D)$	(b) $f_0(0,B,1,D)$	(c) $f_0(1,B,0,D)$	(d) $f_0(1,B,1,D)$
$= \overline{B}D$	$= \overline{B}D$	$= \overline{B} + \overline{D}$	$= \overline{B}D$

$$f(A,B,C,D) = \overline{A}\,\overline{C}(\overline{B}D) + \overline{A}C(\overline{B}D) + A\overline{C}(\overline{B} + \overline{D}) + AC \cdot \overline{B}D$$

ii. Expanding variables A, D

B C f_0	B C f_1	B C f_2	B C f_3
— — —	0 0 1	0 0 1	0 0 1
	0 1 1	1 0 1	0 1 1
(a) $f_0(0,B,C,0)$	(b) $f_0(0,B,C,1)$	(c) $f_0(1,B,C,0)$	(d) $f_0(1,B,C,1)$
$= 0$	$= \overline{B}$	$= \overline{C}$	$= \overline{B}$

$$f(A,B,C,D) = \overline{A}\,\overline{D} \cdot 0 + \overline{A}D \cdot \overline{B} + A\overline{D} \cdot \overline{C} + AD \cdot \overline{B}$$

iii. Expanding variables C, D

A B f_0	A B f_1	A B f_2	A B f_3
1 0 1	0 0 1	— — —	0 0 1
1 1 1	1 0 1		1 0 1
(a) $f_0(A,B,0,0)$	(b) $f_0(A,B,0,1)$	(c) $f_0(A,B,1,0)$	(d) $f_0(A,B,1,1)$
$= A$	$= \overline{B}$	$= 0$	$= \overline{B}$

$$f(A,B,C,D) = \overline{C}\,\overline{D} \cdot A + \overline{C}D \cdot \overline{B} + C\overline{D} \cdot 0 + CD \cdot \overline{B}$$

TABLE 4.7
Expansion of the function around select variables

FIGURE 4.57

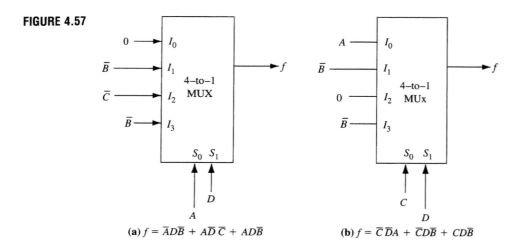

(a) $f = \overline{A}D\overline{B} + A\overline{D}\,\overline{C} + AD\overline{B}$
(b) $f = \overline{C}\,\overline{D}A + \overline{C}D\overline{B} + CD\overline{B}$

4.10 LOGIC DESIGN USING MSI CHIPS □ 119

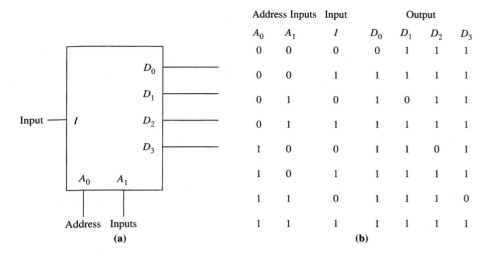

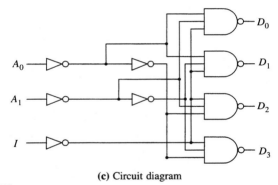

(c) Circuit diagram

FIGURE 4.58
1-to-4 demultiplexer

Figure 4.58b. The implementation of the demultiplexer using NANDs and inverters is shown in Figure 4.58c.

A **decoder** produces a unique output corresponding to each input pattern. If the input line in Figure 4.58c is set to logic 0 (i.e., $I = 0$), the 1-to-4 demultiplexer will act as a decoder. For example, $A_0 = 0$, $A_1 = 0$ will give an output of 0 on line D_0; lines D_1, D_2, and D_3 will be at logic 1. Similarly, $A_0 = 0$, $A_1 = 1$ will give an output of 0 on line D_1; $A_0 = 1$, $A_1 = 0$ will give an output of 0 on line D_2; and $A_0 = 1$, $A_1 = 1$ will give an output of 0 on line D_3. Thus, the 1-to-4 demultiplexer can function as a 2-to-4 decoder, allowing each of the four possible combinations of the input signals A_0 and A_1 to appear on the selected output line. Decoders are available as catalog devices in dual 2-to-4 and single 3-to-8 or 4-to-16 devices. Furthermore, 4-to-10 decoders for decoding BCD numbers are also available.

A decoder of any size can be made up by interconnecting several small decoders. For example, a 6-to-64 decoder can be constructed from four 4-to-16 decoders and a 2-to-4 decoder as shown in Figure 4.59. The 2-to-4 decoder enables one of the four 4-to-

FIGURE 4.59
6-to-64 decoding circuit

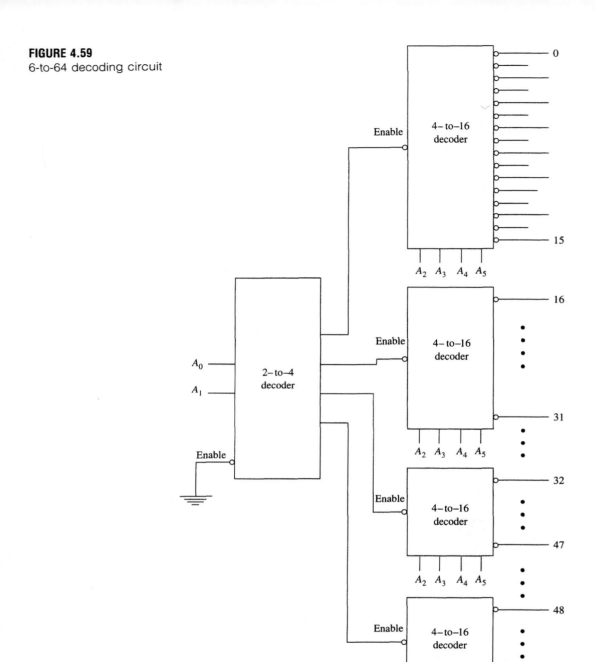

16 decoders, depending on the address bits A_0 and A_1. Address bits A_2, A_3, A_4, and A_5 determine which output of the enabled 4-to-16 decoder goes low. We thus have one of the 64 outputs going low, as selected by the 6-bit address.

Like a multiplexer, a decoder can also be used to implement any arbitrary Boolean functions. As indicated in the previous section, a decoder generates all the product terms of its input variables. Thus, by connecting the outputs of the decoder corresponding to the canonical sum-of-products expression to an output NAND gate, any Boolean function can be realized.

EXAMPLE 4.24

Let us implement the following function using a 3-to-8 decoder (SN74LS138)

$$f_1(A,B,C) = AB + \overline{A}\,\overline{B}\,\overline{C}$$
$$f_2(A,B,C) = \overline{A}B + A\overline{B}$$

First of all the functions must be expressed in canonical form,

$$f_1(A,B,C) = \underbrace{ABC}_{7} + \underbrace{AB\overline{C}}_{6} + \underbrace{\overline{A}\,\overline{B}\,\overline{C}}_{0}$$

$$f_2(A,B,C) = \underbrace{\overline{A}BC}_{3} + \underbrace{\overline{A}B\overline{C}}_{2} + \underbrace{A\overline{B}\,\overline{C}}_{4} + \underbrace{A\overline{B}C}_{5}$$

The input variables A, B, and C are connected to address inputs of the decoder; the outputs of the decoder corresponding to the minterms of the given functions are then fed to the inputs of the two NAND gates. The resulting circuit is shown in Figure 4.60.

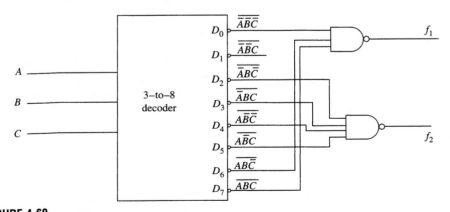

FIGURE 4.60
Implementation of $f = AB + \overline{A}\,\overline{B}\,\overline{C}$ and $f = \overline{A}B + A\overline{B}$ using a 3-to-8 decoder and NAND gates

4.11 COMBINATIONAL CIRCUIT DESIGN USING PLDs

Programmable logic devices, popularly known as PLDs, have a general architecture which a user can customize by inducing physical changes in selected parts. Thus, these devices

FIGURE 4.61
Generic PLD structure

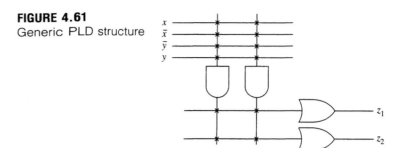

can be configured to be application specific by utilizing their user programmable features. The actual programming of such devices can be done by a piece of equipment called PLD Programmer. The process takes only a few seconds.

Most PLDs consist of an AND array followed by an OR array. The inputs to a PLD enter the AND array through the buffers, which generate both the true and the complement of the input signals. Each gate in the AND array generates a minterm of the input variables; a minterm is known as a product line in the programmable logic device nomenclature. The device outputs are produced by summing the product terms in the array of OR gates.

Figure 4.61 shows the generic structure of a PLD with two inputs and two outputs. As can be seen from the diagram, there are programmable connections between the input lines and the product lines, as well as between product lines and sum lines; such connections are known as **crosspoints**.

In PLDs manufactured using bipolar technology the crosspoints are implemented using **fuses**, which break open when current flowing through them exceeds a certain limit. A major drawback of PLDs based on bipolar technology is that the programming of the fuses is irreversible (i.e., a device cannot be reprogrammed). On the other hand, PLDs manufactured using CMOS technology use memory cells as crosspoints. These cells are reprogrammable; hence, the function of an already programmed device can be altered by merely changing the contents of the memory cells.

The AND-OR structured programmable logic devices can be grouped into three basic types:

Programmable read-only memory (PROM)

Field-programmable logic array (FPLA)

Programmable array logic (PAL)

In a PROM, the AND array is fixed and the OR array is programmable. In an FPLA, both arrays are programmable. A PAL device is the mirror image of the PROM; its AND array is programmable, while the OR array is fixed.

PROM. Traditionally PROMs have been used to store software in microprocessor-based systems. However, they can also be used to implement logic functions. The major advantage of PROM-based logic functions is that there is no need to employ any of the conventional minimization techniques. The input variables in a combinational function are used as inputs to a PROM with the required output values being stored in the location corresponding to the input combination. It will be obvious that a PROM can be used for realizing multi-output combinational circuits, each bit in the PROM output corresponding to a particular output value of the combinational circuit.

To illustrate, we implement a three-output combinational function of four variables (w,x,y,z). A 1 programmed in the PROM represents the presence of a minterm in the function, and a 0 its absence. Thus, the input combination 1001 applied to the PROM will be interpreted as the minterm $w\bar{x}\bar{y}z$; in the output of the OR gate driven by the minterm, the 1's correspond to functions containing that minterm and 0's to functions that do not.

$$f_1(w,x,y,z) = \bar{w}\bar{x}\bar{y}\bar{z} + \bar{w}\bar{x}yz + wxyz + \bar{w}\bar{x}\bar{y}\bar{z}$$
$$f_2(w,x,y,z) = w\bar{x}y\bar{z} + wxyz + \bar{w}x\bar{y}z$$
$$f_3(w,x,y,z) = \bar{w}xy\bar{z} + w\bar{x}\bar{y}z$$

These functions can be implemented by programming a PROM as shown:

w	x	y	z	f_1	f_2	f_3
0	0	0	0	1	0	0
0	0	0	1	0	0	0
0	0	1	0	0	0	0
0	0	1	1	1	0	0
0	1	0	0	0	0	0
0	1	0	1	1	1	0
0	1	1	0	0	0	1
0	1	1	1	0	0	0
1	0	0	0	0	0	0
1	0	0	1	0	0	1
1	0	1	0	0	1	0
1	0	1	1	0	0	0
1	1	0	0	0	0	0
1	1	0	1	0	0	0
1	1	1	0	0	0	0
1	1	1	1	1	1	0

Note that in employing PROMs to implement logic functions, the minterms are programmed directly from the truth table for each of the functions; the minimization process is bypassed, since this does not result in any savings. In fact if an expression is already reduced it must be expanded to its canonical form in order to properly specify the PROM program.

■ **EXAMPLE 4.25**

Let us implement a multiplier of two 2-bit numbers a_1a_0 and b_1b_0 using a PROM. Since the multiplier uses two 2-bit numbers, a PROM with 4 inputs is needed. The product of two such numbers will also contain 4 ($= 2 + 2$) bits, so the PROM will require 4 output lines. Each address to the PROM will consist of 4 bits, and the content corresponding to the address will be a 4-bit number that is the product of the first and the second halves of the address bits. The program table for the PROM is

b_1	b_0	a_1	a_0	o_8	o_4	o_2	o_1
0	0	0	0	0	0	0	0
0	0	0	1	0	0	0	0
0	0	1	0	0	0	0	0
0	0	1	1	0	0	0	0
0	1	0	0	0	0	0	0
0	1	0	1	0	0	0	1
0	1	1	0	0	0	1	0
0	1	1	1	0	0	1	1
1	0	0	0	0	0	0	0
1	0	0	1	0	0	1	0
1	0	1	0	0	1	0	0
1	0	1	1	0	1	1	0
1	1	0	0	0	0	0	0
1	1	0	1	0	0	1	1
1	1	1	0	0	1	1	0
1	1	1	1	1	0	0	1	∎

FPLA. The FPLA structure offers a high level of flexibility because both the AND array and the OR array are programmable. Since the AND array is programmable, FPLAs do not suffer from the limitation of PROMs that the AND array must provide all possible input combinations. Since both the arrays are user programmable, it is possible for an OR gate to access any number of product terms. Moreover, all OR gates can access the same product term(s) simultaneously. In other words, product sharing does not require any additional resources in an FPLA architecture. It should be noted that an FPLA is a special type of PLA (generic name). In FPLAs every connection is made through a fuse at every intersection point; the undesired connections can be removed later by blowing the fuses. Alternatively, the desired connections can be made during the chip fabrication according to a particular interconnection pattern; these types of PLAs are mostly embedded in VLSI chips.

FPLAs are configured by the number of input variables (I), output functions (O), and product terms (P): $I \times P \times O$. For example, Signetics PLS100 is a $16 \times 48 \times 8$ device (i.e., it has 16 input lines, 48 product terms, and 8 OR gates). Figure 4.62 shows the logic diagram of the device. The 16 inputs are each internally buffered and also inverted so as to provide true or complemented values, so there are $2 \times 16 = 32$ inputs to the 48 AND gates (32×48 input arrays). Product terms are formed by appropriate ANDing of any combination of the 16 input variables and their complements. Each of the 8 OR gates can provide a sum of any or all of the 48 product terms. In addition each output function can be individually programmed true or complement. This is accomplished by an EX-OR gate on each output, one input of which is connected to the ground via a fuse so that the sum-of-products function is not inverted (Fig. 4.63). If the fuse is blown, this input is forced to be at logic 1, thus providing the complement of the sum-of-products function. The programmable inversion has the added advantage in that if the complement of a sum-of-products function is simpler to realize than the original output function, then the complemented function may be programmed into the AND array and an inversion of it is obtained at the output.

FIGURE 4.62
Logic diagram of PLS100

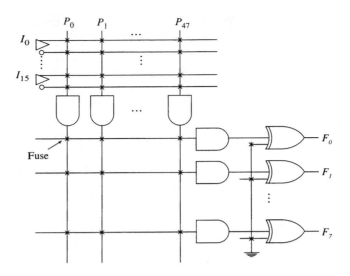

FIGURE 4.63
Programmable output inversion

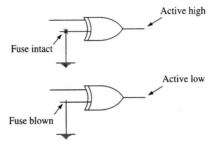

■ **EXAMPLE 4.26**

Let us consider the implementation of the following Boolean expressions using an FPLA

$$f_1(w,x,y,z) = \overline{w}xyz + w\overline{y}\overline{z} + \overline{x}yz$$
$$f_2(w,x,y,z) = w + x + y$$
$$f_3(w,x,y,z) = \overline{w}x + w\overline{x} + \overline{y}z + y\overline{z}$$

Expression f_2 can be rewritten as

$$f_2(w,x,y,z) = \overline{\overline{w}\,\overline{x}\,\overline{y}}$$

which requires only a single product term. Figure 4.64 shows the appropriately programmed logic diagram for the FPLA. ■

■ **EXAMPLE 4.27**

Let us use an FPLA to implement a circuit that receives 8-bit input patterns and indicates whether such a pattern contains all 0's, a single 1, or multiple 1's. We will implement the

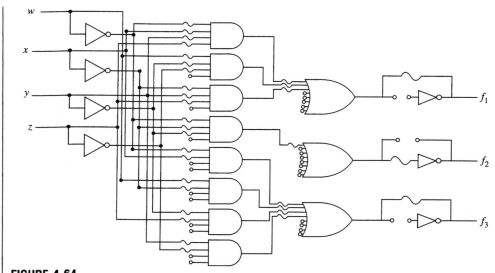

FIGURE 4.64
Logic diagram of the programmed FPLA

circuit using a PLS153 (18 inputs, 10 outputs, and 42 product terms). We need to use 8 of the device's inputs and 3 of the outputs. Let us identify the inputs as $a, b, c, d, e, f, g,$ and h, and the three outputs as W, X, and Y, which correspond to an all 0's input pattern, a single 1 in a pattern, or multiple 1's in a pattern respectively. Figure 4.65 shows the truth table of the circuit. After minimization, the Boolean expressions for the circuit can be rewritten as

$$W = \bar{a}\bar{b}\bar{c}\bar{d}\bar{e}\bar{f}\bar{g}\bar{h}$$
$$X = (a + b + c + d + e + f + g + h)(\bar{e}\bar{f} + \bar{a}\bar{b}\bar{c}\bar{d}\bar{g}\bar{h})(\bar{g}\bar{h} + \bar{a}\bar{b}\bar{c}\bar{d})$$
$$(\bar{b}\bar{c} + \bar{a}\bar{d})(\bar{g} + \bar{h})(\bar{e} + \bar{f})(\bar{a} + \bar{d})(\bar{b} + \bar{c})$$
$$Y = (e + f)(a + b + c + d + g + h) + (g + h)(a + b + c + d)$$
$$+ (b + c)(a + d) + gh + ef + ad + bc$$

These expressions can be converted into sum-of-products form having 37 unique product terms. Since the number of product terms does not exceed the maximum available in PLS153, these expressions can be implemented using this device.

a	b	c	d	e	f	g	h	W	X	Y
0	0	0	0	0	0	0	0	1	0	0
1	0	0	0	0	0	0	0	0	1	0
0	1	0	0	0	0	0	0	0	1	0
0	0	1	0	0	0	0	0	0	1	0
0	0	0	1	0	0	0	0	0	1	0
0	0	0	0	1	0	0	0	0	1	0
0	0	0	0	0	1	0	0	0	1	0

0	0	0	0	0	0	1	0	0	1	0
0	0	0	0	0	0	0	1	0	1	0
←———— All other combinations ————→								0	0	1

FIGURE 4.65
Truth table

A major problem in the FPLA implementation of a combinational logic circuit is to ensure that the number of product terms in the Boolean expressions for the circuit will not exceed the number of product terms available in the FPLA. Therefore, minimization of multi-output Boolean functions is extremely important in the FPLA implementation of combinational circuits (see Sec. 4.6).

To illustrate, let us consider the truth table of a circuit shown in Figure 4.66.* This circuit has 5 inputs and 3 outputs. By using the two-level minimizer ESPRESSO, the following equations are obtained:

$$W = abde + abcd + bcde + acde + abce$$

$$X = a\bar{b}\bar{c}d\bar{e} + \bar{a}\bar{b}\bar{c}de + a\bar{b}c\bar{d}\bar{e} + \bar{a}\bar{b}cd\bar{e} + \bar{a}\bar{b}cd\bar{e} + a\bar{b}cd\bar{e} + ab\bar{c}d\bar{e} + \bar{a}b\bar{c}de$$
$$+ a\bar{b}cd\bar{e} + \bar{a}bcd\bar{e} + \bar{a}bcd\bar{e} + abcde + ab\bar{c}\bar{d}e + a\bar{b}\bar{c}de + abc\bar{d}\bar{e} + \bar{a}b\bar{c}de$$

$$Y = a\bar{b}\bar{d}e + ab\bar{c}\bar{d}e + a\bar{b}\bar{c}de + ab\bar{c}\bar{e} + \bar{a}\bar{c}de + a\bar{b}d\bar{e} + a\bar{b}c\bar{e} + \bar{a}cd\bar{e} + \bar{a}b\bar{d}e$$
$$+ abc\bar{d}\bar{e} + \bar{a}\bar{b}cde + \bar{a}cd\bar{e} + \bar{a}bd\bar{e} + \bar{a}bc\bar{e}$$

Output expressions X and Y have four shared product terms, $ab\bar{c}\bar{d}e$, $a\bar{b}\bar{c}de$, $abc\bar{d}\bar{e}$, and $\bar{a}\bar{b}cde$. Thus, the total number of unique product terms needed to implement the circuit is 31. Note that multi-output minimization often leads to sharing of product terms, which may not be obvious from the truth table. Since the capacity of a PLS153 is 42 product terms, the circuit under consideration easily fits into the device. In general, determining whether a design will fit into a particular PLD requires sophisticated software capability.

FIGURE 4.66
A truth table

a	b	c	d	e	W	X	Y
1	—	1	1	1	1	—	—
1	1	—	1	1	1	—	—
1	1	1	1	—	1	—	—
1	1	1	—	1	1	—	—
—	1	1	1	1	1	—	—
0	1	—	0	1	—	—	1
—	0	1	1	0	—	—	1
0	0	1	—	1	—	—	1
1	—	0	0	1	—	—	1
1	—	1	0	0	—	—	1
1	1	0	—	0	—	—	1
0	1	1	—	0	—	—	1
1	0	0	1	—	—	—	1
0	—	0	1	1	—	—	1
—	1	0	1	0	—	—	1

---

*Circuit rd53, an MCNC (Microelectronics Center of North Carolina) benchmark circuit.

FIGURE 4.66
(Continued)

—	0	1	0	1	—	—	1	
0	1	1	1	0	—	1	—	
0	0	0	1	0	—	1	—	
0	1	0	0	0	—	1	—	
1	1	1	1	1	—	1	—	
0	0	1	0	0	—	1	—	
0	0	1	1	1	—	1	—	
1	1	1	0	0	—	1	—	
1	1	0	1	0	—	1	—	
0	1	1	0	1	—	1	—	
0	1	0	1	1	—	1	—	
1	0	1	1	0	—	1	—	
1	0	0	0	0	—	1	—	
1	1	0	0	1	—	1	—	
0	0	0	0	1	—	1	—	
1	0	1	0	1	—	1	—	
1	0	0	1	1	—	1	—	

PAL. The basic PAL architecture is exactly opposite to that of a PROM. It is comprised of a programmable AND array and a fixed OR array. The programmability in the AND array removes one of the main deficiencies in PROMs—that the AND plane must be large enough to produce product terms corresponding to all possible input combinations. Thus, as in FPLAs, only the desired input combinations have to be programmed. Moreover, logic minimization techniques can be employed to further reduce the required number of product terms. However, since the OR array is not programmable, only a fixed number of product terms, typically eight, can drive a specific OR gate.

PAL devices were introduced by Monolithic Memories, now part of Advanced Micro Devices Inc. The first family of PAL devices was aimed at substitutions of SSI/MSI chips, rather than replacing complex blocks of logic on a circuit board. Some of these devices have feedback paths from the outputs back to the AND array. Some devices have programmable I/O pins (i.e., the pins can be configured to function either as inputs or outputs).

PAL devices that are specifically aimed for implementing combinational logic functions include

10H8	10L8	16P8
12H6	10L6	16L8
14H4	14L4	18P8
16H2	16L2	20C1
16C1	16H8	20L2

The first two digits in a device number specify the number of inputs, the last digit the number of outputs. Letters **H** and **L** indicate that the corresponding device has an active-high or active-low output respectively. The letter **C** denotes that both the true output and its complement are available. The letter **P** indicates that each output in a device can be programmed active-high or active-low.

The devices with active-high outputs can implement AND-OR logic, and the devices with active-low outputs can implement AND-NOR logic. Figures 4.67a and b show

FIGURE 4.67
Structure of active-high/active-low output PALs

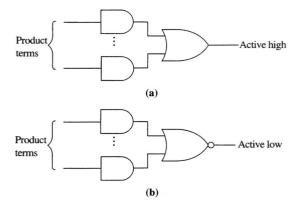

the logic for a cell from an active-high and an active-low PAL respectively. Any combinational function represented by NAND-NAND, OR-NAND, or NOR-OR logic can be replaced by a PAL with active-high output. Similarly, a function in NAND-AND, OR-AND, or NOR-NOR form can be realized by a PAL with active-low outputs.

■ **EXAMPLE 4.28**

Let us illustrate the application of combinational PAL devices in logic design by using such a device to implement a circuit with four inputs (a,b,c,d) and two outputs (X,Y) [11]. One output (X) is high when the majority of the inputs are high and low at other times. The remaining output (Y) is high only during a tie (i.e., when two inputs are high and two are low). Figure 4.68 shows the truth table for the circuit.

FIGURE 4.68
Truth table for a 4-input and 2-output combinational logic circuit

a	b	c	d	X	Y
0	0	0	0	0	0
0	0	0	1	0	0
0	0	1	0	0	0
0	0	1	1	0	1
0	1	0	0	0	0
0	1	0	1	0	1
0	1	1	0	0	1
0	1	1	1	1	0
1	0	0	0	0	0
1	0	0	1	0	1
1	0	1	0	0	1
1	0	1	1	1	0
1	1	0	0	0	1
1	1	0	1	1	0
1	1	1	0	1	0
1	1	1	1	1	0

The output expressions for X and Y can be minimized by using Karnaugh maps as shown in Figure 4.69. From the Karnaugh map for X we get

$$X = abd + abc + bcd + acd \tag{4.5}$$

The expression for X can also be represented as

$$X = \overline{\bar{c}\bar{d} + \bar{a}\bar{b} + \bar{b}\bar{c}d + \bar{a}c\bar{d} + \bar{b}cd + \bar{a}\bar{c}d} \tag{4.6}$$

Similarly,

$$Y = \bar{a}\bar{b}cd + \bar{a}b\bar{c}d + \bar{a}bc\bar{d} + ab\bar{c}\bar{d} + a\bar{b}\bar{c}d + \bar{a}bc\bar{d} \tag{4.7}$$

and also

$$Y = \overline{\bar{a}\bar{b}\bar{c} + \bar{a}\bar{b}cd + \bar{a}b\bar{c}\bar{d} + \bar{a}bcd + ab\bar{c}d + acd + abc\bar{d} + a\bar{b}\bar{c}d} \tag{4.8}$$

FIGURE 4.69
Karnaugh maps for output expressions X and Y

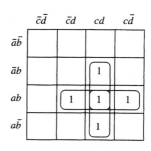
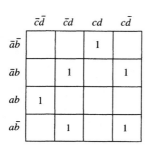

It is possible to use either an active-high or an active-low output PAL to implement the expressions for X and Y. If we wish to use an active-high PAL, then Equations (4.5) and (4.7) will have to be employed, whereas Equations (4.6) and (4.8) will be required for implementation using an active-low PAL. Figures 4.70a and b show the implementation of the expressions in a PAL16H2 and PAL16L2 device respectively.

FIGURE 4.70
(a) Active-high PAL;
(b) Active-low PAL

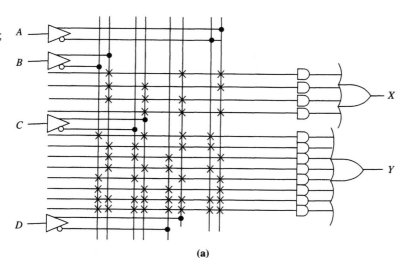

(a)

FIGURE 4.70
(Continued)

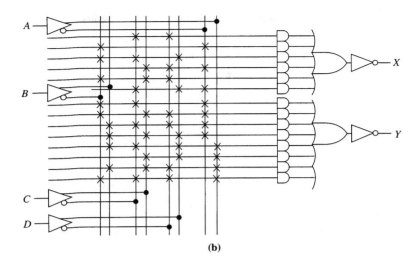

(b)

*ABEL

The design process using PLDs requires two distinct steps:

i. Specification of the function that a desired PLD should perform. Truth tables and state diagrams are often most convenient.
ii. Generation of a **fuse map** for the selected PLD and programming the device. A fuse map represents the cells or fuses in a device (e.g., the 1's and 0's represent the blown and intact fuses respectively).

Several software packages are available for translating the function specifications of a circuit to the corresponding fuse map, which can then be programmed into the selected PLD. Many PLD manufacturers provide software tools to develop and simulate designs based on their proprietary chips. In addition, several programming languages that support all types of PLDs are also available: **ABEL** from Data I/O Inc., **CUPL** from Logical Devices Inc., and **PLDesigner** from Minc Inc. All of these languages have similar features, but vary in syntax. In the following section, we consider the basic features of ABEL and show how it can be used.

ABEL uses a hardware description language (HDL) to describe a logic design. The design description is then used by the ABEL compiler as a source file to generate the object file, also known as the **JEDEC** (Joint Electronic Device Engineering Council) file. The JEDEC file contains the fuse map of the original logic design and is downloaded to a PLD programmer to program the specified PLD.

ABEL uses several intermediate tasks in sequence to convert a source file into a JEDEC file. The following is an overview of what happens during each task. The output file generated by a task is used as an input file by its succeeding task. The program module used for performing a task is identified within the parenthesis.

1. **Compile (ahdl2pla)** checks the source file for correct syntax, and translates the design description into a PLA format. The source file optionally includes test vectors that

show the expected output responses for various input patterns. These test vectors are stored in a separate file during the compilation process.
2. **Simulate equations (plasim)** simulates the logic design using the PLA file and the test vector file.
3. **Optimize (plaopt)** uses the two-level minimizer ESPRESSO to optimize the PLA file.
4. **Partmap (fuseasm)** creates the JEDEC file for the PLD selected from the device library and generates the design documentation.
5. **Simulate JEDEC (jedsim)** simulates the operation of the selected device which is assumed to be programmed with the information in the JEDEC file, using the test vector file.

A source file in ABEL is written in the hardware description language, ABEL-HDL. Typically a source file in the ABEL has the following format:

module name
title 'string of characters'
device identifier **device** device type
pin declaration
constant declaration
equations (or **truth_table** or **state_diagram**)
equations, truth table, or state diagram representation of the circuit
end

Equations are written using either of the following set of operations; however, if alternate operations are to be used, the keyword **@alternate** must be included before the pin declaration.

ABEL-HDL operator	*Description*	*Alternate Operator*
!	NOT	/
&	AND	*
#	OR	+
$	XOR	:+:
!$	XNOR	:*:

■ **EXAMPLE 4.29**

Let us write the ABEL source file for a combinational circuit that has 4 inputs (a,b,c,d) and 2 outputs (u,v); the output expressions for u and v are as follows:

$$u = ab\bar{c} + ad + b\bar{c}\bar{d} + ac\bar{d}$$
$$v = ab + \bar{a}d + bd + \bar{c}d + ac$$

Assuming the circuit is to be implemented using a PAL16H2, the ABEL source file for the circuit is as shown in Figure 4.71. The **plaopt** program in ABEL reduces the output expressions to

$$u = b\bar{c}\bar{d} + ad + ac$$
$$v = ac + ab + d$$

The JEDEC file for the PAL16H2 corresponding to the above expressions is shown in

4.11 COMBINATIONAL CIRCUIT DESIGN USING PLDs □ 133

Figure 4.72. Rows L0000, L0032, L0064, L0256, L0288, and L0320 in the JEDEC file correspond to rows 0, 32, 64, 256, 288, and 320 respectively in the logic diagram of PAL16H2 shown in Figure 4.74. A 1 in an L row indicates that the fuse at the intersection of the row and the corresponding column is blown, whereas a 0 indicates that the fuse is intact. Rows V0001, ..., V0014 represent the status of the input and the output pins of the device for each of the 14 test vectors. The letter N in columns 10 and 20 of a row indicates that these pins are ground and V_{CC} respectively. Figure 4.73 shows the outputs produced by the JEDEC file in response to selected test vectors.

```
                                ┌── module name
        module comb1
        title 'example of combinational circuit design using PLD'

        chip1 device 'p16h2';
        a,b,c,d pin 1,2,3,4;   ⎫
device identifier u,v pin 15,16;  ⎬ ── pin declaration
        H=1;                      ⎫
        L=0;     ── constant declaration
        X=.x.;                    ⎭
                  ── don't care
```

```
equations
  u=a&b&!c#a&d#b&!c&!d#a&c&!d;
  v=a&b#!a&d#b&d#!c&d#a&c;

test_vectors

  ([a,b,c,d] ->[u,v])
   [L,L,L,L] ->[L,L];
   [X,X,L,H] ->[L,H];
   [L,L,H,L] ->[L,L];
   [X,X,H,H] ->[L,H];
   [X,H,L,L] ->[H,L];
   [L,H,H,L] ->[L,L];
   [L,X,X,H] ->[L,H];
   [H,L,L,L] ->[L,L];
   [H,L,L,H] ->[H,H];
   [H,L,H,L] ->[H,H];
   [H,X,X,H] ->[H,H];
   [H,H,L,X] ->[H,H];
   [H,X,H,L] ->[H,H];
   [H,H,H,H] ->[H,H];

end comb 1
```
FIGURE 4.71 ABEL source file

FIGURE 4.72
JEDEC file

ABEL 4.01 Data I/O Corp. JEDEC file for: P16H2 V9.0
Created on: Tue Jul 13 16:16:13 1993

```
example of combinational circuit design using PLD
*
QP20* QF512* QV14* F0*
X0*
NOTE Table of pin names and numbers*
NOTE PINS a:1 b:2 c:3 d:4 u:15 v:16*
L0000   11010111111111111111111111111111*
L0032   11111111011111111111111111111111*
L0064   01011111111111111111111111111111*
L0256   01111011101111111111111111111111*
L0288   11011111011111111111111111111111*
L0320   11010111111111111111111111111111*
V0001   0 0 0 0 X X X X X N X X X X L L X X X N *
V0002   X X 0 1 X X X X X N X X X X L H X X X N *
V0003   0 0 1 0 X X X X X N X X X X L L X X N *
V0004   X X 1 1 X X X X X N X X X X L H X X X N *
V0005   X 1 0 0 X X X X X N X X X X H L X X X N *
V0006   0 1 1 0 X X X X X N X X X X L L X X X N *
V0007   0 X X 1 X X X X X N X X X X L H X X X N *
V0008   1 0 0 0 X X X X X N X X X X L L X X X N *
V0009   1 0 0 1 X X X X X N X X X X H H X X X N *
V00010  1 0 1 0 X X X X X N X X X X H H X X N *
V00011  1 X X 1 X X X X X N X X X X H H X X X N *
V00012  1 1 0 X X X X X X N X X X X H H X X X N *
V00013  1 X 1 0 X X X X X N X X X X H H X X X N *
V00014  1 1 1 1 X X X X X N X X X X H H X X X N *
```

FIGURE 4.73
Simulation results

```
Simulate ABEL 4.03 Date Tue Jul 13 16:16:41 1993
Fuse file: 'comb1.tt1' Vector file: 'comb1.tmv'
Part: 'PLA'
example of combinational circuit design using PLD

          a b c d    u v

V0001    0 0 0 0    L L
V0002    0 0 0 1    L H
V0003    0 0 1 0    L L
V0004    0 0 1 1    L H
V0005    0 1 0 0    H L
V0006    0 1 1 0    L L
V0007    0 0 0 1    L H
V0008    1 0 0 0    L L
V0009    1 0 0 1    H H
V0010    1 0 1 0    H H
V0011    1 0 0 1    H H
V0012    1 1 0 0    H H
V0013    1 0 1 0    H H
V0014    1 1 1 1    H H
14 out of 14 vectors passed.
```

4.11 COMBINATIONAL CIRCUIT DESIGN USING PLDs 135

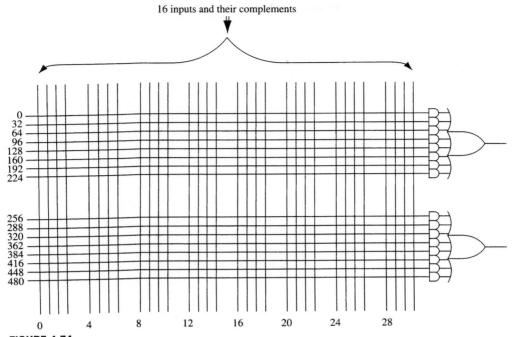

FIGURE 4.74
Logic diagram of PAL16H2; Input pins: 1, 2, 3, 4, 5, 6, 7, 8, 9, 11, 12, 13, 14, 17, 18, 19; Output pins: 15, 16

EXAMPLE 4.30

Let us consider a dual 8-to-1 multiplexer that is to be implemented using a single PAL. Let the dual inputs be x_0, x_1, x_2, x_3, x_4, x_5, x_6, and x_7, and y_0, y_1, y_2, y_3, y_4, y_5, y_6, and y_7. Also let the control signals, common to both multiplexers, be a, b, and c. The outputs of the multiplexers are z_0 and z_1. It will be clear from this specification that the selected PAL device needs to have the following inputs,

$$x_0 - x_7, y_0 - y_7, a, b, c$$

and two outputs,

$$z_1, z_0$$

The output expressions for the multiplexer are as follows:

$$z_1 = \bar{a}\bar{b}\bar{c} \cdot x_0 + \bar{a}\bar{b}c \cdot x_1 + \bar{a}b\bar{c} \cdot x_2 + \bar{a}bc \cdot x_3 + a\bar{b}\bar{c} \cdot x_4 + a\bar{b}c \cdot x_5 + ab\bar{c} \cdot x_6 + abc \cdot x_7$$

$$z_0 = \bar{a}\bar{b}\bar{c} \cdot y_0 + \bar{a}\bar{b}c \cdot y_1 + \bar{a}b\bar{c} \cdot y_2 + \bar{a}bc \cdot y_3 + a\bar{b}\bar{c} \cdot y_4 + a\bar{b}c \cdot y_5 + ab\bar{c} \cdot y_6 + abc \cdot y_7$$

Since there are eight product terms in z_1 and z_0, and each output of PAL20L2 can gener-

```
.module trial9
title 'dual-multiplexer implementation using PAL20L2'
bjdu          device 'p2012';

a,b,c                           pin 2,3,4;
x0,x1,x2,x3,x4,x5,x6,x7         pin 5,6,7,8,9,10,11,13;
y0,y1,y2,y3,y4,y5,y6,y7         pin 14,15,16,17,20,21,22,23;
z0,z1                           pin 18,19 is type 'invert';
X=.x.;

equations

z0 = (!a & !b & !c & x0
    # !a & !b &  c & x1
    # !a &  b & !c & x2
    # !a &  b &  c & x3
    #  a & !b & !c & x4
    #  a & !b &  c & x5
    #  a &  b & !c & x6
    #  a &  b &  c & x7);
z1 = (!a & !b & !c & y0
    # !a & !b &  c & y1
    # !a &  b & !c & y2
    # !a &  b &  c & y3
    #  a & !b & !c & y4
    #  a & !b &  c & y5
    #  a &  b & !c & y6
    #  a &  b &  c & y7);

test_vectors
  ([a, b, c, x0,x1,x2,x3,x4,x5,x6,x7,y0,y1,y2,y3,y4,y5,y6,y7]->[z1, z0])
   [0, 0, 1, X, 0, X, X, X, X, X, X, 1, X, X, X, X, X, X] ->[1,   0];
   [0, 1, 0, X, X, 1, X, X, X, X, X, 0, X, X, X, X, X, X] ->[0,   1];
   [1, 0, 0, X, X, X, X, 0, X, X, X, X, X, X, X, 0, X, X] ->[0,   0];
   [1, 1, 1, X, X, X, X, X, X, X, 1, X, X, X, X, X, X, 1] ->[1,   1];
end
```

FIGURE 4.75
ABEL source file

ate the sum of exactly eight product terms, these expressions can be implemented using a single PAL20L2. The ABEL source file for the dual multiplexer is shown in Figure 4.75. As mentioned previously, there is an inverter in a 20L2 between an output pin and the OR gate that generates the sum-of-products expression. Therefore, if the outputs of the multiplexer are to be active-high, the sum-of-products expression for the multiplexer outputs z_0 and z_1 must be inverted using DeMorgan's theorem. The ABEL compiler automatically does this inversion while generating the programming file for 20L2. ∎

EXERCISES

1. Derive the truth tables for the following functions:
   a. $f(a,b,c) = ac + ab + bc$
   b. $f(a,b,c,d) = (a\bar{c} + b\bar{d})(\bar{a}b\bar{c} + a\bar{d})$
   c. $f(a,b,c,d) = a \oplus b \oplus c + \bar{a}\bar{c}\bar{d}$

2. Derive the minterm and the maxterm list form of the Boolean functions specified by the following truth tables

   a.
a	b	c	f(a,b,c)
0	0	0	0
0	0	1	1
0	1	0	1
0	1	1	0
1	0	0	1
1	0	1	1
1	1	0	0
1	1	1	0

   b.
a	b	c	d	f(a,b,c,d)
0	0	0	0	1
0	0	0	1	1
0	0	1	1	1
0	1	1	1	1
1	0	0	0	1
1	1	0	0	1
1	0	1	1	1
All other combinations				0

3. Derive the product-of-sums form of the following sum-of-products Boolean expressions:
   a. $f(a,b,c) = \bar{a}\bar{b}\bar{c} + a\bar{b}c + ab\bar{c} + abc$
   b. $f(a,b,c,d) = ac + ad + ab + c\bar{d}$
   c. $f(a,b,c,d) = \bar{a}\bar{b}\bar{c} + \bar{a}bcd + a\bar{c}$

4. Derive the canonical product-of-sums form of the following functions:
   a. $f(a,b,c) = (a + \bar{b})(\bar{a} + \bar{c})(b + c)$
   b. $f(a,b,c,d) = (a + \bar{b} + d)(b + \bar{c} + \bar{d})(\bar{a} + c + \bar{d})(b + c + d)$
   c. $f(a,b,c,d,e) = (a + \bar{b} + c + \bar{e})(b + \bar{c} + d + \bar{e})(\bar{a} + c + \bar{d} + e)$

5. Derive the canonical sum-of-products form of the following functions:
   a. $f(a,b,c,d) = ac\bar{d} + a\bar{b}c + \bar{b}cd + \bar{a}d + ab$
   b. $f(a,b,c,d) = \bar{a}b + a\bar{c} + ad$
   c. $f(a,b,c,d,e) = \bar{a}bc\bar{e} + b\bar{d}e + ab d\bar{e} + \bar{a}\bar{b}\bar{c}e + \bar{a}de$

6. Minimize the following functions using Karnaugh maps:
   a. $f(a,b,c,d) = \bar{a}\bar{b}\bar{c}\bar{d} + ab\bar{c}\bar{d} + a\bar{b}\bar{c}\bar{d} + abcd + \bar{a}\bar{b}c\bar{d} + ab\bar{c}d + \bar{a}bcd$
   b. $f(a,b,c,d) = \Sigma m(1,3,4,6,9,11,13,15)$
   c. $f(a,b,c,d) = \Pi M(3,6,7,9,11,12,13,14,15)$
   d. $f(a,b,c,d,e) = \Sigma m(4,8,10,15,17,20,22,26) + d(2,3,12,21,27)$

7. Derive the set of prime implicants for the following functions using Quine-McCluskey method. In each case identify the essential prime implicants if there are any.
   a. $f(a,b,c,d) = \Sigma m(0,1,2,3,7,9,12,13,14,15,22,23,29,31)$
   b. $f(a,b,c,d) = \Sigma m(5,6,7,10,14,15) + d(9,11)$
   c. $f(a,b,c,d) = \Sigma m(1,7,9,11,13,21,24,25,30,31) + d(0,2,6,8,15,17,22,28,29)$

8. The prime implicant chart for two Boolean functions are shown in (a) and (b). Obtain the minimal sum-of-products expression for each case.

   a.

	0	4	5	6	11	13	15
$PI_1 = \bar{a}\bar{c}$	X	X	X				
$PI_2 = \bar{c}d$			X			X	
$PI_3 = \bar{b}d$					X		
$PI_4 = ad$					X	X	X
$PI_5 = \bar{a}b\bar{d}$		X		X			

b.

	0	1	4	5	6	7	9	11	15
$PI_1 = \bar{a}\bar{c}$	X	X	X	X					
$PI_2 = \bar{a}b$			X	X	X	X			
$PI_3 = ac$								X	X
$PI_4 = bc$					X	X			X
$PI_5 = a\bar{b}d$							X	X	

9. Implement the function $f(w,x,y,z) = (w + \bar{x})(w + \bar{x} + \bar{y})(y + \bar{z})$ using NOR-NOR logic.

10. Design a combinational circuit to generate the parity bit for digits coded in BCD code. The circuit should also have an additional output that produces an error signal if a non-BCD digit is input to the circuit. Realize the circuit using NAND-NAND logic.

11. Determine the Boolean function for the circuit shown. Obtain an equivalent circuit with fewer NOR gates. (Assume only 2-input or 3-input NOR gates.)

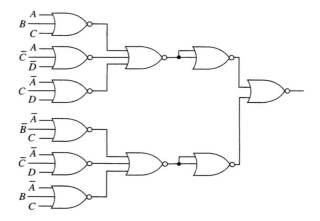

12. A computing system consists of four processors P_1, P_2, P_3, and P_4, and four blocks of memories M_1, M_2, M_3, and M_4. Processor P_1 is allowed to use memory blocks M_1 and M_4 only. Memory block M_2 can be used only by P_2 and P_3. Processor P_3 can use all blocks of memories. All processors can use block M_4. Design a circuit that will produce an active high output only if a processor uses an appropriate memory block.

13. Show how two 2-to-1 multiplexers can be used to implement a half-adder.

14. The function $f(w,x,y,z) = \Sigma m(3,5,6,7,9,10,11,13)$ is to be implemented using a multiplexer. Select an appropriate multiplexer, and show how its inputs can be selected to implement the given function.

15. Each of the four members of a "jukebox jury" is provided with a pushbutton; the button is pressed if the record played is a "hit" and is not pressed if the record played is a "miss." Design a digital circuit that lights a lamp when a majority of the jury thinks the record is a hit. Implement the circuit using a miltiplexer.

16. Assuming you have any number of 8-to-1 multiplexers and a single 4-to-1 MUX, design a 32-to-1 MUX.

17. Design a multiple output logic circuit whose input is BCD data and whose outputs are

   w: detects inputs, that are divisible by 3

x: detects number greater than or equal to 4

Implement the circuit using an appropriate decoder.

18. A combinational circuit has six inputs $x_i (i = 1, \ldots, 6)$ and six outputs $z_i (i = 1, \ldots, 6)$. An output z_j is to be 1 if and only if x_j is 1 and each $x_i = 0$ for all $i < j$. Implement the circuit using decoder(s) and the minimum number of gates.

19. A combinational circuit is to be designed to control a 7-segment display of decimal digits. The inputs to the circuit are BCD codes. The seven outputs correspond to the segments that are activated to display a given decimal digit.
    a. Develop a truth table for the circuit.
    b. Derive simplified sum-of-products and product-of-sums expressions.
    c. Implement the expressions using either NAND or NOR gates as appropriate.

20. Given the following Boolean expression

$$f(A,B,C) = AB\overline{C} + \overline{A}BC + A\overline{B}C + \overline{A}\,\overline{B}C$$

   a. Develop an equivalent expression using NAND functions only, and draw the logic diagram.
   b. Develop an equivalent expression using NOR functions only, and draw the logic diagram.

21. A certain "democratic" country is ruled by a family of four members (A, B, C, and D). A has 35 votes, B has 40 votes, C has 15 votes, and D has 10 votes. Any decision taken by the family is based on its receiving at least 60% of the total number of votes. Design a circuit that will produce an output of 1 if a certain motion is approved by the family.

22. In a digital system, a circuit is required that will compare two 3-bit binary numbers, $X = x_2,x_1,x_0$ and $Y = y_2,y_1,y_0$, and generate separate outputs corresponding to the conditions $X = Y$, $X > Y$, and $X < Y$. Implement the circuit using NAND gates only.

23. A combinational circuit that will generate the square of all the combinations of a 3-bit binary number is to be implemented using an FPLA. Show the program table for implementing the circuit using the format shown in the text.

24. Implement the following Boolean functions using a PROM

$$f_1(w,x,y,z) = wxy + \overline{w}\,\overline{x}y$$
$$f_2(w,x,y,z) = \overline{w} + \overline{x} + y + z$$
$$f_3(w,x,y,z) = w + \overline{x} + \overline{y}z + wz$$
$$f_4(w,x,y,z) = wyz + w\overline{y}\,\overline{z} + \overline{x}yz + \overline{w}xz$$

25. A 6-to-64 decoder is to be implemented using 3-to-8 decoders only. Show the block diagram of the 6-to-64 decoder.

26. Design a combinational circuit to generate a parity (even) bit for digits coded in 5421 code. Also, provide an error output if the input to the circuit is not a 5421 code.

27. Implement the following functions using NAND gates having a maximum fan-in of three.

$$f(A,B,C,D) = \Sigma m(0,1,3,7,8,12) + d(5,10,13,14)$$
$$f(A,B,C,D) = AB(C + D) + CD(A + B)$$

28. Implement the following function using NOR gates having a maximum fan-in of three.

$$f(A,B,C,D,E) = \overline{A}(B + C + DE)(\overline{B} + CD + \overline{A}E)$$
$$f(A,B,C,D) = \overline{A}B + \overline{B}CD + A\overline{B}\,\overline{D}$$

29. Prove that
   a. $A + B = A \oplus B \oplus AB$
   b. $A \oplus B \odot C = \overline{(A \oplus B \oplus C)}$

30. It is required to design a lighting for a room such that the lights may be switched on or off from any one of three switch points. Implement the circuit using the minimum number of EX-OR gates.

31. Derive the kernels and cokernels of the following function:

$$f(a,b,c,d,e,f,g) = adf + aef + bdf + bef + cdf + cef + g$$

32. Implement the logic circuit represented by the following Boolean functions in a multilevel form*:

$$f_1(a,b,c,d,e,f,g) = ac + ade + bc + bde$$
$$f_2(a,b,c,d,e,f,g) = afg + bfg + efg$$

References

1. R. Burgoons, "Improve your Karnaugh mapping skills," *Electronic Design*, No. 26, Dec. 21, 1972, pp. 54–56.

2. R. K. Brayton, G. D. Hachtel, C. T. McMullen, and A. L. Sangiovanni-Vincentelli, *Logic Minimization Algorithms for VLSI Design*, Kluwer, 1984.

3. SIS: A system for sequential circuit synthesis, Electronic Research Laboratory Memorandum No. UCB/ERLM92/41, May 4, 1992, University of California, Berkeley.

4. R. K. Brayton, G. D. Hachtel, and A. L. Sangiovanni-Vincentelli, "Multilevel logic synthesis," *Proc. IEEE*, February 1990, pp. 264–300.

5. R. Rudell, *Logic Synthesis for VLSI Design*, Ph.D Thesis, Univ. of California, Berkeley, 1989.

6. S. Hurst, "More powerful logic will simplify design philosophy," *Electronic Engineering*, February 1976, pp. 53–55.

7. P. K. Lala, *Digital Systems Design using PLDs*, Prentice Hall, 1990.

8. B. A. Twicker, "Synthesize logic with exclusive-OR ICs," *Electronic Engineering*, July 1970, pp. 52–53.

9. P. K. Lala, *Fault Tolerant and Fault Testable Hardware Design*, Prentice Hall, 1985.

10. A. Pal, "An algorithm for optimal logic design using multiplexers," *IEEE Trans. on Computers*, August 1986, pp. 755–757.

11. F. Cave and D. Terell, *Digital Technology with Microprocessors*, Reston, 1981.

Further Reading

1. R. H. Katz, *Contemporary Logic Design*, Benjamin/Cummings, 1994.

2. F. J. Hill and G. R. Peterson, *Computer Aided Logic Design with Emphasis on VLSI*, Wiley, 1993

---

*MCNC (Microelectronics Center of North Carolina) Technical Report TR87-15.

5
Fundamental Concepts of Sequential Logic

5.1 INTRODUCTION

Combinational logic refers to circuits whose output is strictly dependent on the present value of the inputs. As soon as input values are changed, the information regarding the previous inputs is lost; in other words, combinational logic circuits have no memory. In many applications, information regarding input values at a certain instant of time is needed at some future time. Circuits whose output depends not only on the present values of the input but also on the past values of the inputs are known as **sequential logic circuits**. The mathematical model of a sequential circuit is usually referred to as a **sequential machine** or a **finite state machine**. A general model of a sequential circuit is shown in Figure 5.1. As can be seen in the diagram, sequential circuits are basically combinational circuits with the additional property of **memory** (to remember past inputs). The combinational part of the circuit receives two sets of input signals: **secondary** (coming from the memory) and **primary** (coming from the circuit environment). The particular combination of secondary input variables at a given time is called the **present state** of the circuit; the secondary input variables are also known as **state variables**. If there are m secondary input variables in a sequential circuit, then the circuit can be in any one of 2^m different present states.

The outputs of the combinational part of the circuit are divided into two sets. The primary outputs are available to control operations in the circuit environment, whereas the secondary outputs are used to specify the **next state** to be assumed by the memory. The number of secondary output variables, often called the **excitation variables**, depends on the type of memory element used.

5.2 SYNCHRONOUS AND ASYNCHRONOUS OPERATION

Sequential logic circuits can be categorized into two classes: **synchronous** and **asynchronous**. In synchronous circuits internal states change at discrete instants of time under the control of a synchronizing pulse, called the **clock**. The clock is generally some form of square wave as illustrated in Figure 5.2. The **on-time** is defined as the time the

FIGURE 5.1
General model of a sequential logic circuit

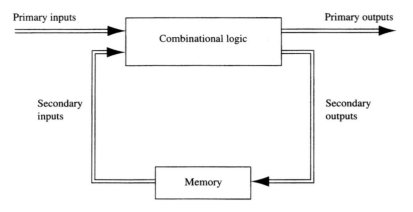

wave is in the 1 state; the **off-time** is defined as the time the wave is in the 0 state. The **duty cycle** is defined as

$$\text{Duty cycle} = \frac{\text{On time}}{\text{Period}} \text{ (expressed as a percentage)}$$

State transitions in synchronous sequential circuits are made to take place at times when the clock is making a transition from 0 to 1 or from 1 to 0. The 0-to-1 transition is called the **positive edge** or the **rising edge** of the clock signal, whereas the 1-to-0 transition is called the **negative edge** or the **falling edge** of the clock signal (as shown in Figure 5.2). Between successive clock pulses there is no change in the information stored in memory. Synchronous sequential circuits are also known as **clocked sequential circuits.**

In asynchronous sequential circuits the transition from one state to another is initiated by the change in the primary inputs; there is no external synchronization. Since state transitions do not have to occur at specific instants of time, asynchronous circuits can operate at their own speed. The memory portion of asynchronous circuits is usually implemented by feedback among logic gates. Thus, asynchronous circuits can be regarded as combinational circuits with feedback. Because of the difference in the delays through various signal paths, such circuits can give rise to transient conditions during the change of inputs or state variables. Hence, such circuits have to be designed in a special way in order to ensure that their operations are not affected by transient conditions.

FIGURE 5.2
Clock signal

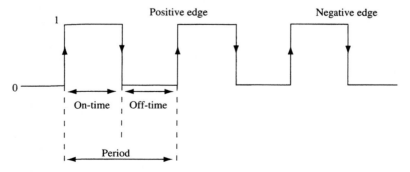

5.3 LATCHES

As can be seen from Figure 5.1, the memory unit is an essential part of a sequential circuit. This unit usually consists of a series of **latches**. A latch is a logic circuit with two outputs, which are the complement of each other. A basic latch can be constructed by cross-coupling two NOR gates as shown in Figure 5.3. The two inputs are labeled **Set** and **Reset**. The outputs Q and $\overline{Q}$ are always the complement of each other during normal operation. Let us assume both inputs are at 0 initially, output Q is at 1, and output $\overline{Q}$ is at 0. Thus, the output of gate G_1 is at 1. Since the output of gate G_1 is fed back to the input of gate G_2, the output of G_2 will be 0. The circuit is therefore stable with Q at 1 and $\overline{Q}$ at 0, as was originally assumed.

If the Reset input is now taken to 1, the output of G_1 will change to 0. Both the inputs of G_2 are at 0, so its output will change to 1. The circuit is now stable with $Q = 0$ and $\overline{Q} = 1$. The circuit remains in this stable state even if the Reset input is changed back to 0. If the Set input is now taken to 1, the output of G_2 (i.e., $\overline{Q}$) will be at 0. Since both inputs of G_1 are now at 0, its output (i.e., Q) will be at 1. The circuit remains stable with $Q = 1$ and $\overline{Q} = 0$ even when the Set input returns to 0.

The input combination Set = 1 and Reset = 1 is not allowed because both Q and $\overline{Q}$ go to 0 in this case, which violates the condition that Q and $\overline{Q}$ should be complements of each other. Furthermore, when the Set and Reset inputs are returned to 0, an ambiguous situation arises. For example, if the propagation delay of G_1 is lower than that of G_2, then the output of G_1 (i.e., Q) will change to 1 first. This in turn will make the output of G_2 (i.e., $\overline{Q}$) = 0. On the other hand, if the output of G_2 changes to 1 first, then the output of G_1 will be forced to 0. In other words, it is impossible to predict the output. Therefore, the Set = 1 and Reset = 1 combination is avoided during the operation of this type of latch. The behavior of the latch circuit can be represented by the truth table of Figure 5.4a. The cross-coupled NOR latch is generally known as an *SR* (Set-Reset) latch. The logic symbol used to represent the *SR* latch is shown in Figure 5.4b.

An SR latch can also be constructed by cross-coupling NAND gates as shown in Figure 5.5a. The circuit operates in the same manner as the NOR latch, but it does have a few subtle differences. Unlike the NOR latch, the NAND latch inputs are normally 1 and must be changed to 0 to change the output. An ambiguous output results when both the Set and the Reset inputs are at 0. Figure 5.5b shows the truth table of the NAND latch. The logic symbol for the NAND latch is shown in Figure 5.5c; the circles denote that the latch responds to 0 on its inputs (i.e., it has active-low inputs).

In many applications a latch has to be set or reset in synchronization with a control signal. Figure 5.6a shows how the NAND latch of Figure 5.5a has been modified to incorporate a control input, which is usually driven by a clock. The resulting circuit is

FIGURE 5.3
Cross-coupled NOR flip-flop

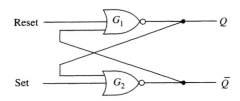

FIGURE 5.4

Set	Reset	Q	$\bar{Q}$
0	0	No change	
0	1	0	1
1	0	1	0
1	1	Ambiguous	

(a) Truth table for cross-coupled NOR latch (b) Logic symbol

known as a **gated latch** or **clocked latch**. As long as the control is at 0 in Figure 5.6a, the outputs of gates G_3 ad G_4 will be 1 and the latch will not change state. When the enable input changes to 1, the set and reset inputs affect the latch. Thus, if Set = 1 and Reset = 0, the output of G_3 is 0 and that of G_4 is 1, Q goes to 1, and $\bar{Q}$ goes to 0. If Set = 0 and Reset = 1, Q goes to 0. The truth table for the latch is constructed as shown in Figure 5.6b. Note that Q_t is the present state of the latch, and Q_{t+1} is the next state. The next state of the latch depends on the present state of the latch and the present value of the input. These types of latches are often called **transparent** because the output changes (after the latch propagation delay) as the inputs change, if the enable input is high. The gated NOR latch and its truth table are shown in Figures 5.7a and b respectively.

It is possible for *SR* latches to have more than two inputs. Figure 5.8 represents an *SR* latch constructed from two 3-input NOR gates. The Q output will be 1 if any of the Set inputs is at 1 and the Reset inputs are at 0. The Q output goes to 0 if any of the Reset inputs is at 1 and all the Set inputs are at 0. As before, Set and Reset inputs are not simultaneously allowed to be at 1. A normal *SR* latch can be converted to another type of latch, known as a *D* latch, by generating Reset = $\overline{\text{Set}}$ (using an additional inverter).

FIGURE 5.5

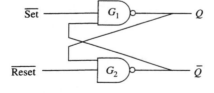

Set	Reset	Q	$\bar{Q}$
0	0	Ambiguous	
0	1	1	0
1	0	0	1
1	1	No change	

(a) NAND latch

(b) Truth table

(c) Logic symbol for NAND latch

5.4 FLIP-FLOPS

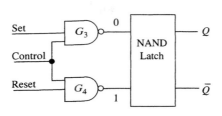

(a) Gated NAND latch with enable input

Control	Set	Reset	G_3	G_4	Q_{t+1}
0	—	—	1	1	Q_t
1	0	0	1	1	Q_t
1	0	1	1	0	0
1	1	0	0	1	1
1	1	1	0	0	Ambiguous

(b) Truth table

FIGURE 5.6

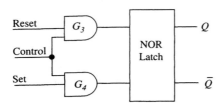

(a) Gated NOR latch with enable input

Control	Set	Reset	G_3	G_4	Q_{t+1}
0	—	—	0	0	Q_t
1	0	0	0	0	Q_t
1	0	1	1	0	0
1	1	0	0	1	1
1	1	1	1	1	Ambiguous

(b) Truth table

FIGURE 5.7

FIGURE 5.8
Three-input NOR SR latch

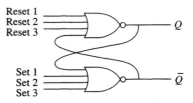

Figure 5.9a shows a D latch constructed from an SR NAND latch. As can be seen from the truth table of the latch (Fig. 5.9b), the output Q follows the D input if the control line is at 1. With the control line at 0, Q holds the value of D prior to the 1-to-0 transition of the control line. The D latch has the advantage over the SR latch that no input combination produces ambiguous output, so no input has to be avoided.

5.4 FLIP-FLOPS

A flip-flop, like a gated latch, possesses two stable states, but unlike a gated latch, transitions in a flip-flop are activated by the edge rather than the level of the clock pulse on the control input. Figure 5.10a shows timing diagrams for a D latch and a D flip-flop.

A flip-flop is often called a bistable element because when its Q output is at logic 1, the $\overline{Q}$ output is at logic 0 or vice-versa. However, it is also possible for a flip-flop to

FIGURE 5.9

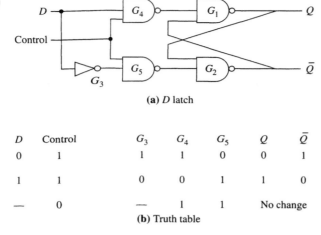

(a) *D* latch

D	Control	G_3	G_4	G_5	Q	$\bar{Q}$
0	1	1	1	0	0	1
1	1	0	0	1	1	0
—	0	—	1	1	No change	

(b) Truth table

be in a **metastable** state. Metastability implies that the output of a flip-flop is undefined (i.e., neither 0 nor 1) for a certain period of time. This phenomenon occurs if the data input to a flip-flop does not satisfy the specified **setup time** and **hold time** with respect to the clock. The setup time is the period of time the data must be stable at the input of a flip-flop before the flip-flop is triggered by the clock edge. The period of time the data must remain stable after the flip-flop has been triggered is known as the hold time. Figure 5.10b illustrates the meaning of setup time and hold time. In order to guarantee that valid data is produced at the output of a flip-flop after a maximum clock-to-output delay time (i.e., the time from the rising edge of the clock to the time valid data is available on the output), the input data must not violate the specified setup and hold times. Otherwise, the output will be in a metastable state for a time greater than the maximum clock-to-output delay time. The indeterminate logic value produced by a flip-flop while it is in a metastable state can result in an unpredictable circuit behavior.

D Flip Flop

The input data to a *D* flip-flop is transferred to the output and held there on a $0 \rightarrow 1$ transition of the clock pulse (see Fig. 5.10a). In other words, the *D* flip-flop is triggered on the positive edge of the clock pulse. A positive edge–triggered *D* flip-flop can be implemented by modifying the NAND latch configuration as shown in Figure 5.11a. In addition to the *D* input, the flip-flop has a pair of direct clock-independent (asynchronous) inputs, $\overline{\text{Reset}}$ and $\overline{\text{Set}}$. When the $\overline{\text{Reset}}$ input is at logic 0, the *Q* output of the flip-flop goes to 0. On the other hand, when the $\overline{\text{Set}}$ input is at logic 0, the *Q* output goes to 1. The $\overline{\text{Set}}$ and $\overline{\text{Reset}}$ inputs are not allowed to be at 0 simultaneously because this will make both *Q* and $\bar{Q}$ outputs of the flip-flop to go to 1; in other words, the condition that $\bar{Q}$ must always be the complement of *Q* in a flip-flop is violated.

5.4 FLIP-FLOPS □ 147

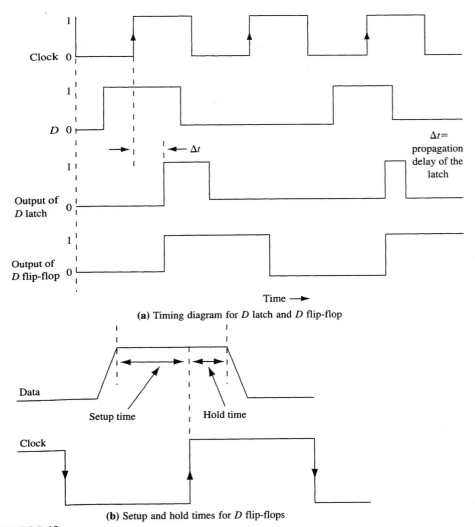

FIGURE 5.10

Figure 5.11b shows the timing diagram of the D flip-flop. As can be seen from the diagram, the D flip-flop transfers the input value at D to the Q output on the positive edge of the clock pulse only when both $\overline{\text{Set}}$ and $\overline{\text{Reset}}$ inputs are at 1. Table 5.1 summarizes the operation of the D flip-flop. The logic symbol for the D flip-flop is shown in Figure 5.12a. The small circles on the Set and Reset inputs mean that when the Set input is driven to 0, the flip-flop is preset to 1, and when the Reset input is at 0, the flip-flop is reset to 0. The characteristics of the D flip-flop can be represented by the following equation (Fig. 5.12b).

$$Q_{t+1} = D_t$$

In other words, the next state of a D flip-flop corresponds to the data input applied at time t and is independent of the present state of the flip-flop.

JK Flip-Flop

The JK flip-flop has similar functions to an SR latch, with J equivalent to Set and K to Reset. In addition, if both J and K are set to 1, the outputs complement when the device is clocked. Thus, a JK flip-flop does not have any invalid input combinations. Figure

FIGURE 5.11

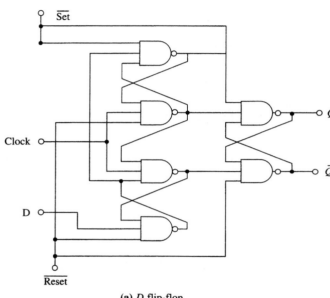

(a) D flip-flop

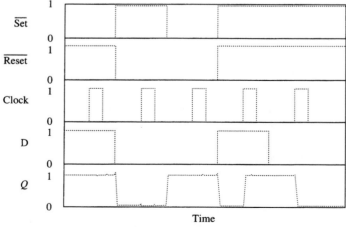

(b) Timing diagram of D flip-flop

FIGURE 5.12

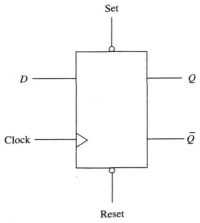

(a) Logic symbol for *D* flip-flop

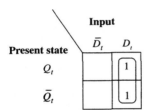

(b) Karnaugh map for deriving characteristic equation

5.13a shows the implementation of a positive-edge-triggered *JK* flip-flop. The analysis of the circuit operation is similar to that of the *D* flip-flop. The logic symbol and the truth table for the flip-flop are shown in Figures 5.13a and b respectively. The characteristic equation for the *JK* flip-flop can be derived from its truth table (Fig. 5.13c) and is given by

$$Q_{t+1} = J_t \overline{Q}_t + \overline{K}_t Q_t$$

TABLE 5.1
Function table of the *D* flip-flop

Set	Reset	D	Clock	Q	$\overline{Q}$
0	1	—	—	1	0
1	0	—	—	0	1
0	0	—	—	1	1
1	1	0	0→1	0	1
1	1	1	0→1	1	0

FIGURE 5.13

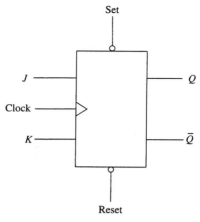

(a) Logic symbol for *JK* flip-flop

Present state	Inputs			Next state
Q_t	J_t	K_t	Clock	Q_{t+1}
0	0	0	0 → 1	0
0	0	1	0 → 1	0
0	1	0	0 → 1	1
0	1	1	0 → 1	1
1	0	0	0 → 1	1
1	0	1	0 → 1	0
1	1	0	0 → 1	1
1	1	1	0 → 1	0

(b) Truth table

	$\bar{J}_t\bar{K}_t$	$\bar{J}_tK_t$	J_tK_t	$J_t\bar{K}_t$
$\bar{Q}_t$			1	1
Q_t	1			1

(c) Karnaugh map for deriving characteristic equation

The characteristic equation indicates that a *JK* flip-flop can also be implemented using a *D* flip-flop. This is shown in Figure 5.14.

The *JK* flip-flops are often used in place of *SR* latches, as these devices are not commonly available as commercial parts. Figure 5.15a shows the proper connections of a *JK*

FIGURE 5.14
Implementation of a JK flip-flop using an edge-triggered D flip-flop

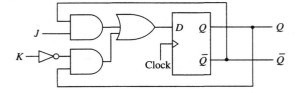

FIGURE 5.15

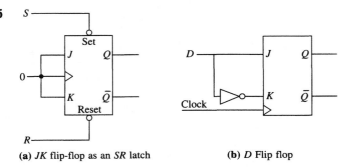

(a) JK flip-flop as an SR latch (b) D Flip flop

flip-flop in order to operate as a *SR* latch. A *JK* flip-flop can also operate as a *D* flip-flop; the configuration used to achieve this is shown in Figure 5.15b. In this case $J = D$ and $K = \overline{D}$; thus, from the *JK* flip-flop characteristic equation,

$$Q_{t+1} = D_t \cdot \overline{Q}_t + \overline{\overline{D}_t} Q_t$$
$$= D_t$$

T Flip-Flop

The *T* flip-flop, known as a **toggle** or a **trigger** flip-flop, has a single input line. The symbol for the *T* flip-flop is shown in Figure 5.16a. If $T = 1$ when the clock pulse changes from 0 to 1, the flip-flop assumes the complement of its present state; if $T = 0$, the flip-flop does not change state. The truth table of the flip-flop is shown in Figure 5.16b. The characteristic equation for the flip-flop can be derived from its truth table, as shown in Figure 5.16c.

$$Q_{t+1} = Q_t \overline{T} + \overline{Q}_t T$$

Thus, a *T* flip-flop can be considered as a single-input version of *JK* flip-flop. A *JK* flip-flop can be configured as shown in Figure 5.17 to operate as a *T* flip-flop. Alternatively, a *T* flip-flop can also be derived from a *D* flip-flop (Fig. 5.18). The *D* input is driven by an exclusive-OR gate, which in turn is fed by the *Q* output of the flip-flop and the *T* input line, as dictated by the characteristic equation of the *T* flip-flop. It should be noted that *T* flip-flops are not available as commercial parts; they are constructed from *JK* or *D* flip-flops.

5.5 TIMING IN SYNCHRONOUS SEQUENTIAL CIRCUITS

As mentioned previously, in a synchronous sequential circuit the state transition (i.e., the change in the outputs of the flip-flops) occurs in synchronization with a pulse. In posi-

FIGURE 5.16

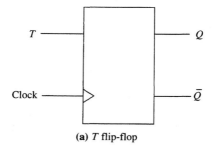

(a) T flip-flop

Present state	Inputs		Next state
Q_t	T	Clock	Q_{t+1}
0	0	0 → 1	0
0	1	0 → 1	1
1	0	0 → 1	1
1	1	0 → 1	0

(b) Truth table

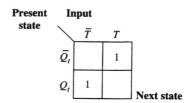

(c) Karnaugh map for deriving characteristic equation

tive-edge-triggered flip-flops, the delay between the positive-going transition on the clock input and the changes on the outputs is specified as propagation delay, t_{pd}. Usually the delays are different for the two possible directions of output change and are specified as

t_{PLH}: the delay between the transition midpoint of the clock signal and the transition midpoint of the output, where the output is changing from low to high. This delay is also referred to as **rise delay**.

t_{PHL}: the delay between the transition midpoint of the clock signal and the transition midpoint of the output, where the output is changing from high to low. This delay is also referred to as **fall delay**.

Figure 5.19 illustrates these propagation delays. The values of t_{PLH} and t_{PHL} are gener-

FIGURE 5.17
Construction of a T flip-flop from a JK flip-flop

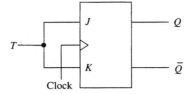

FIGURE 5.18
Construction of a T flip-flop from a D flip-flop

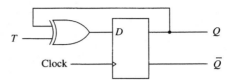

ally not the same. The manufacturers' data sheets usually specify typical and maximum value; the minimum value must obviously be nonzero.

The clock frequency f_{max}, which determines the maximum speed at which a synchronous sequential circuit can reliably operate, is related to the minimum clock period T_{min} by

$$f_{max} = \frac{1}{T_{min}}$$

where T_{min} = minimum setup time for flip-flops
+ minimum hold-time for flip-flops
+ maximum gate propagation delay
+ maximum flip-flop propagation delay

Figure 5.20 shows a very simple synchronous sequential circuit, designed using two-level NAND gates and two flip-flops. Assuming the gates have propagation delays of 4 ns, the propagation delay of the flip-flops is 10 ns, their setup time is 3 ns, and the hold

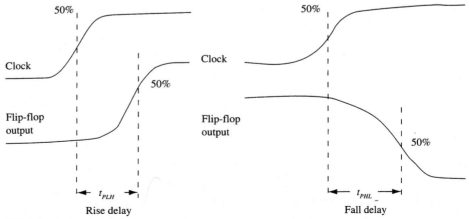

FIGURE 5.19
Flip-flop propagation delay

FIGURE 5.20
A sequential circuit

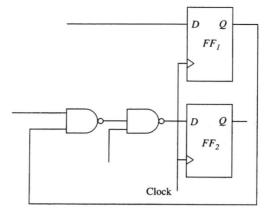

time is 1 ns, we first determine T_{min}:

$$T_{min} = 3 + 1 + 2 \times 4 + 10 = 22 \text{ ns}$$

Thus, the maximum clock frequency is

$$f_{max} = \frac{1}{22 \times 10^{-9}} = 45.5 \text{ MHz}$$

One additional consideration that has to be taken into account in sequential circuits is the **clock skew**. So far, we have assumed that all the flip-flops in the circuit are triggered simultaneously. However, in a large circuit this assumption is rarely true, so it is more realistic to assume that the clock signal appears at the clock inputs of the various flip-flops at different times. This is due to the delays in the conducting paths between the clock generator and the flip-flops as well as the delay variations between different clock buffers. Hence, the flip-flops are triggered at different times, resulting in incorrect circuit operation.

For example, in the circuit of Figure 5.20 FF_1 drives FF_2 through two gates with delay t_g ($= 8$ ns by previous assumption). If the clock signal arrives at FF_2 later than FF_1, this delay must be less than

$$\begin{aligned}\Delta t_{max} &= t_{pd}(FF_1) + t_g + t_s(FF_2) \\ &= 10 + 8 + 3 \\ &= 21 \text{ ns}\end{aligned}$$

for the circuit to operate properly. If the clock pulse arrives later than Δt_{max}, then the new state of FF_1 will be clocked into FF_2.

5.6 STATE TABLES AND STATE DIAGRAMS

In Section 5.1 we examined a general model for sequential circuits. In this model the effect of all previous inputs on the outputs is represented by a state of the circuit. Thus, the output of the circuit at any time depends upon its current state and the input; these also determine the next state of the circuit. The relationship that exists among primary input variables, present state variables, next state variables, and output variables can be specified by either the **state table** or the **state diagram**. In the state table representation of a sequential circuit, the columns of the table correspond to the primary inputs and the rows correspond to the present state of the circuit. The entries in the table are the next state and the output associated with each combination of inputs and present states. As an example consider the sequential circuit of Figure 5.21. It has one input x, one output z, and two state variables $y_1 y_2$ (thus having four possible present states 00, 01, 10, 11). The behavior of the circuit is determined by the following equations:

$$Z = xy_1$$
$$Y_1 = \bar{x} + y_1$$
$$Y_2 = x\bar{y}_2 + \bar{x}\bar{y}_1$$

These equations can be used to form the state table. Suppose the present state (i.e., $y_1 y_2$) = 00 and input $x = 0$. Under these conditions, we get $Z = 0$, $Y_1 = 1$, and $Y_2 = 1$. Thus, the next state of the circuit $Y_1 Y_2 = 11$, and this will be the present state after the clock pulse has been applied. The output of the circuit corresponding to the present state $y_1 y_2 = 00$ and $x = 1$ is $Z = 0$. This data is entered into the state table as shown in Figure 5.22a. Normally each combination of present state variables is replaced by a letter; Figure 5.22b is derived from Figure 5.22a by replacing states 00, 01, 11, and 10 by A, B, C, and D respectively. The output and the next state entries corresponding to other present states and the input are derived in a similar manner. In general, an m-input, n-state machine will have n rows and one column for each of the 2^m combinations of inputs. The next state and output corresponding to a present state and an input combination are entered at their intersection in the table.

FIGURE 5.21
A sequential circuit

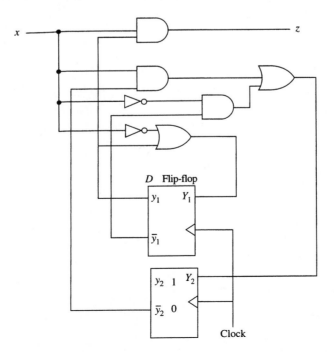

Present State	Input	
$y_1 y_2$	$x = 0$	$x = 1$
0 0	11, 0	01, 0
0 1	11, 0	00, 0
1 1	10, 0	10, 1
1 0	10, 0	11, 1
	Next state, output	
(a)		

Present State	Input	
	$x = 0$	$x = 1$
A	C,0	B,0
B	C,0	A,0
C	D,0	D,1
D	D,0	C,1
(b)		

FIGURE 5.22

A sequential circuit can also be represented by a **state diagram** or a **state transition graph**. A state diagram is a directed graph with each node corresponding to a state in the circuit and each edge representing a state transition. The input that causes a state transition and the output that corresponds to the present state/input combination are given beside each edge. A slash separates the input from the output. The state diagram for the sequential circuit of Figure 5.22 is shown in Figure 5.23; states 00, 01, 11, and 10 are denoted by the letters A, B, C, and D respectively. For example, the edge from B to A and the associated label 1/0 indicate that if B is in the present state and the input is 1, then the next state is A and the output is 0.

Both state tables and state diagrams can be used to define the operation of sequential circuits; they provide exactly the same information. However, in general state diagrams are used to represent the overall circuit operation, whereas the state table is employed for the actual circuit design.

FIGURE 5.23
State diagram

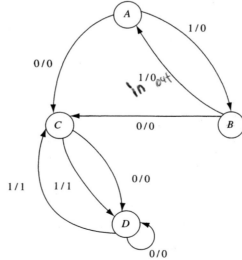

FIGURE 5.24
Moore model of sequential circuits

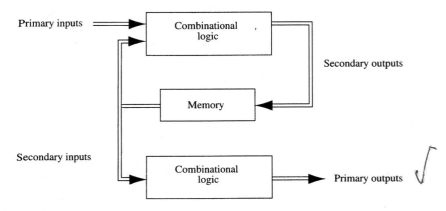

5.7 MEALY AND MOORE MODELS

So far, we have considered sequential circuits in which the output at any time depends on the present state and the input, and these also determine the next state. This particular model of sequential circuits is known as the **Mealy model** (Fig. 5.1). In an alternative model, called the **Moore model**, the next state depends on the present state and the input, but the output depends only on the present state. Figure 5.24 illustrates the Moore model of the sequential circuit.

In the state table representation of Moore-type sequential circuits, the rows of the table correspond to present states and the columns to input combinations as in Mealy-type circuits. The entries in the table for the input/present state combinations are the next states associated with each combination of inputs and present states; there is a separate output column with the entry corresponding to each row (i.e., present state) in the table. An example of such a table is shown in Figure 5.25a. The state diagram for the circuit is shown in Figure 5.25b. Since each state has a unique output, the output can be associated with

Present state	Input		Output
	$x=0$	$x=1$	
A	B	D	1
B	C	A	0
C	C	D	0
D	B	D	0
	Next state		

(a) State table

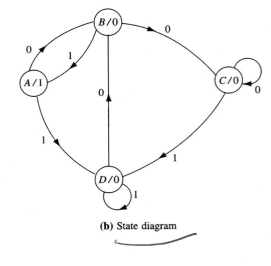

(b) State diagram

FIGURE 5.25

the state. Thus the state and the output are labeled within the node, separated by a slash. An edge corresponds to a state transition, and the input causing a transition is given beside the edge.

A sequential circuit can be represented either by a Moore model or by a Mealy model, and conversion from one model to the other is always possible. Let us first consider the conversion of the Mealy model of a sequential circuit to an equivalent Moore model [1].

EXAMPLE 5.1

The state table of a Mealy-type sequential circuit is shown in Figure 5.26. If in the Mealy-type circuit, a next state entry S is always associated with the same output Z, then S will be associated with the output Z in the state table of the equivalent Moore-type circuit. For example, in Figure 5.26, next state entries B, D, and E are always associated with outputs 0, 0, and 1 respectively. Hence, in the equivalent Moore-type circuit shown in Figure 5.27, B is associated with an output of 0, D with 0, and E with 1.

FIGURE 5.26
A Mealy-type sequential circuit

Present State	Input	
	$x = 0$	$x = 1$
A	B,0	A,1
B	D,0	C,1
C	B,0	C,0
D	E,1	E,1
E	A,0	B,0

If a state S is the next state entry for several states in the Mealy-type circuit and is associated with n different outputs, then in the state table of the equivalent Moore-type circuit, state S is replaced by n different states. The next state entries corresponding to each of theses new states are identical to the next state entries for S in the Mealy-type cir-

FIGURE 5.27
State table of equivalent Moore-type circuit

Present State	Input		Output
	$x = 0$	$x = 1$	
A,0	B	A,1	0
A,1	B	A,1	1
B	D	C,1	0
C,0	B	C,0	0
C,1	B	C,0	1
D	E	E	0
E	A,0	B	1

cuit. For instance, A as a next state entry is associated with both outputs 0 and 1 in Figure 5.26. Hence, A is replaced by the two states $A,0$ and $A,1$ in Figure 5.27, $A,0$ being associated with output 0 and $A,1$ being associated with output 1. The next state entries for both $A,0$ and $A,1$ are B when $x = 0$, and $A,1$ when $x = 1$. The reason for $A,1$ being the next state entry when $x = 1$ is because in Figure 5.26, the next state for A when $x = 1$ is A and the associated output entry is 1. Similarly state C in Figure 5.26 is replaced by two states $C,0$ and $C,1$ in Figure 5.27. ∎

Let us now illustrate the conversion of a Moore-type circuit to an equivalent Mealy-type circuit.

■ **EXAMPLE 5.2**

The state table of a Moore-type circuit is shown in Figure 5.28. In converting a Moore-type circuit to a Mealy-type circuit, if a state S is associated with an output Z, then the output associated with the next state S in the state table of the Mealy-type circuit will be Z. The state table of a Mealy-type circuit is derived from Figure 5.28 as shown in Figure 5.29.

FIGURE 5.28
A Moore-type circuit

Present State	Input		Output
	$x = 0$	$x = 1$	
A	B	C	0
B	A	C	0
C	D	E	1
D	C	B	0
E	A	F	1
F	A	E	1
	Next State		

FIGURE 5.29
State table of the equivalent Mealy-type circuit

Present State	Input	
	$x = 0$	$x = 1$
A	B,0	C,1
B	A,0	C,1
C	D,0	E,1
D	C,1	B,0
E	A,0	F,1
F	A,0	E,1
	Next State, Output	

∎

5.8 ANALYSIS OF SYNCHRONOUS SEQUENTIAL CIRCUITS

In the previous section we discussed several models for sequential circuits. Such circuits are used to perform many different functions. This section covers the analysis of these circuits. As will be seen, the analysis of the behavior of such circuits identifies the processes that are required to synthesize them. We shall consider the sequential circuit shown in Figure 5.30. The circuit has one input x and one output z. Two JK flip-flops are used as memory elements that define the four possible states of the circuit, $y_1 y_2 = 00, 01, 10$, or 11. The equations that describe the circuit's operation, known as the **design equations**, can be derived directly from Figure 5.30.

$$J_1 = x + y_2 \qquad J_2 = xy_1$$
$$K_1 = x + \bar{y}_2 \qquad K_2 = \bar{x}$$
$$z = y_1 \cdot y_2$$

The characteristic equation for JK flip-flops was derived in Section 5.4. It is repeated here:

$$Q_{t+1} = J\bar{Q}_t + \bar{K}Q_t$$

where Q_t and Q_{t+1} are respectively the present and the next state of a flip-flop. By substituting J_1, K_1 and J_2, K_2 in this equation, the next state functions of the two flip-flops are obtained:

$$(y_1)_{t+1} = (x + y_2)\bar{y}_1 + \overline{(x + \bar{y}_2)}y_1$$
$$= x\bar{y}_1 + \bar{y}_1 y_2 + \bar{x}y_1 y_2$$
$$= x\bar{y}_1 + \bar{x}y_2$$
$$(y_2)_{t+1} = xy_1\bar{y}_2 + \bar{x} \cdot y_2$$
$$= xy_1\bar{y}_2 + \bar{x}y_2$$
$$= xy_1 + \bar{x}y_2$$

It is now possible to construct a table (Table 5.2) that gives the next state values of the flip-flops for given present state values and for a given input. This form of the state table is known as a **transition table**. The two output columns of the table result from the interpretation of the output equation $Z = y_1 \cdot y_2$. Thus the output of the circuit is 1 when the present state of the circuit is $y_1 y_2 = 11$, irrespective of the input value. Replacing $y_1 y_2 = 00, 01, 10, 11$ by A, B, C, and D respectively we can derive the state table (Table 5.3) of the circuit from its transition table.

FIGURE 5.30
An example of a sequential circuit

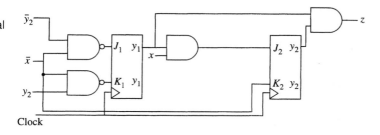

TABLE 5.2
Transition table for the circuit of Figure 5.30

Present State		Input		Output	
y_1	y_2	$x = 0$	$x = 1$	Z (when $x = 0$)	Z (when $x = 1$)
0	0	00	10	0	0
0	1	10	11	0	0
1	0	00	01	0	0
1	1	10	01	1	1
		Next State			

TABLE 5.3
State table of the circuit of Figure 5.30

Present State	Input	
	$x = 0$	$x = 1$
A	A,0	C,0
B	C,0	D,0
C	A,0	B,0
D	C,1	B,1

The state diagram of the circuit can be derived from its state table and is shown in Figure 5.31.

FIGURE 5.31
State diagram of the circuit of Figure 5.30

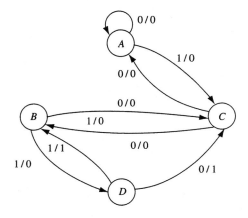

EXERCISES

1. A D flip-flop is connected as shown. Determine the output of the flip-flop for ten clock pulses assuming it has initially been reset. What function does this configuration perform?

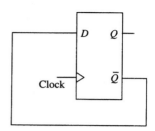

2. Modify a JK flip-flop such that when both J and K inputs are at logic 0, the flip-flop is reset.
3. Modify a T flip-flop such that it functions as a JK flip-flop.
4. Assume a D flip-flop with a separate set (S) and reset (R) inputs. How can this flip-flop be configured such that its output will be set to logic 1, when both S and R inputs are high simultaneously?
5. Modify a T flip-flop such that it functions as a D flip-flop.
6. A sequential circuit uses two D flip-flops as memory elements. The behavior of the circuit is described by the following equations:

$$Y_1 = y_1 + \bar{x}y_2$$
$$Y_2 = x\bar{y}_1 + \bar{x}y_2$$
$$z = \bar{x}y_1 y_2 + x\bar{y}_1 \bar{y}_2$$

Draw the state diagram of the circuit.

7. Derive the transition table for the following circuit.

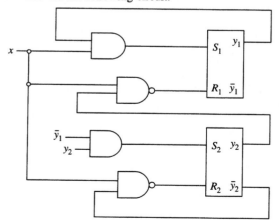

8. For the circuit shown, fill in the values for the J and K inputs, and the output values in the table.

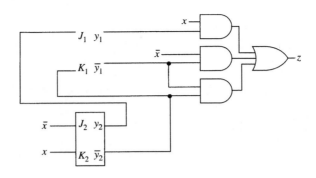

	x = 0		x = 1		z
y_1y_2	J_1,K_1	J_2,K_2	J_1,K_1	J_2,K_2	$x = 0, x = 1$
0 0					
0 1					
1 1					
1 0					

9. Derive the state tables for the following circuits:

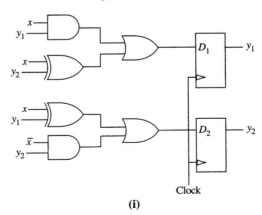

(i)

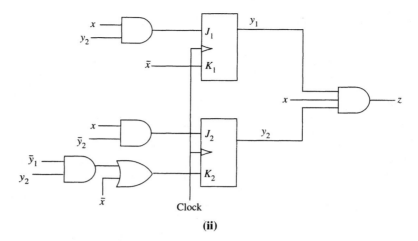

(ii)

10. Convert the following state tables for Mealy-type sequential circuits into those of Moore-type circuits.

	$x = 0$	$x = 1$
A	B,0	A,0
B	B,0	C,0
C	D,0	A,0
D	B,0	C,1

	$x = 0$	$x = 1$
A	B,0	C,1
B	A,0	E,0
C	E,1	D,0
D	B,1	A,1
E	C,0	D,1

11. Convert the following state tables for Moore-type sequential circuits into those of Mealy-type circuits.

	$x = 0$	$x = 1$	z
A	B	A	1
B	B	C	0
C	D	A	1
D	B	C	0

	$x = 0$	$x = 1$	z
A	B	A	0
B	D	C	1
C	B	C	1
D	A	E	0
E	E	D	0

References

1. C. L. Sheng, *Introduction to Switching Logic*, International Text Book Co., 1972.

6
Synchronous Sequential Circuits

6.1 INTRODUCTION

The analysis of a synchronous sequential circuit in Section 5.7 identified the major steps required for synthesizing such circuits; these steps have to be executed in sequence as shown in Figure 6.1. This chapter discusses each step individually and demonstrates that collectively these steps constitute a design procedure for implementing arbitrary synchronous sequential circuits. Henceforth, unless indicated otherwise, a sequential circuit will mean a synchronous sequential circuit.

The purpose of the first step in Figure 6.1 is to provide a precise definition of the intended behavior of the circuit to be designed. This definition should not constrain the means by which the circuit achieves the desired behavior, but it should completely define

Specify the problem.
↓
Derive the state diagram or the state table.
↓
Reduce the number of states.
↓
Choose a state assignment.
↓
Construct the transition table.
↓
Develop the design equations.
↓
Implement the circuit.

FIGURE 6.1
Design procedure for sequential circuits

the external characteristics of the circuit such that it is possible to verify the circuit's behavior from its specification. In other words, the specification should define a **black box,** whose behavior is known but whose internal construction is unknown.

In the second step of the design process, the specification of the circuit is expressed in terms of the states of the circuit. No formal procedure is available that can be used to derive state diagrams or tables. In fact, this step may be considered as the most difficult part of the design process, and only with experience can a logic designer acquire the skill to describe the state-to-state behavior of a sequential circuit. The third step is generally known as **state minimization** and consists of removing the equivalent states (if there are any) from the state table derived in the second step. This results in a state table with fewer states, which often leads to the simplification of the logic needed to realize the state table.

In the fourth step a unique binary code is assigned to each state; this is known as **state assignment.** The problem is to assign codes to different states such that more economic logic realization than that obtained by arbitrary assignment can be achieved. In the fifth step, the type of flip-flops to be used is decided and the Boolean logic equations (known as **excitation equations**) are derived for the flip-flops from the transition table. This table is also used to derive the output logic equations. Finally, the logic diagram of the sequential circuit is drawn using the chosen flip-flops and the logic equations derived in the fifth step. In the following sections we shall examine each of these steps in detail.

6.2 PROBLEM DEFINITION OF SEQUENTIAL CIRCUITS

As mentioned before, the development of state diagrams/state tables from the original specification of a circuit is mainly an intuitive process, and is heavily dependent on past experience. Either a Moore model or a Mealy model can be used to represent a sequential circuit; however, in practice the Mealy model is often preferred because it is more general.

■ EXAMPLE 6.1

Let us derive the state diagram for a synchronous sequential circuit required to recognize the 4-bit sequence 1101 and to produce an output 1 whenever the sequence occurs in a continuous serial input. For example, if the input sequence is 0101101101011, the output sequence is 000000100100.

We assume an initial state A where the circuit waits to receive the first input symbol:

At this state the circuit can receive either a 1 or 0. There is no change in state if 0 is received (indicated by a self-loop). If a 1 is applied, the circuit goes to a new state B with an output 0:

If while in state B, a 1 is received (i.e., the sequence 11), the circuit changes to state C; on the other hand, a 0 input takes the circuit back to state A:

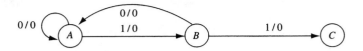

When in state C, if a 1 is received the circuit remains in the same state. The circuit moves to a new state D if a 0 is applied (i.e., the sequence is 110).

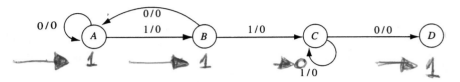

The next input symbol will be the fourth bit of the 4-bit sequence; therefore, the circuit must decide whether or not the sequence is the one to be recognized. If a 1 is applied, the sequence is correct and the circuit changes to B, giving the required output. However, if a 0 is received when the circuit is in state D, it returns to state A to await the start of another sequence. This completes the derivation of the state diagram for the sequential circuit. Figure 6.2 shows the state diagram.

It is now possible to construct a state table with the aid of this state diagram. Table 6.1 shows the resulting state table.

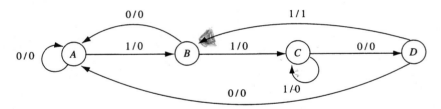

FIGURE 6.2
State diagram for 1101-sequence detector

TABLE 6.1
State table for the 1101-sequence detector

Present State	Input	
	$x = 0$	$x = 1$
A	A,0	B,0
B	A,0	C,0
C	D,0	C,0
D	A,0	B,1
	Next State, Output	

■ **EXAMPLE 6.2**

Let us derive the state diagram and the state table for a sequential circuit that has a single input x and a single output z. It examines incoming serial data in consecutive sequences

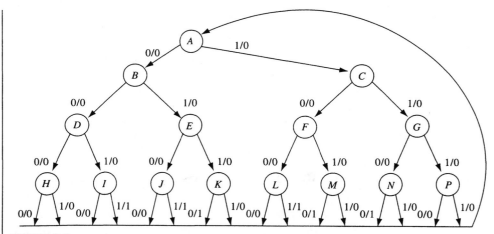

FIGURE 6.3
State diagram for a 2-out-of-4 code detector

of four bits. The output of the circuit is 1 if and only if an input sequence is a 2-out-of-4 code word (i.e., there are exactly two 1's in a 4-bit sequence).

Assume the initial state is A. Since the circuit has only a single serial input, each state in the state diagram will have two transition edges, one corresponding to input 0 and the other corresponding to input 1. Besides, an input sequence consists of 4 bits, so we must go back to the initial state after 4 bits have been examined. Figure 6.3 shows the complete state diagram for the desired circuit. The information in the diagram is transferred to the state table (Table 6.2).

TABLE 6.2
State table for a 2-out-of-4 code detector

	Input	
Present State	$x = 0$	$x = 1$
A	B,0	C,0
B	D,0	E,0
C	F,0	G,0
D	H,0	I,0
E	J,0	K,0
F	L,0	M,0
G	N,0	P,0
H	A,0	A,0
I	A,0	A,1
J	A,0	A,1
K	A,1	A,0
L	A,0	A,1
M	A,1	A,0
N	A,1	A,0
P	A,0	A,0

As can be seen from Figure 6.3, each combination of 4 bits has been taken into consideration while deriving the state diagram. This has produced quite a few redundant states in the state diagram/state stable; a state is **redundant** if its function can be served by another state in the circuit. In the following section, various techniques available for determining redundant states are considered.

6.3 STATE REDUCTION

The number of states in a sequential circuit is closely related to the complexity of the resulting circuit. It is therefore desirable to know when two or more states play identical roles (i.e., are equivalent in all respects). The process of eliminating the equivalent or redundant states from a state table/diagram is known as **state reduction.** It corresponds to the process of minimization of logic functions in combinational circuit design.

Let us first consider an intuitive approach for state reduction. Table 6.3 shows the state table of an arbitrary sequential circuit. It can be seen from the table that present states A and F both have the same next states, B (when $x = 0$) and C (when $x = 1$). They also produce the same outputs 1 (when $x = 0$) and 0 (when $x = 1$). It can be reasoned that if the next states are the same, the outputs produced to any subsequent inputs will also be the same. Thus one of the states, A or F, can be removed from the state table. For example, if we remove row F from Table 6.3 and replace all F's by A's in the columns, the state table is modified as in Table 6.4. It is apparent from Table 6.4 that states B and E

TABLE 6.3
A state table

Present State	Input	
	$x = 0$	$x = 1$
A	B,1	C,0
B	F,0	D,0
C	D,1	E,1
D	F,0	E,1
E	A,0	D,0
F	B,1	C,0
	Next State, Output	

TABLE 6.4
State F removed

Present State	Input	
	$x = 0$	$x = 1$
A	B,1	C,0
B	A,0	D,0
C	D,1	E,1
D	A,0	E,1
E	A,0	D,0
	Next State, Output	

are equivalent. Removing E and replacing E's by B's results in the reduced table shown in Table 6.5. The removal of the equivalent states has reduced the number of states in the circuit from six to four. Note that in the original state table (Table 6.3) states B and E are not equivalent, because the next states for B and E when $x = 0$ were different. Thus, two states may be equivalent even though their next states are not the same, provided there is an equivalence between the unlike states. Two states are considered to be **equivalent** if and only if for every input sequence the circuit produces the same output sequence irrespective of which one of the two states is the starting state. State equivalence is a mathematical equivalence relationship. This, if states A and B are equivalent and states B and C are equivalent, then A is equivalent to C; the three states form a set of equivalent states. If no two states in a circuit are equivalent, then the circuit is reduced.

Partitioning

The equivalent sets of states in a sequential circuit can be determined by using a procedure based on partitioning. The first step is to partition the set of states of a circuit into a number of blocks so that all states in a block have identical output for each possible input. Consider for example Table 6.6. The output produced for each of the states A, C, and E is 0 for both $x = 0$ and $x = 1$. The outputs associated with the inputs 0 and 1 are respectively 1 and 0 for each of the states B, D, and F. Hence, the first partition Π_1 for the circuit is

$$\Pi_1 = (ACE)(BDF)$$

TABLE 6.5
Reduced state table

Present State	Input	
	$x = 0$	$x = 1$
A	B,1	C,0
B	A,0	D,0
C	D,1	B,1
D	A,0	B,1
	Next State, Output	

TABLE 6.6

Present State	Input	
	$x = 0$	$x = 1$
A	B,0	A,0
B	D,1	D,0
C	A,0	C,0
D	B,1	F,0
E	B,0	E,0
F	D,1	E,0
	Next State, Output	

The next step of the procedure is to derive a partition Π_2 by placing two states in the same block if for each input value their next states lie in a common block of Π_1. In the example of Table 6.6 the next states for A, C, and E (i.e., states in the first block of Π_1) corresponding to $x = 0$ are B, A, and B respectively. Since A and B are in different blocks of Π_1, partition Π_2 must separate C from A and E. For $x = 1$ the next states A, C, and E lie in the same block. In the second block of Π_1, the next states for B, D, and F with $x = 0$ belong to the same block of Π_1. However, for $x = 1$ the next state of F lies in a different block of Π_1 than the next states of B and D. Hence the block (BDF) is split into blocks $(F)(BD)$. Thus, partition Π_2 is

$$\Pi_2 = (C)(AE)(F)(BD)$$

Partition Π_3 can be formed in a similar manner. The next states for A and E lie in the same blocks of Π_2 for both $x = 0$ and $x = 1$, so block (AE) cannot be separated. However, the next states for B and D with $x = 1$ lie in different blocks of Π_2, so block (BD) must be split into blocks $(B)(D)$. Therefore

$$\Pi_3 = (C)(AE)(F)(B)(D)$$

The next partition, Π_4, is derived from Π_3 in the same way and is given by

$$\Pi_4 = (C)(AE)(F)(B)(D)$$

Since Π_3 and Π_4 are identical, all subsequent partitions Π_5, Π_6, ... will also be identical to Π_3. Therefore, if a partition Π_{k+1} is identical to its predecessor partition Π_k, the partitioning process is terminated, and partition Π_k is said to be an **equivalence partition**. All states belonging to a block in the equivalence partition are equivalent. For the example under consideration, Π_3 is the equivalence partition and states A and E are equivalent. The original state table (Table 6.6) can be reduced by eliminating row E and replacing each E by an A (Table 6.7).

■ **EXAMPLE 6.3**

Let us consider the state table of a circuit shown in Table 6.8a. The partitions for the state table are

$$\Pi_1 = (AFG)\,(BCDE)$$
$$\Pi_2 = (AFG)\,(BD)\,(CE)$$

TABLE 6.7
Reduced state table

Present State	Input	
	$x = 0$	$x = 1$
A	B,0	A,0
B	D,1	D,0
C	A,0	C,0
D	B,1	F,0
F	D,1	A,0
	Next State, Output	

172 CHAPTER 6 / SYNCHRONOUS SEQUENTIAL CIRCUITS

$$\Pi_3 = (AFG)(BD)(CE)$$

Thus, Π_2 is the equivalence partition. Assuming $(AFG) = \alpha$, $(BD) = \beta$, and $(CE) = \gamma$, the reduced state table will be as shown in Table 6.8b. ∎

Implication Table

An alternative method for finding equivalent states in a sequential circuit is based on an **implication table,** which shows the necessary conditions or **implications** that exist between all possible equivalent pairs of states. We shall consider the state table of Table 6.6 to explain the method.

The first step is to form a table with the rows consisting of all but the first states and the columns consisting of all states except the last. The resulting table has as many

TABLE 6.8

Present State	Input	
	$x = 0$	$x = 1$
A	F,0	D,1
B	C,1	F,1
C	F,1	B,1
D	E,1	G,1
E	A,1	D,1
F	G,0	B,1
G	A,0	D,1

Next State, Output

(a) A state table

Present State	Input	
	$x = 0$	$x = 1$
α	α,0	β,1
β	γ,1	α,1
γ	α,1	β,1

Next State, Output

(b) Reduced state table

FIGURE 6.4

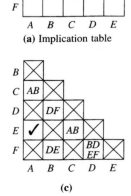

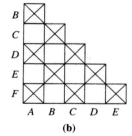

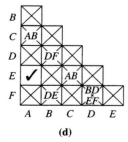

cells as there are permissible state pairs. Figure 6.4a shows the implicant table for our example. Next we consider whether a state pair in the implicant table is equivalent or not; a state pair cannot possibly be equivalent if the states have different outputs. A cross is placed in a cell of the implicant table if the corresponding state pair has differing outputs (Fig. 6.4b). The non-equivalent state pairs are called **incompatibles.** The vacant cells must now be completed.

Each vacant cell is filled with the required state pairs whose equivalence implies the equivalence of the state pair that defines the vacant cell. For example, consider the cell corresponding to the state pair AC. We enter into AC the state pair AB, which must be equivalent in order for A and C to be equivalent (Fig. 6.4c). A check is inserted in a cell if the corresponding state pair is equivalent. For example, in Figure 6.4c the cell defined by the state pair AE has a check, indicating that the states A and E are equivalent. When the table is completed it is examined column by column, starting from the extreme right-hand column, to determine whether any other cells should be crossed out.

In Figure 6.4c the first cell to be considered is the one defined by D and F; it contains the pair BD and EF. Since the cell defined by E and F was already crossed out, it follows that any state pair whose equivalence is implied by the equivalence of E and F must also be crossed out. Hence, the cell corresponding to D and F is crossed out (Fig. 6.4d). The procedure is repeated until no further cells can be crossed out. The state pairs corresponding to the cells that have not been crossed out are the equivalent states. The only equivalent state pair in Figure 6.4d is AE. Thus, the equivalence partition Π is

$$\Pi = (AE)(B)(C)(D)(F)$$

Note that this equivalence partition is identical to the one derived earlier by partitioning.

■ | EXAMPLE 6.4

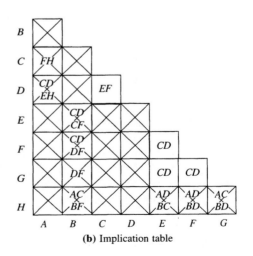

Present state	Input	
	$x = 0$	$x = 1$
A	D,0	H,1
B	F,1	C,1
C	D,0	F,1
D	C,0	E,1
E	C,1	D,1
F	D,1	D,1
G	D,1	C,1
H	B,1	A,1
	Next state, Output	

(a) State table

(b) Implication table

FIGURE 6.5

Let us consider the application of the implication table in deriving the equivalence partition for the state table shown in Figure 6.5a. The corresponding implication table is shown in Figure 6.5b. As can be seen from Figure 6.5b, the equivalence partition is

FIGURE 6.5
(Continued)

Present State	Input x = 0	x = 1
α	α,0	β,1
β	α,1	α,1
γ	α,0	ω,1
δ	β,1	α,1
ω	δ,1	γ,1

Next State, Output

(c) Reduced state table

$$\Pi = (CD)(EF)(EG)(FG)(A)(B)(H)$$

By using the transitivity relationship the state pairs (EF), (EG), and (FG) can be grouped into a set of states (EFG). Thus

$$\Pi = (CD)(EFG)(A)(B)(H)$$

Assigning $(CD) = \alpha$, $(EFG) = \beta$, $(A) = \gamma$, $(B) = \delta$, and $(H) = \omega$, the reduced state table can be derived as shown in Figure 6.5c. ■

6.4 DERIVATION OF DESIGN EQUATIONS

Once the reduced state table has been obtained, the next step in the design process is to encode the states in binary form. This is known as **state assignment.** A state assignment must allocate a unique binary combination to each state. In order to obtain a distinct binary combination for each state of an *n*-state circuit, we need *s* secondary input variables such that $s = \lceil \log_2 n \rceil$ (i.e., $s \geq \log_2 n$). Each secondary variable is generated by a flip-flop. Thus, the number of flip-flops required to implement an *n*-state sequential circuit is $\lceil \log_2 n \rceil$.

The flip-flops in a sequential circuit are **excited** to take on the various states in proper sequence as required by the state table of the circuit. Suppose the current content (i.e., the present state) of a *D* flip-flop is 0. To change the content of the flip-flop from 0 to 1, its input must be set to 1. In other words, the *D* flip-flop must be excited to 1 in order to make a transition from the present state 0 to next state 1. The present state–to–next state transition input requirement of any flip-flop can be derived from its excitation table (Table 6.9). For example, if the present state of a *JK* flip-flop is 0 and it has to be changed to 1, then the *J* input should be set to 1, while the *K* input can be either 0 or 1, i.e., a don't care (—), because it does not affect the next state of the flip-flop. As we shall see shortly, the information contained in the excitation table is necessary to obtain the design equation for each flip-flop used in a circuit.

■ **EXAMPLE 6.5**

Let us derive the design equations for the sequence detector circuit of Example 6.1. The state table for the circuit is shown in Table 6.10. Next, we select the state assignment. The man-

TABLE 6.9
Excitation table for flip-flops

Present State	Next State	D Flip-Flop	JK Flip-Flop		T Flip-Flop
Q_t	Q_{t+1}	D	J	K	T
0	0	0	0	—	0
0	1	1	1	—	1
1	0	0	—	1	1
1	1	1	—	0	0

ner in which the binary combinations are assigned to the states of a circuit has a considerable effect on the complexity of the combinational logic necessary to implement the design equations. We consider some general rules for finding reasonably good state assignments in Section 6.5; a systematic technique that yields optimum state assignments will also be presented. The state assignment for the sequence detector circuit is arbitrarily chosen as follows:

$$A = 00$$
$$B = 01$$
$$C = 10$$
$$D = 11$$

Since there are four states A, B, C, and D, a minimum of two state variables is required to represent them; consequently, two flip-flops are needed.

We may choose any type of flip-flop to implement the memory portion of a sequential circuit. Let us use D flip-flops for this example. Before we can derive the excitation equations for the D flip-flops, the transition table corresponding to the state assignment has to be derived from the state table. The entries in the transition table (Table 6.11) represent the next states of the D flip-flops for each combination of present state and input value. Since the next state value of a D flip-flop is the same as the excitation input, the transition table entries in effect specify the required excitation of the D flip-flops. Karnaugh maps can now be plotted for each of the flip-flop excitation inputs. These are shown in Figures 6.6a and b; the positions of rows 10 and 11 are swapped in these maps in order to satisfy the requirements of a Karnaugh map. The Karnaugh map for output Z is shown in Figure 6.6c. Hence, the Boolean equations needed to implement the sequence detector circuit are as follows:

$$D_1 = y_1\bar{y}_2 + x\bar{y}_1 y_2$$
$$D_2 = \bar{x}y_1\bar{y}_2 + xy_1 y_2 + x\bar{y}_1\bar{y}_2$$
$$Z = xy_1 y_2$$

The logic diagram for the completed design is shown in Figure 6.7.

TABLE 6.10
Reduced state table

Present State	Input $x = 0$	Input $x = 1$
A	A,0	B,0
B	A,0	C,0
C	D,0	C,0
D	A,0	B,1
	Next State, Output	

TABLE 6.11
Transition table derived from the reduced state table of Table 6.10

Present State $y_1 y_2$	Input $x = 0$	Input $x = 1$
$A \rightarrow 00$	00,0	01,0
$B \rightarrow 01$	00,0	10,0
$C \rightarrow 10$	11,0	10,0
$D \rightarrow 11$	00,0	01,1

y_1 y_2	Input $x = 0$	Input $x = 1$
0 0	0	0
0 1	0	1
1 1	0	0
1 0	1	1

(a) Karnaugh map for D_1

y_1 y_2	Input $x = 0$	Input $x = 1$
0 0	0	1
0 1	0	0
1 1	0	1
1 0	1	0

(b) Karnaugh map for D_2

y_1 y_2	Input $x = 0$	Input $x = 1$
0 0	0	0
0 1	0	0
1 1	0	1
1 0	0	0

(c) Karnaugh map for output Z

FIGURE 6.6

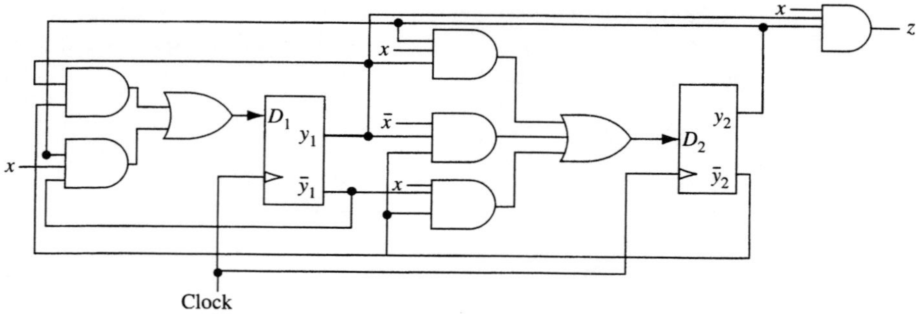

FIGURE 6.7
Logic diagram of the sequence detector circuit ■

6.4 DERIVATION OF DESIGN EQUATIONS   177

EXAMPLE 6.6

Let us derive the design equations for the sequential circuit specified in Table 6.12. We will use *JK* flip-flops to realize the circuit. By choosing the state assignment

$$A = 00$$
$$B = 01$$
$$C = 10$$
$$D = 11$$

the transition table shown in Table 6.13 is obtained.

Next we derive the excitation equations for the two flip-flops from the Karnaugh maps of their *J* and *K* inputs (Fig. 6.8); the maps are formed by applying the excitation table of *JK* flip-flops to the transition table of the circuit. The Karnaugh map for the output function is shown in Figure 6.9.

The design equations for the sequential circuit can be derived from the Karnaugh maps:

$$J_1 = \bar{x}_1 + x_2 \qquad K_1 = 1$$
$$J_2 = x_2 \qquad K_2 = \bar{x}_2$$
$$Z = x_1 \cdot y_2 + x_1 \cdot \bar{x}_2 + \bar{x}_2 \cdot y_2$$

The actual implementation of the circuit is shown in Figure 6.10.

The Karnaugh maps for *T* flip-flop realization of this sequential circuit can be derived from Table 6.9; these are shown in Figure 6.11. The output map remains the same as shown in Figure 6.9. The design equations obtained from Figures 6.9 and 6.11 are as follows:

$$T_1 = \bar{x}_1 + x_2 + y_1$$
$$T_2 = \bar{x}_2 \cdot y_2 + x_2 \cdot \bar{y}_2$$
$$Z = x_1 \cdot y_2 + x_1 \cdot \bar{x}_2 + \bar{x}_2 \cdot y_2$$

TABLE 6.12
State table of a sequential circuit

		Input x_1x_2			
Present State		00	01	10	11
A		C,0	D,0	A,1	D,0
B		C,1	D,0	A,1	D,1
C		A,0	B,0	A,1	B,0
D		A,1	B,0	A,1	B,1
				Next State, Output	

TABLE 6.13
Transition table

		Input x_1x_2			
y_1y_2		00	01	10	11
00		10,0	11,0	00,1	11,0
01		10,1	11,0	00,1	11,1
10		00,0	01,0	00,1	01,0
11		00,1	01,0	00,1	01,1

FIGURE 6.8
Karnaugh maps for JK flip-flop realization

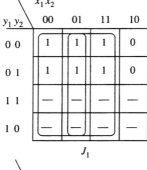

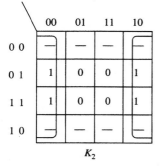

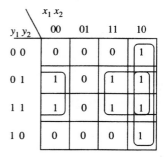

FIGURE 6.9
Karnaugh map for output Z

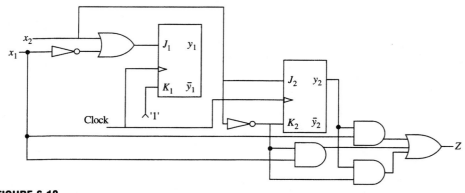

FIGURE 6.10
Realization of the sequential circuit specified in Table 6.12

FIGURE 6.11
Karnaugh maps for T flip-flop realization

$y_1 y_2 \backslash x_1 x_2$	00	01	11	10
00	1	1	1	0
01	1	1	1	0
11	1	1	1	1
10	1	1	1	1

$y_1 y_2 \backslash x_1 x_2$	00	01	11	10
00	0	1	1	0
01	1	0	0	1
11	1	0	0	1
10	0	1	1	0

The corresponding circuit is shown in Figure 6.12. ■

As we saw in the previous examples, a sequential circuit with n states requires $\lceil \log_2 n \rceil$ flip-flops. However, there are occasions when the number of states in a sequential circuit is fewer than the maximum number that can be specified with $\lceil \log_2 n \rceil$ flip-flops. For example, a sequential circuit with five states requires three flip-flops, but three flip-flops can specify up to eight states, so there are three unused or invalid states in the circuit. Normally, when power is turned on, the flip-flops in a sequential circuit can settle in any state, including one of the invalid states. In that case it is necessary to ensure that the circuit goes to a valid or specified state with the fewest number of clock pulses (a circuit changes state only after the application of a clock pulse). Once the circuit goes to a valid state, it can continue to operate as required.

■ **EXAMPLE 6.7**

Let us design the sequential circuit specified by Table 6.14. The circuit has six states, so three flip-flops will be needed to implement the circuit. However, three flip-flops can specify eight states, so there are two invalid states in the circuit. In order to make sure the circuit is transferred to a valid state (i.e., $A, B, \ldots, F$) from an invalid state with one clock pulse, it will be necessary to augment the state table as shown in Table 6.15. The choice

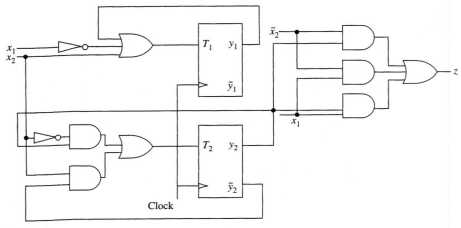

FIGURE 6.12
T flip-flop implementation

TABLE 6.14
State table of a sequential circuit

Present State	Input	
	$x = 0$	$x = 1$
A	A,0	B,0
B	B,0	C,0
C	C,0	D,0
D	D,0	E,0
E	E,0	F,0
F	F,0	A,1

TABLE 6.15
Augmented state table

	Present State	Input	
		$x = 0$	$x = 1$
Valid States	A	A,0	B,0
	B	B,0	C,0
	C	C,0	D,0
	D	D,0	E,0
	E	E,0	F,0
	F	F,0	A,1
Invalid States	G	E,0	B,0
	H	D,0	A,0

of next states for the invalid states G and H are arbitrary in this case; in practice, the next states are selected such that a minimal increase in the circuitry is needed to implement the augmented state table as compared to the original state table. The state assignment for the circuit is arbitrarily chosen as follows:

$$A = 000$$
$$B = 001$$
$$C = 010$$
$$D = 011$$
$$E = 100$$
$$F = 101$$
$$G = 110$$
$$H = 111$$

The resulting transition table is shown in Table 6.16. The Karnaugh maps for a *JK* flip-flop realization of the circuit are shown in Figure 6.13. The output map is shown in Figure 6.14. The design equations obtained from Figures 6.13 and 6.14 are

6.4 DERIVATION OF DESIGN EQUATIONS — 181

TABLE 6.16
Transition table

Present State $y_1y_2y_3$	Input	
	$x = 0$	$x = 1$
000	000,0	001,0
001	001,0	010,0
010	010,0	011,0
011	011,0	100,0
100	100,0	101,0
101	101,0	000,1
110	100,—	001,—
111	011,—	000,—

FIGURE 6.13
Karnaugh maps for JK flip-flop realization

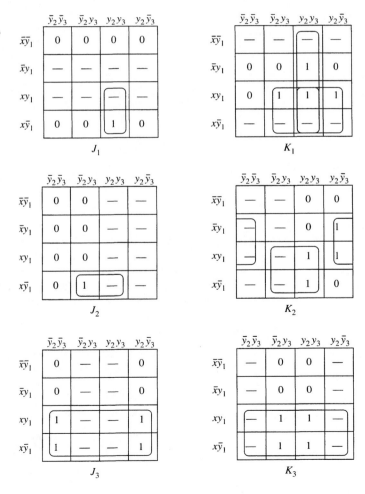

FIGURE 6.14
Karnaugh map for output Z

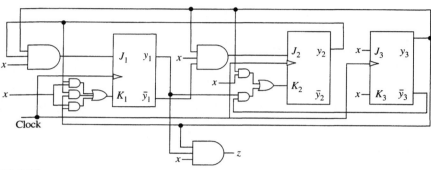

FIGURE 6.15

$$J_1 = x \cdot y_2 \cdot y_3 \qquad K_1 = x \cdot y_2 + xy_3 + y_2 y_3$$
$$J_2 = x \cdot \bar{y}_1 \cdot y_3 \qquad K_2 = x \cdot y_3 + y_1 \bar{y}_3$$
$$J_3 = x \qquad\qquad K_3 = x$$
$$Z = x \cdot y_1 \cdot y_3$$

The circuit implementation is shown in Figure 6.15. ∎

6.5 STATE ASSIGNMENT

So far in all the design problems we have considered, an arbitrary state assignment has been adopted. For example, in the 1101 sequence detector circuit designed in Figure 6.7 the state assignment selected was

$$A = 00, B = 01, C = 10, D = 11$$

However, a different state assignment may be chosen, and this will lead to a different set of design equations. For example, if we choose the following assignment

$$A = 00, B = 11, C = 01, D = 10$$

then the design equations can be derived as shown in Figure 6.16, and are given by

$$D_1 = x \cdot \bar{y}_2 + \bar{x} \cdot \bar{y}_1 \cdot y_2$$
$$D_2 = x$$
$$Z = x \cdot y_1 \cdot \bar{y}_2$$

FIGURE 6.16

$y_1 y_2$ \	$x=0$	$x=1$
0 0	00,0	11,0
1 1	00,0	01,0
0 1	10,0	01,0
1 0	00,0	11,1

(a) Transition table

$y_1 y_2$ \	$x=0$	$x=1$
0 0	0	1
0 1	1	0
1 1	0	0
1 0	0	1

(b) Karnaugh map for D_1

$y_1 y_2$ \	$x=0$	$x=1$
0 0	0	1
0 1	0	1
1 1	0	1
1 0	0	1

(c) Karnaugh map for D_2

$y_1 y_2$ \	$x=0$	$x=1$
0 0	0	0
0 1	0	0
1 1	0	0
1 0	0	1

(d) Karnaugh map for Z

Alternatively, the state assignment

$$A = 00, B = 10, C = 11, D = 01$$

will result in the following design equations (derived as shown in Fig. 6.17)

$$D_1 = x \qquad D_2 = y_1 \cdot y_2 + x \cdot y_1$$
$$Z = x \cdot \bar{y}_1 \cdot y_2$$

Either of these assignments will lead to a simpler circuit for the 1101 sequence detector circuit than that obtained by choosing the assignment in Example 6.5. This can be seen from Table 6.17, which shows a comparison of the number of gates required to implement the circuit for each of the three assignments. In fact the third assignment turns out to be the best; as we shall see later, this is not just a happy coincidence.

As we saw in the preceding example, the criterion for a good state assignment is that it should result in simpler design equations. The problem associated with state assignment, therefore, is to select the state variables such that the complexity of the combinational logic required to control the memory elements of the sequential circuit is minimized. However, the number of possible state assignment increases very rapidly with the number of states of the sequential circuit. If a circuit has n states, $s = \lceil \log_2 n \rceil$ state variables are needed for an assignment; thus 2^s combinations of state variables are available. The first state of the circuit can be allocated any one of the 2^s combinations, the second state can be allocated any one of the remaining $2^s - 1$ combinations, etc.; hence the nth

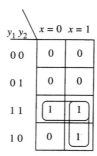

FIGURE 6.17

(a) Transition table

$y_1 y_2$	$x=0$	$x=1$
A= 0 0	00, 0	10, 0
D= 0 1	00, 0	10, 1
C= 1 1	01, 0	11, 0
B= 1 0	00, 0	11, 1

(b) Karnaugh map for D_1

$y_1 y_2$	$x=0$	$x=1$
0 0	0	1
0 1	0	1
1 1	0	1
1 0	0	1

(c) Karnaugh map for D_2

$y_1 y_2$	$x=0$	$x=1$
0 0	0	0
0 1	0	0
1 1	1	1
1 0	0	1

(d) Karnaugh map for Z

$y_1 y_2$	$x=0$	$x=1$
0 0	0	0
0 1	0	1
1 1	0	0
1 0	0	0

TABLE 6.17
Gate comparison for three state assignments

No.	State Assignment	No. of 2-Input ANDs	No. of 3-Input ANDs	No. of Inverters	No. of 2-Input ORs	No. of 3-Input ORs	Total No. of Gates
1	$A = 00, B = 01, C = 10, D = 11$	1	4	1	1	1	8
2	$A = 00, B = 11, C = 01, D = 10$	1	2	1	1	0	5
3	$A = 00, B = 10, C = 11, D = 01$	2	1	0	1	0	4

state of the circuit can be assigned any one of the $2^s - n + 1$ combinations of state variables. Thus there are

$$2^s \cdot (2^s - 1) \cdot \ldots \cdot (2^s - n + 1) = \frac{2^s!}{(2^s - n)!}$$

ways of assigning 2^s combinations of state variables to the n states. The state variables can be permuted in $s!$ ways. In addition, each state variable can be complemented, so the set of state variables s can be complemented in 2^s ways. Therefore, the number of unique state assignments is

TABLE 6.18
Number of state assignments

n	s	No. of State Assignments
2	1	1
3	2	3
4	2	3
5	3	140
6	3	420
7	3	840
8	3	840
9	4	10,810,800
10	4	75,675,600

$$\frac{2^s!}{(2^s - n)!\, s!2^s} = \frac{(2^s - 1)!}{(2^s - n)!\, s!}$$

Table 6.18 shows the number of state assignments for different values of n. Thus, even for a circuit with six states the number of possible state assignments to be considered is 420, and it rapidly rises to more than ten million for a circuit with nine states! Since the number of possible state assignments grows profusely with the number of internal states, it is almost impossible to try all possible assignments in order to select the one that leads to the simplest design equations. However, rather than using exhaustive evaluation, one may follow two simple rules that often result in good state assignments:

1. Assign adjacent codes (i.e., differing in one bit) to states with the same next state in a column.
2. Assign adjacent codes to states that are the next states of the same present state.

If there is any conflict in the adjacencies obtained by using these rules, then the adjacencies obtained from the first rule take precedence.

EXAMPLE 6.8

Let us apply the rules for state assignment to the four-state sequential circuit specified by Table 6.10. Using the first rule, states A and B should be given adjacent assignments because both of them go to state A for $x = 0$. Similarly, state pairs (B,D), (A,D), and (B,C) should be given adjacent assignments; the pair (A,D) appears twice. The application of the second rule shows that A and B should be given adjacent assignments because they are the next states of the present state A. For similar reasons state pairs (A,C) and (C,D) should be adjacent, with (A,B) appearing twice. ∎

The plotting of the three state assignments for the sequential circuit (Table 6.10) on two-variable Karnaugh maps is illustrated in Figure 6.18. It can be seen from the Karnaugh map for assignment III that it satisfies most of the adjacencies and hence produces a better result than the other assignments. It should be noted that although assignment II

FIGURE 6.18
Comparison of the state assignments for the circuit defined by Table 6.10

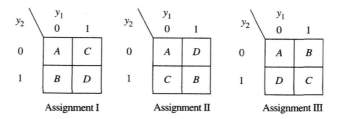

satisfies the same number of adjacencies as assignment III, it does not fulfill the adjacency requirement for the state pair (A,B) as determined by rule 1.

*State Assignment Based on Partitioning

An alternative way of obtaining good state assignments for sequential circuits is to select assignments such that each state variable corresponding to the next state depends on as few state variables of a present state as possible. In other words, if the next state variables depend on small subsets of the present state variables, the resulting excitation equations for the sequential circuit are considerably simplified. Hartmanis [1] has shown that a sequential circuit can have a state assignment with reduced dependency if there exists a partition with **substitution property** on the states of the circuit.

As defined in Chapter 3, a partition Π on a set of elements S is a collection of disjoint subsets of S such that their set union is S. The subsets of Π are called the **blocks** of S. A partition Π on a set of states S of a sequential circuit is said to have the **substitution property** (*SP*) if any two states belonging to a block of Π, under the same input combinations, move to next states that again belong to a common block of Π. This common block may or may not be the same block containing the two original states. For example, the following partitions on the states of the sequential circuit described by Table 6.19 have substitution properties:

$$\Pi_1 = (ABF)(CDE)$$
$$\Pi_2 = (AF)(CE)(BD)$$

The partitions with substitution properties for a given sequential circuit can be determined as follows:

TABLE 6.19
A state table

Present State	Input	
	$x = 0$	$x = 1$
A	C,0	A,1
B	D,1	F,1
C	A,0	B,0
D	B,1	F,0
E	F,0	B,0
F	E,0	F,1

i. Identify any two distinct states S_1 and S_2.
ii. Identify the pairs of states S_{1K} and S_{2K} to which S_1 and S_2 move if we apply the Kth input, $K = 1, 2, \ldots, m$.
iii. To this set of states, add those pairs that can be identified by the transitive law—i.e., if S_i and S_j are identified and S_j and S_k are also identified, then we have to identify S_i and S_k.
iv. Repeat the process, looking up the new identifications induced by the new pairs.

If after x steps, the $(x + 1)$th step does not yield any new identifications, a partition Π with the substitution property is obtained on the set of states of the circuit. If a nontrivial partition with the substitution property does not exist, then the process stops after identifying all states of the circuit. For a machine with n states, it is necessary to try $n(n - 1)/2$ distinct pairs of states before deciding whether or not a partition with the substitution property exists.

EXAMPLE 6.9

Let us apply the procedure for obtaining all partitions with the substitution property to the sequential circuit specified by Table 6.20. For this circuit we must consider $5(5 - 1)/2 = 10$ distinct pairs of states in order to determine all partitions with the substitution property. We start with the state pair (A,B). From Table 6.20 we see that when $x = 0$, both states A and B go to state B. The $x = 1$ column entries show that when $x = 1$, A goes to D and B goes to E. Since the pairs of states (A,B) and (D,E) are disjoint, it is not necessary to add any new state pairs because of the transitive law. This step may be represented as

$$(A,B) \rightarrow (B,B)(D,E)$$

Here the arrow signifies "implies" or "requires." Note that requirements such as (B,B) (i.e., B must be in the same block as B) are always satisfied and need not require further consideration.

Since the pairs of states (A,B) and (D,E) are disjoint, it is not necessary to add new state pairs because of the transitive law. States D and E have to be identified next. From the state table we see that D and E go to C when $x = 0$ and to A when $x = 1$:

$$(A,B) \rightarrow (D,E) \rightarrow (A,A)(C,C)$$

Since there are no more state pairs to be identified, the process is computed and we get the following partition

TABLE 6.20
State table for a sequential circuit

Present State	Input	
	$x = 0$	$x = 1$
A	B,0	D,1
B	B,1	E,0
C	B,0	D,0
D	C,1	A,0
E	C,0	A,1
	Next State, Output	

$$(A,B) \to (D,E) \to (A,A)(C,C) \equiv (A,B)(D,E)(C) \equiv \Pi_1$$

Continuing in the same manner for the other state pairs, we obtain the following partitions

$$(A,C) \to (B,B)(D,D) \equiv (A,C)(B)(D)(E) \equiv \Pi_2$$
$$(A,D) \to (B,C) \to (D,E) \equiv (A,D,E)(B,C) \equiv \Pi_3$$
$$(A,E) \to (B,C)(A,D) \to (D,E) \equiv (A,D,E)(B,C) \equiv \Pi_3$$
$$(B,C) \to (D,E) \to (C,C)(A,A) \equiv (A)(B,C)(D,E) \equiv \Pi_4$$
$$(B,D) \to (B,C)(A,E) \to (D,E)(A,D) \equiv (A,B,C,D,E) \equiv \Pi(I)$$
$$(B,E) \to (B,C)(A,E) \to (D,E)(A,D) \equiv (A,B,C,D,E) \equiv \Pi(I)$$
$$(C,D) \to (B,C)(A,D) \to (D,E) \equiv (A,B,C,D,E) \equiv \Pi(I)$$
$$(C,E) \to (B,C)(A,D) \to (D,E) \equiv (A,B,C,D,E) \equiv \Pi(I)$$
$$(D,E) \to (C,C)(A,A) = (A)(C)(D,E)(B) \equiv \Pi_5$$

Having obtained the partitions with the substitution property (Π_1, Π_2, Π_3, Π_4, and Π_5), we must next find the sum (lowest upper bound) and the product (greatest lower bound) of every pair of partitions.

$$\Pi_1 + \Pi_2 = (A,B,C)(D,E) = \Pi_6 \quad \text{(a new partition)}$$
$$\Pi_1 \cdot \Pi_2 = (A)(B)(C)(D)(E) = \Pi(0)$$
$$\Pi_1 + \Pi_3 = (A,B,C,D,E) = \Pi(I)$$
$$\Pi_1 \cdot \Pi_3 = (B)(A)(D,E)(C) = \Pi_5$$
$$\Pi_2 + \Pi_3 = (A,B,C,D,E) = \Pi(I)$$
$$\Pi_2 \cdot \Pi_3 = (A)(B)(C)(D)(E) = \Pi(0)$$
$$\Pi_1 + \Pi_4 = (A,B,C)(D,E) = \Pi_6$$
$$\Pi_1 \cdot \Pi_4 = (A)(B)(C)(D,E) = \Pi_5$$
$$\Pi_1 + \Pi_5 = (A,B)(C)(D,E) = \Pi_6$$
$$\Pi_1 \cdot \Pi_5 = (A)(B)(C)(D,E) = \Pi_5$$
$$\Pi_2 + \Pi_4 = (A,B,C)(D,E) = \Pi_6$$
$$\Pi_2 \cdot \Pi_4 = (A)(B)(C)(D)(E) = \Pi(0)$$
$$\Pi_2 + \Pi_5 = (A,C)(B)(D,E) = \Pi_2$$
$$\Pi_2 \cdot \Pi_5 = (A)(B)(C)(D)(E) = \Pi(0)$$
$$\Pi_3 + \Pi_4 = (A,D,E)(B,C) = \Pi_3$$
$$\Pi_3 \cdot \Pi_4 = (A)(D,E)(B,C) = \Pi_4$$
$$\Pi_3 + \Pi_5 = (A,D,E)(B,C) = \Pi_3$$
$$\Pi_3 \cdot \Pi_5 = (A)(B)(C)(D,E) = \Pi_5$$
$$\Pi_4 + \Pi_5 = (A,B,C)(D,E) = \Pi_4$$
$$\Pi_4 \cdot \Pi_5 = (A)(B)(C)(D,E) = \Pi_5$$

Thus, the complete set of partitions with the substitution property for the example is Π_1, Π_2, Π_3, Π_4, Π_5, and Π_6. ∎

Next we consider how to use these partitions in deriving a state assignment for the example circuit. A sequential circuit with n states has a binary variable assignment of length $s\ (= \lceil \log_2 n \rceil)$, which can be split into two parts such that the first k variables $1 \leq k \leq s$, and the last $(s - k)$ variables can be computed independently, if and only if there exist two nontrivial partitions Π_a and Π_b with the substitution property that satisfy the following conditions

(i) $\quad \Pi_a \cdot \Pi_b = \Pi(0)$
(ii) $\quad \lceil \log_2 \#(\Pi_a) \rceil + \lceil \log_2 \#(\Pi_b) \rceil = s$

where $\#\Pi_i$ denotes the number of blocks or subsets in Π_i. The sequential circuit under consideration has five states, so we need three binary digits to represent these states. Partitions Π_1 and Π_2 (and also Π_2 and Π_3) satisfy the first condition. However, only the partition pair, Π_2 and Π_3, satisfies the second condition,

$$\begin{aligned} s &= \lceil \log_2 \#(\Pi_2) \rceil + \lceil \log_2 \#(\Pi_3) \rceil \\ &= \lceil \log_2 4 \rceil + \lceil \log_2 2 \rceil \\ &= 2 + 1 = 3 \end{aligned}$$

Therefore, the assignment is made such that
a. The secondary variables y_1 and y_2 distinguish the blocks of Π_2.
b. The secondary variable y_3 distinguishes the blocks of Π_3.

	Blocks	$y_1 y_2$		Blocks	y_3
Π_2	(A,C)	00	Π_3	(A,D,E)	0
	(B)	01		(B,C)	1
	(D)	10			
	(E)	11			

The resulting transition table for the sequential circuit is shown in Table 6.21. By utilizing the don't care conditions resulting from the three unused binary combinations 010, 101, and 111, the simplified design equations for the D flip-flop implementation of the circuit are as follows:

$$\begin{aligned} D_1 &= x\bar{y}_1 \\ D_2 &= \bar{x} \cdot \bar{y}_1 + \bar{y}_1 \cdot y_2 \\ D_3 &= \bar{x} \\ Z &= \bar{x}\bar{y}_1 \cdot y_2 + \bar{x} \cdot y_1 \cdot \bar{y}_2 + x \cdot y_1 \cdot y_2 + x \cdot \bar{y}_1 \cdot \bar{y}_3 \end{aligned}$$

It can be seen that the dependence of the next state variables on present state variables is reduced because of the choice of partition pairs. D_1 is dependent upon y_1 alone, D_2 is dependent upon both y_1 and y_2, and D_3 is not dependent on any of the present state variables.

*Computer-Aided Techniques for State Assignment

As indicated earlier, the goal of state assignment is to optimally assign binary codes to the states of a sequential circuit such that the resulting circuit is minimized. For all but the sim-

TABLE 6.21
Transition table

Present States $y_1y_2y_3$	Input $x=0$	$x=1$
A → 000	011,0	100,1
B → 011	011,1	110,0
C → 001	011,0	100,0
D → 100	001,1	000,0
E → 110	001,0	000,1

Next State, Output

plest of sequential circuits, computer-aided-design techniques have to be used to derive optimal circuits. Two efficient state assignment algorithms that have been incorporated in the University of California's CAD tool SIS for logic synthesis are NOVA [2] and JEDI [3]. The state assignment based on NOVA results in a two-level implementation of sequential circuits. On the other hand, JEDI produces multilevel implementation of sequential circuits. However, none of these techniques guarantees optimal state assignments.

In order to use either NOVA or JEDI in the SIS system, a file corresponding to the state table/diagram of a sequential circuit has to be created using a special format. One such format is KISS2. In KISS2 the following information is included at the beginning of a file:

.i	number of inputs
.o	number of outputs
.p	number of transition terms
.s	number of states
.r	reset state (the initial state of the sequential circuit)

Then the present state/next state transitions are specified in the following manner:

<input> <present state> <next state> <output>

The reset state is the symbolic name of the initial state of a sequential circuit. The input (output) is a pattern consisting of as many bits as there are number of inputs (outputs) in the circuit; each bit is assigned a value from {0, 1, or –}, where – indicates that the status of the bit is unknown.

To illustrate, let us specify the state diagram (Fig. 6.19) in KISS2 format. As can be seen from the state diagram, the circuit has seven states, three inputs, and one output. The corresponding file in KISS2 format, `Seq1.kiss2`, is

```
.i     3
.o     2
.s     7
.p     15[1]
.r     s0
1--s0  s001
0--s0  s110
-0-s1  s201
```

6.5 STATE ASSIGNMENT 191

```
-1-s₁    s₆10
0--s₂    s₂01
1--s₂    s₃10
1--s₃    s₃01
0--s₃    s₄10
-00s₄    s₂01
-01s₄    s₅00
-1-s₄    s₆11
0--s₅    s₅00
1--s₅    s₀01
0--s₆    s₆11
1--s₆    s₀10
```

[1]Note that there are 15 present state-to-next state transition terms; the inputs are identified as p, c, and f, and the outputs as z_0 and z_1 in deriving the next state and output expressions.

The file is then read into the SIS system:

sis > read_kiss seq1.kiss2

The binary code for each state can then be derived by using the state assignment technique NOVA (or JEDI):

sis > state_assign Nova
         or
sis > state_assign Jedi

The Boolean expressions generated by these techniques are then minimized using the command

sis > source script.rugged

The actual minimized expressions are obtained by using

sis > write_eqn

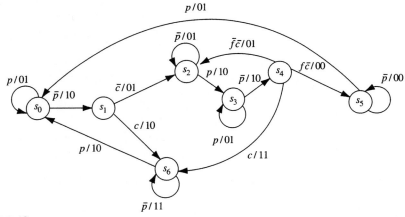

FIGURE 6.19
State diagram

The binary code for each state can be obtained by using the command

```
sis > write_blif
```

BLIF (Berkeley logic interchange format) is an alternative format for representing a logic circuit in textual form.

The `write_blif` command creates the BLIF format of the design equations generated by the `state_assign` command. The state encodings are part of the BLIF format representation. For example, state assignment NOVA generates the following design equations after minimization for output

$$z_0 = py_0y_1y_2 + \bar{p}\bar{y}_1(y_0 + y_2) + c\bar{y}_0y_1 + c(y_0 + \bar{y}_2) + \bar{y}_0y_1\bar{y}_2$$
$$z_1 = z_0\bar{y}_0y_1y_2 + \bar{f}\bar{y}_0y_1y_2 + \bar{p}z_0\bar{y}_0\bar{y}_2 + p\bar{y}_1 + z_0y_0$$

and for next state

$$Y_0 = z_0\bar{y}_0\bar{y}_1 + y_1(y_0\bar{y}_2 + \bar{z}_0z_1) + z_1y_0$$
$$Y_1 = z_0\bar{y}_1 + z_1y_1 + y_0\bar{y}_2$$
$$Y_2 = p\bar{y}_0\bar{y}_1 + y_0y_2 + \bar{z}_0z_1$$

The binary codes assigned to the state are

	y_0	y_1	y_2
s_0	0	0	1
s_1	1	1	0
s_2	1	1	1
s_3	1	0	1
s_4	0	1	1
s_5	0	0	0
s_6	0	1	0

State assignment JEDI generates the following design equations for the same sequential circuit: next state,

$$Y_0 = y_2(y_0 + \bar{y}_1) + p\bar{y}_0y_2 + p\bar{y}_1$$
$$Y_1 = \bar{c}fy_1\bar{y}_2 + \bar{p}\bar{y}_0y_1y_2 + \bar{p}y_0\bar{y}_1$$
$$Y_2 = y_0(z_0 + Y_1) + \bar{y}_1y_2 + \bar{y}_0y_2$$

and output,

$$z_0 = cy_1(y_0 + \bar{y}_2) + p\bar{y}_0\bar{y}_1 + \bar{p}y_0\bar{y}_1 + \bar{y}_0\bar{y}_1y_2$$
$$z_1 = \bar{y}_1\bar{Y}_0\bar{Y}_1 + y_1y_2\bar{Y}_1 + y_0\bar{z}_0 + \bar{z}_0\bar{Y}_1$$

The binary codes assigned to the states are

	y_0	y_1	y_2
s_0	1	0	1
s_1	1	1	1

s_2	0	0	0
s_3	1	0	0
s_4	0	1	0
s_5	0	1	1
s_6	0	0	1

Note that the design equations generated by NOVA and JEDI for the given sequential circuit consist of 56 and 49 literals respectively.

State Assignment Based on 1-Hot Encoding

One straightforward approach for encoding the state of a sequential circuit is to assign a 1-out-of-n code to each state, where n is the number of states in the circuit. In such an n-bit code word only one bit is 1 (**hot**), and the rest of the bits are 0's. The state assignment based on 1-out-of-n code is also known as 1-hot encoding. Let us illustrate the 1-hot encoding for the sequential circuit (Fig. 6.19) considered in the previous section. The states are assigned codes as follows:

	y_0	y_1	y_2	y_3	y_4	y_5	y_6
s_0	1	0	0	0	0	0	0
s_1	0	1	0	0	0	0	0
s_2	0	0	1	0	0	0	0
s_3	0	0	0	0	1	0	0
s_4	0	0	0	0	0	1	0
s_5	0	0	0	0	0	0	1
s_6	0	0	0	1	0	0	0

The minimized next state and output expressions corresponding to this assignment are

$$Y_0 = p\bar{y}_1\bar{y}_2\bar{y}_4\bar{y}_5 + py_3$$
$$Y_1 = \bar{p}y_0$$
$$Y_2 = \bar{c}\bar{f}y_5 + \bar{p}y_2 + \bar{c}y_1$$
$$Y_3 = c\bar{y}_0\bar{y}_2\bar{y}_3\bar{y}_4\bar{y}_6 + \bar{p}y_3$$
$$Y_4 = p(y_2 + y_4)$$
$$Y_5 = \bar{p}y_4$$
$$Y_6 = \bar{c}fy_5 + \bar{p}y_6$$
$$z_0 = py_3 + py_2 + Y_1 + Y_3 + Y_5$$
$$z_1 = y_4Y_4 + y_3\bar{Y}_0 + \bar{y}_3Y_0 + cy_6 + Y_2$$

The SIS system can be used to derive the output and the next state expressions corresponding to 1-hot encoding by using the command `one_hot`.

The main disadvantage of 1-hot encoding is that the resulting sequential circuit uses significantly more flip-flops than the minimum number required. The advantage of this approach is that a state can be identified without encoding, and the next state and output logic expressions are relatively straightforward. However, the complexity of the circuit depends on how the 1-hot code words are assigned to the states. The determination of code assignment for minimal circuitry is not a trivial task.

*State Assignment Using *m*-out-of-*n* Code

An alternative for encoding the states of a sequential machine is to use m-out-of-n code. As discussed in Chapter 2, an m-out-of-n code has m 1's and $(m-n)$ 0's, with a Hamming distance of $2d$ ($d = 1, 2, \ldots, \lfloor n/2 \rfloor$) between code words. Let us consider how to select the m and n values for representing the states of a sequential circuit. The n represents the number of flip-flops required. Note that unlike in 1-hot encoding, where the number of flip-flops is equal to the number of states in the sequential circuit, in the m-out-of-n encoding the minimum value of n is selected such that together with properly chosen value of m, the number of code words will be sufficient to uniquely represent each state. Table 6.22 shows the values of n and m needed for encoding different number of states.

Let us implement the sequential circuit of Figure 6.20 using m-out-of-n codes for state assignment. There are seven states in the circuit, so we can select $m = 2$ and $n = 5$ (i.e., 2-out-of-5 code). Since there are 10 possible code words, we arbitrarily choose seven of these for state encoding:

State	$y_0 y_1 y_2 y_3 y_4$
A	1 1 0 0 0
B	1 0 0 1 0
C	0 1 1 0 0
D	0 1 0 1 0
E	1 0 0 0 1
F	1 0 1 0 0
G	0 1 0 0 1

TABLE 6.22
Selection of n and m values

No of States	n	m
4–6	4	2
7–10	5	2
11–20	6	3
21–35	7	3
36–70	8	4
71–126	9	4

FIGURE 6.20
State table

	$x = 0$	$x = 1$
A	F,100	D,100
B	E,100	C,100
C	E,100	G,100
D	F,100	F,010
E	A,010	B,010
F	A,001	B,001
G	E,100	F,010

The next state expressions resulting from the above assignment are

$$Y_0 = \overline{Y}_1 Y_4 + \overline{y}_1 \overline{Y}_2 + \overline{Y}_1 Y_2 + \text{int}$$
$$\text{int} = x\overline{y}_0 \overline{y}_2$$
$$Y_1 = \overline{y}_1 \overline{Y}_3 \overline{Y}_4 + \overline{y}_1 \overline{y}_3 \overline{Y}_3 + xY_4 + y_1 Y_3$$
$$Y_2 = \overline{x} y_1 \overline{y}_2 \overline{y}_4 + x\overline{y}_1 y_3 + \text{int}$$
$$Y_3 = x\overline{y}_1 \overline{y}_3 + xy_0 y_1$$
$$Y_4 = \overline{y}_0 \overline{y}_3 \overline{\text{int}} + x\overline{y}_0 \overline{\text{int}} + \overline{x} \overline{y}_1 y_3$$
$$z_0 = \overline{x} Y_0 \overline{Y}_1 + xY_1 + Y_4$$
$$z_1 = \overline{y}_1 y_4 + \text{int}$$
$$z_2 = \overline{y}_1 y_2$$

The advantage of using m-out-of-n-code for state assignment is that if there is a fault (see Chapter 11) in the next state logic, the circuit may move to an erroneous state, which will be identified by a non–code word. If a dedicated circuit is incorporated to check whether the outputs of the memory elements are code words, then an erroneous state can be easily detected. It is also possible to encode the outputs of a sequential circuit using m-out-of-n code or Berger code such that if a non–code word output is produced, a fault is assumed to be present in the output logic and/or in the next state logic.

6.6 INCOMPLETELY SPECIFIED SEQUENTIAL CIRCUITS

So far we have considered sequential circuits whose state tables do not contain don't cares. In other words all next state and/or output entries are specified. However, in practice it is very likely that some input combinations will not be applied to a sequential circuit, so next states and outputs corresponding to these inputs are of no consequence and hence the corresponding entries in the state table of the circuit are don't cares. A state table with one or more don't care entries is known as an **incompletely specified state table**. A sequential circuit with an incompletely specified state table is called an **incompletely specified sequential circuit**. A don't care entry in an incompletely specified state table is usually denoted by a dash (–). As in completely specified sequential circuits, reduction of states in an incompletely specified state table is needed for efficient implementation of the corresponding sequential circuit.

One approach to reducing the number of states in an incompletely specified sequential circuit will be to specify entries for the unspecified next states and outputs such that some of the states become equivalent. However, considerable trial and error will be required to determine the best way to fill in the unspecified entries, especially when their number is large.

We shall consider a systematic procedure in this section to minimize the number of states in an incompletely specified state table. The procedure is based on the concept of compatibility. Two states S_i and S_j are said to be **compatible** if in response to an applicable input sequence, the same output sequence is produced irrespective of whether S_i or S_j is the initial state; if S_i and S_j are not compatible, then they are said to be **incompatible**. An **applicable input sequence** always leads to a specified next state for each bit of the sequence. For example, in the incompletely specified table (Table 6.23), the output sequences produced in response to the input sequence 1100 for starting states B and D are the same when both are specified:

Input	1	1	0	0
State	$B \to B \to B \to C \to D$			
Output		0	0	1	0
State	$D \to B \to B \to C \to D$			
Output		–	0	1	0

A set of states for which every pair of states is compatible is called a **compatibility class**. A **maximal compatible** is a compatibility class that is not a subset of any other compatibility class. For Table 6.23, {AD} is not a maximal compatible because it is a proper subset of {ABD}, whereas {ABD} is a maximal compatible. The implication table (discussed in Sec. 6.3) can be used to determine all pairs of compatible states in an incompletely specified state table. Figure 6.21 shows the implication table for the state table shown in Table 6.23.

From the implication table the compatible pairs are

$$(AB)(AC)(AD)(AE)(BD)(CD)(CE)$$

The maximal compatibles are found by combining these state pairs into larger groups using the following procedure [4]:

Step 1: List the pairwise compatibles, if any, for the rightmost column of the implication table. If the rightmost column does not have a compatible pair, then move to the next column on the left and check whether it has a compatible pair. Continue this process until a column containing a compatible pair is found.

TABLE 6.23
An unspecified state table

Present State	Input	
	$x = 0$	$x = 1$
A	A,–	–,–
B	C,1	B,0
C	D,0	–,1
D	–,–	B,–
E	A,0	C,1

Next State, Output

FIGURE 6.21
Implication table

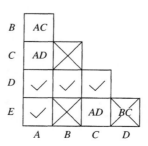

Step 2: Proceed to the next column on the left. If the state to which this column corresponds is compatible with all members of a previously determined compatible class, then add this state to the class, thereby forming a larger class. If the state to which this column corresponds is compatible only with a subset of the compatible class, then form a new class consisting of the subset and this state. Finally, list all the compatible pairs that are not included in an already formed compatible class.

Step 3: Repeat the above step until all columns in the implication table have been considered. The compatibility classes remaining are the set of maximal compatibles.

Applying this procedure to the implication table of Figure 6.21 yields the following sequence of compatibility classes:

$$\begin{array}{ll} \text{Column } D & (\ ) \\ \text{Column } C & (CD)(CE) \\ \text{Column } B & (CD)(CE)(BD) \\ \text{Column } D & (CD)(CE)(BD)(AB)(AC)(AD)(AE) \end{array}$$

Thus, there are three maximal compatibles—$C_1' = \{ABD\}$, $C_2' = \{ACD\}$, and $C_3' = \{ACE\}$. Notice that the sets of maximal compatibles are similar to the blocks of an equivalence partition of a completely specified state table. However, the blocks of an equivalence partition are disjoint, whereas the maximal compatibles are not necessarily so, because they can have common states.

In order to determine the reduced table for an incompletely specified machine we must select a set of maximal compatibles that satisfy the **covering** and **closure** conditions:

1. A set of maximal compatibles **covers** an incompletely specified sequential circuit if each state of the circuit is contained in at least one of the maximal compatibles.
2. A set of maximal compatibles is **closed** if for every compatible contained in the set, the next states corresponding to the states in the compatible for all possible input combinations are also contained in a maximal compatible of the set.

The maximal compatibles derived here cover all the states of Table 6.23. Maximum compatible C_1' covers states A, B, and D, and C_3' covers states A, C, and E. The resulting set $\{C_1'C_3'\}$ covers all the states, and also satisfies the closure conditions as shown in Table 6.24. The incompletely specified state table can therefore be reduced to a table with two states corresponding to C_1' and C_3'. Denoting C_1' and C_3' by α and β respectively, the reduced state table is shown in Table 6.25.

TABLE 6.24
Verification of the closure condition

Present State	Input $x = 0$	$x = 1$
$C_1' = \{ABD\}$	$C_3', 1$	$C_1', 0$
$C_3' = \{ACE\}$	$C_1', 0$	$C_3', 1$
	Next State, Output	

6.7 ALGORITHMIC STATE MACHINE (ASM)

The Algorithmic State Machine (ASM) format is used to describe state machine behavior in the form of a flowchart. Unlike state diagrams, which become unwieldy except for very small state machines, ASMs provide an unambiguous way of specifying complex state machines. Once the ASM chart of a state machine has been formulated, it is easy to convert it into a state table or a state diagram.

Three primary symbols are used in an ASM chart—rectangle, diamond, and oval. A machine state is represented by a rectangle in ASM charts. The outputs for that state are listed inside the rectangle. The state is given a symbolic name and is placed on either side of the rectangle. The binary code (i.e., state code) assigned to the state is written on top of the state box. Figure 6.22 is an example of a state box. It shows a state with symbolic name S_1, which has output START. The binary code for this state is 010. A diamond in ASM charts represents a conditional path from the state machine's present state to a next state. The input condition that causes the transition from the present state to a next state is written inside the diamond. One branch from the diamond is taken if the input condition is true and the other if the condition is not satisfied. Figure 6.23 shows a decision diamond. An oval in an ASM chart identifies outputs that may result when a machine in a particular state receives different input conditions. In other words, an oval is used to

TABLE 6.25
Reduced state table

Present State	Input	
	$x = 0$	$x = 1$
α	$\beta,1$	$\alpha,0$
β	$\alpha,0$	$\beta,1$

Next State

FIGURE 6.22
ASM state representation

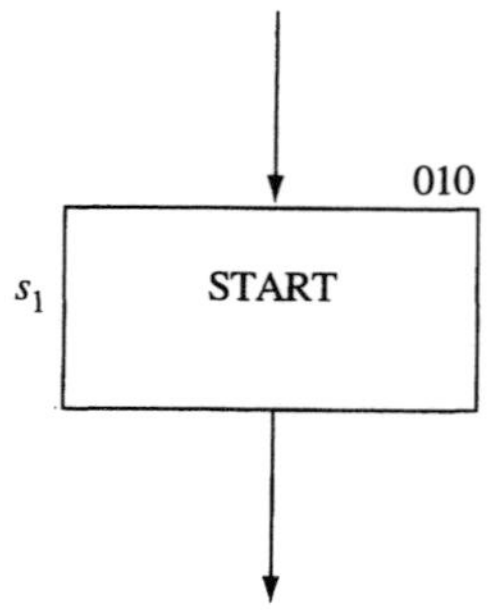

FIGURE 6.23
Decision diamond

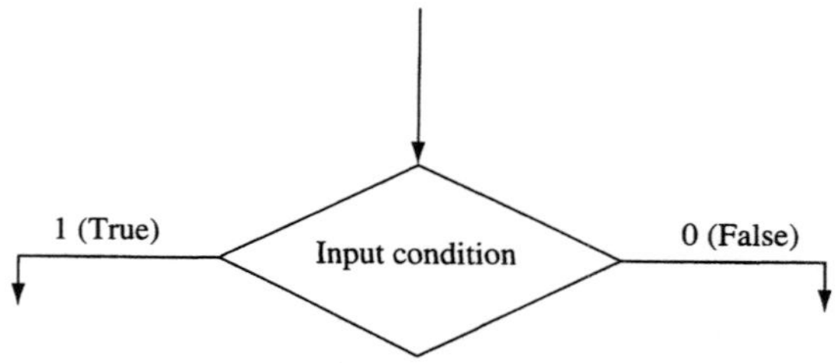

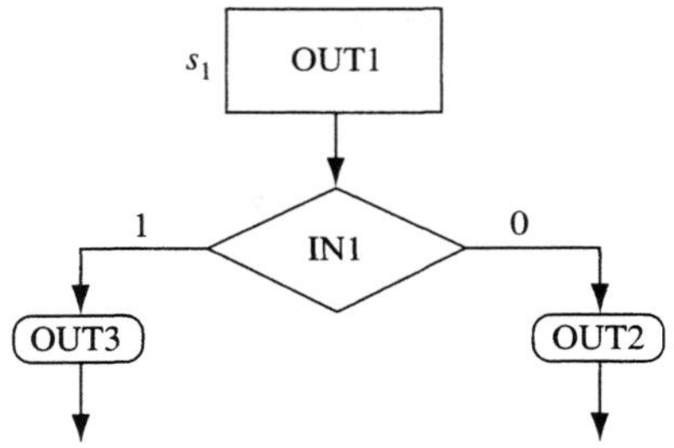

FIGURE 6.24
Input-dependent outputs

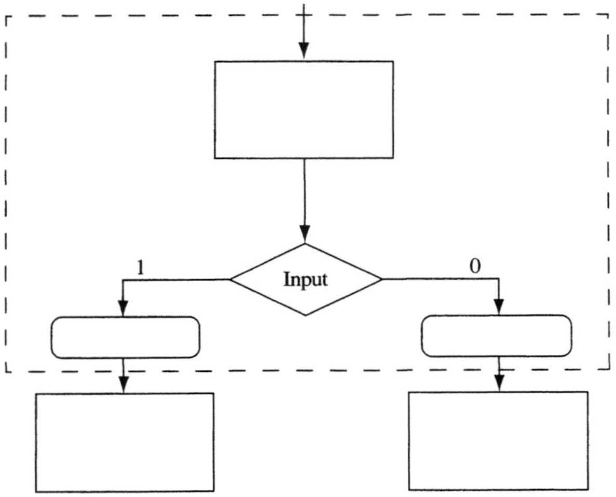

FIGURE 6.25
ASM block

represent input-dependent output (i.e., output associated with a conditional branch). Thus, as shown in Figure 6.24 the input to a conditional output box comes from an exit path of a decision box. The output OUT1 occurs in state S_1 irrespective of the inputs, so it is included in the state box. On the other hand, outputs OUT2 and OUT3 are produced in state S_1 if input IN1 is 0 and 1 respectively. Therefore, OUT2 and OUT3 are listed in conditional output boxes in the exit paths corresponding to IN1 = 0 and 1 respectively.

A state box together with all decision diamonds connected to it forms an ASM block. Figure 6.25 illustrates such an ASM block. Each exit path from an ASM block must be connected to a state box. All decision diamonds and conditional ovals may be shared among the ASM blocks.

■ **EXAMPLE 6.10**

Let us derive the ASM chart for a sequential circuit that generates an output $z = 1$, whenever the input pattern 101 is shifted into the circuit (overlapping patterns are acceptable). The ASM chart for the machine is shown in Figure 6.26. The circuit is initially at state A (000), waiting to receive an input bit at x. If $x = 1$, the circuit produces an output of 0 and remains in state A. Note that the ASM block associated with state A has a decision

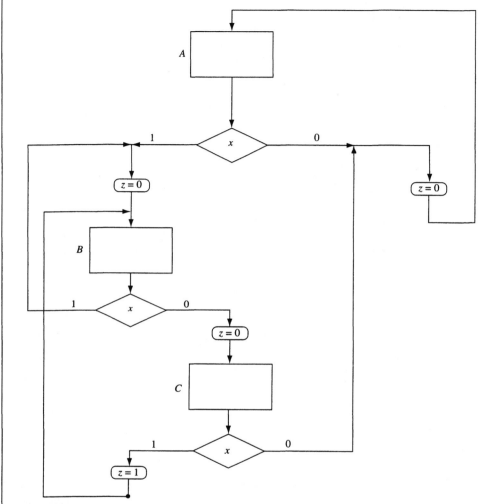

FIGURE 6.26
ASM chart

box for input x and a conditional box for output z. The ASM chart has two additional blocks associated with states B and C. Each block in an ASM chart represents a single clock period. ∎

It should be emphasized that ASM charts are suitable mainly for manual design. Current digital design practices are heavily dependent on using CAD tools for converting design specifications into hardware. Therefore, state machines often are described using a hardware description language (HDL) rather than as flowcharts. The HDL description is then compiled into a state machine implementation using suitable logic modules. In the next section, we show the realization of state machines described in ABEL-HDL, using PLDs.

6.8 SEQUENTIAL LOGIC DESIGN USING PLDs

The PLDs considered in Chapter 4 are combinational devices, their outputs at any instant of time are functions of their inputs at that instant. Although it is possible to implement sequential logic circuits using these devices with memory elements, this results in an increase in the number of packages. Therefore, the incorporation of memory elements within PLDs has significant advantage in that a sequential circuit may be implemented using a single package.

Sequential Circuit Description in ABEL

A state diagram or state table of a sequential circuit is described in ABEL-HDL using the following format:

```
module 'name'
pin assignment
node assignment
equation
state_diagram 'sreg'
State state_symbol 1:
IF-THEN-WITH;
ELSE-WITH;
         :
State state_symbol n:
IF-THEN-WITH;
ELSE-WITH;
test_vectors
end
```

The pin assignment for sequential circuits is done in a similar manner as for combinational circuits. In addition, the characteristics of signals can be defined by the `istype` statement. For example, the outputs of flip-flops in a device can be identified by the following statement

```
pin istype 'reg'
```

Depending upon the types of flip-flops, the string `'reg'` can be more specifically defined as `reg_D`, `reg_JK`, etc. Similarly, a signal can be defined to be combinational by the statement

```
pin istype 'com'
```

The symbols associated with the states of a sequential circuit are assigned binary patterns that define the value of a register known as `'sreg'`, for each state. For example, if a circuit has three states A, B, and C, sreg will have two bits (e.g., q_1, q_0); a possible binary assignment for the states may be as follows:

```
sreg = [q1 q0];
A    = 0 0;
B    = 1 0;
C    = 1 1;
```

Alternatively, the states could be assigned decimal values instead of binary patterns:

```
sreg = [q1 q0];
A = 0; B = 2; C = 3;
```

The identifier `state_diagram` indicates the beginning of the description of a sequential circuit. The string `'sreg'` is listed after `state_diagram`. Each symbolic state of the circuit is listed after the identifier `state`. The next state and the output associated with each state are defined using `IF-THEN-WITH, ELSE-WITH` statements. For example, suppose the current state of a sequential circuit is A and the current output is 0. If input $x = 1$, the circuit goes to state C, generating an output $z = 1$; otherwise, it moves to state B, with $z = 0$. These transitions can be specified as

```
State A:
IF (x) THEN C WITH z = 1;
ELSE B WITH z = 0;
```

A `GOTO` statement can be used to specify an unconditional transition to a particular state.

Equations in sequential circuit descriptions are used mainly to specify clocking of flip-flops and enabling of tristate outputs. For example, the following equation

```
[q1,q0].clk = clock
```

indicates that flip-flops q_1 and q_0 are clocked on the positive edge of the clock signal. The equation

```
[q1,q0].oe = !enab
```

specifies that an active-low input signal `'enab'` is required to control the output enable of flip-flops q_1 and q_0.

Field Programmable Logic Sequencer (FPLS)

FPLS devices are programmable sequential circuits of the Mealy type offered by Signetics Corporation. Figure 6.27 shows the basic organization of one such sequencer, the PLS155. It has 4 internal inputs, 45 product terms, and 4 *JK* flip-flops, which can also be converted to *D* flip-flops. In addition it has 8 bidirectional i/o lines (i.e., these lines can be used as output or additional input lines). It also has a complement array, which consists of a single 45-input OR gate that drives an inverter. The output of the inverter is fed back to the AND gates. Note that the outputs of the flip-flops can be observed by enabling the tristate inverters at the output of the flip-flops. This simplifies identification of the states of a sequential circuit implemented using this device.

Figure 6.28 shows the input and the output connections of the flip-flops in PLS155. If the fuse (identified by *X* in the diagram) is intact, then a *JK* flip-flop is converted into *D* flip-flop provided the NOR gate is disabled, and the output of the OR gate *B* is at logic 0. If the fuse is blown and the outputs of both OR gates *A* and *B* are at logic 1, the *JK* flip-flop functions as a *T* flip-flop.

■ EXAMPLE 6.11

Let us implement the sequential circuit specified in Figure 6.29 [5] using a PLS155. Since there are 8 states in the circuit, 3 flip-flops will be required. The circuit has one input (*a*) and

6.8 SEQUENTIAL LOGIC DESIGN USING PLDs   203

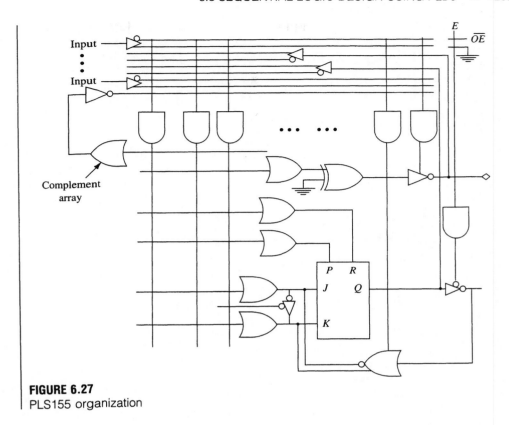

FIGURE 6.27
PLS155 organization

FIGURE 6.28
Flip-flop in PLS155

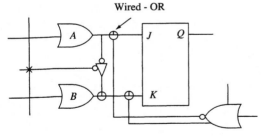

× blown, the inverter is disabled

× intact, the inverter is enabled

FIGURE 6.29
State diagram of a circuit

one output (z). Thus, 1 external input, 1 output, and 3 *JK* flip-flops of PLS155 will be needed to implement the circuit. The clock signal is applied via the clock input (pin 1) of the device. We select the following state assignment for the circuit:

	q_2	q_1	q_1
A	0	0	0
B	0	0	1
C	0	1	1
D	0	1	0
E	1	1	0
F	1	0	0
G	1	0	1
H	1	1	1

The pin assignment in the ABEL description of the circuit will be as follows:

```
clock       pin 1;
a           pin 2;
z           pin 13 istype 'com';
q2,q1,q0    pin 16, 15, 14 istype 'reg_jk'.
sreg        =[q2,q1,q0];
A=0; B=1; C=2; D=2; E=6; F=4; G=5; H=7;
```

If one wishes to use *D* flip-flops rather than *JK* flip-flops as memory of the circuit, the string `'reg_jk'` in the fourth line above has to be replaced by simply `'reg'`. The ABEL compiler will automatically assume *D* flip-flops rather than the actual *JK* flip-flops available in the device. In other words, the reconfiguration of a *JK* flip-flop to a *D* flip-flop is accomplished by the ABEL compiler; the generated fuse map contains the appropriate information necessary to change from the *JK* to the *D* configuration. Figure 6.30 shows the complete ABEL description of the sequential circuit. The resulting equations are

$$q_2.J = \bar{a}\,\bar{q}_0\bar{q}_2 + \bar{q}_0 q_1 \bar{q}_2$$
$$q_2.K = a q_0 q_1 q_2$$
$$q_1.J = a\bar{q}_1\bar{q}_2 + q_0 \bar{q}_1 q_2$$
$$q_1.K = a q_0 q_1 q_2 + \bar{q}_0 q_1 q_2$$
$$q_0.J = \bar{q}_0 \bar{q}_1 q_2 + \bar{a}\,\bar{q}_0 \bar{q}_2 + \bar{q}_0 q_1 \bar{q}_2$$
$$q_0.K = \bar{a} q_0 \bar{q}_2 + \bar{a} q_0 q_1 + q_0 q_1 \bar{q}_2$$
$$z = \overline{(a\bar{q}_0\bar{q}_2 + q_0\bar{q}_1 q_2 + \bar{a} q_0 q_1 + \bar{q}_0 q_1 q_2 + q_0 q_1 \bar{q}_2 + \bar{q}_0 q_1 \bar{q}_2)}$$

FIGURE 6.30
ABEL source file

```
module trial1
title 'sequential circuit design using f155'
    bjdu    device 'f155';

    clock       pin 1;
    a,rst       pin 2,3;
    z           pin 13 istype 'com';
```

FIGURE 6.30
(Continued)

```
            q2,q1,q0     pin 16,15,14 istype 'reg_jk';
            sreg     =   [q2,q1,q0];
            A=0;B=1;C=3;D=2;E=6;F=4;G=5;H=7;
            X=.x.;

        equations
          sreg.clk  = clock;
          sreg.ap   = rst;

        state_diagram sreg
          state A:
          IF (a) THEN E WITH z=1
          ELSE B WITH z=0;

          state B:
          GOTO C WITH z=0;

          state C:
          GOTO D WITH z=0;

          state D:
          GOTO A WITH z=0;

          state E:
          IF (!a) THEN H WITH z=1
          ELSE F WITH z=1;

          state F:
          GOTO G WITH z=0;

          state G:
          GOTO A WITH z=0;

          state H:
          IF (!a) THEN D WITH z=1
          ELSE G WITH z=0;

        test_vectors
          ([clock,rst,a] ->[sreg,z])
           [  .c.,   1,X]  ->[   A,X];
           [  .c.,   0,1]  ->[   E,1];
           [  .c.,   0,0]  ->[   H,1];
           [  .c.,   0,1]  ->[   G,0];
           [  .c.,   0,X]  ->[   A,0];
           [  .c.,   0,0]  ->[   B,0];
           [  .c.,   0,X]  ->[   C,0];
           [  .c.,   0,X]  ->[   D,0];
           [  .c.,   0,X]  ->[   A,0];
           [  .c.,   0,0]  ->[   B,0];
           [  .c.,   0,X]  ->[   C,0];
           [  .c.,   0,0]  ->[   D,0];
```

FIGURE 6.30
(Continued)

```
    [  .c.,   0,X] ->[   A,0];
    [  .c.,   0,1] ->[   E,1];
    [  .c.,   0,0] ->[   H,1];
    [  .c.,   0,1] ->[   G,0];
end
```

Figure 6.31 shows the JEDEC file corresponding to these equations. The JEDEC file is tested to ensure that it represents the circuit specification by applying the test vectors included in the ABEL description. The simulation is done by the program module JEDECSIM in ABEL; the simulation results are shown in Figure 6.32.

FIGURE 6.31
JEDEC file

```
ABEL 4.01 Data I/O Corp. JEDEC file for: F155 V9.0
Created on: Tue Jul 13 13:59:00 1993
sequential circuit design using f155
*
QP20* QF2108* QV16* F0*
X0*
NOTE Table of pin names and numbers*
NOTE PINS clock:1 a:2 rst:3 z:13 q2:16 q1:15 q0:14*
L0000 011111111011101111111111111111111111011111111111111111*
L0054 011111111110101111111111111111111111111111111111110111*
L0108 110111111111111111111111111111111111111111100111111111*
L0162 011111110101011111111111111111111111111111111111101011*
L0216 111111110110011111111111111111111111011111111111110111*
L0270 101111110111101111111111111111111111111111111111111110*
L0324 101111110101111111111111111111111111011111111111111110*
L0378 111111111010011111111111111111111111011111111111111101*
L0432 101111111011101111111111111111111111111111111101111101*
L0486 111111110101101111111111111111111111011111111111111110*
L0540 111111111001011111111111111111111111011111111111111011*
L0594 111111111001101111111111111111111111011111111101111101*
L0648 000000000000000000000000000000000111111111111111111111*
L0702 000000000000000000000000000000000111111111111111111111*
L0756 000000000000000000000000000000000111111111111111111111*
L0810 000000000000000000000000000000000111111111111111111111*
L0864 000000000000000000000000000000000111111111111111111111*
L0918 000000000000000000000000000000000111111111111111111111*
L0972 000000000000000000000000000000000111111111111111111111*
L1026 000000000000000000000000000000000111111111111111111111*
L1080 000000000000000000000000000000000111111111111111111111*
L1134 000000000000000000000000000000000111111111111111111111*
L1188 000000000000000000000000000000000111111111111111111111*
L1242 000000000000000000000000000000000111111111111111111111*
L1296 000000000000000000000000000000000111111111111111111111*
L1350 000000000000000000000000000000000111111111111111111111*
L1404 000000000000000000000000000000000111111111111111111111*
L1458 000000000000000000000000000000000111111111111111111111*
L1512 000000000000000000000000000000000111111111111111111111*
```

FIGURE 6.31
(Continued)

```
L1566 000000000000000000000000000000001111111111111111111*
L1620 000000000000000000000000000000001111111111111111111*
L1674 000000000000000000000000000000001111111111111111111*
L1893 11111111111111111111111111111111*
L2091 1*
L2100 0111*
V0001 CX1XXXXXXXNXXNLLLXXXN*
V0002 C10XXXXXXXNXXHLHHXXXN*
V0003 C00XXXXXXXNXXHHHHXXXN*
V0004 C10XXXXXXXNXXLHLHXXXN*
V0005 CX0XXXXXXXNXXLLLLXXXN*
V0006 C00XXXXXXXNXXLHLLXXXN*
V0007 CX0XXXXXXXNXXLHHLXXXN*
V0008 CX0XXXXXXXNXXLLHLXXXN*
V0009 CX0XXXXXXXNXXLLLLXXXN*
V0010 C00XXXXXXXNXXLHLLXXXN*
V0011 CX0XXXXXXXNXXLHHLXXXN*
V0012 C00XXXXXXXNXXLLHLXXXN*
V0013 CX0XXXXXXXNXXLLLLXXXN*
V0014 C10XXXXXXXNXXHLHHXXXN*
V0015 C00XXXXXXXNXXHHHHXXXN*
V0016 C10XXXXXXXNXXLHLHXXXN*


Simulate ABEL 4.03 Date Tue Jul 13 12:53:11 1993
Fuse file: 'trial1.ttl' Vector file: 'trial1.tmv' Part: 'PLA'
sequential circuit design using f155
        c
        1
      o r
      c s    q q q
      k t a  2 1 0 z
V0001 C 1 0  L L L L
V0002 C 0 1  H H L H
V0003 C 0 0  H H H H
V0004 C 0 1  H L H L
V0005 C 0 0  L L L L
V0006 C 0 0  L L H L
V0007 C 0 0  L H H L
V0008 C 0 0  L H L L
V0009 C 0 0  L L L L
V0010 C 0 0  L L H L
V0011 C 0 0  L H H L
V0012 C 0 0  L H L L
V0013 C 0 0  L L L L
V0014 C 0 1  H H L H
V0015 C 0 0  H H H H
V0016 C 0 1  H L H L
16 out of 16 vectors passed.
```

FIGURE 6.32
Simulation results

Sequential PAL Devices

A number of PAL devices are currently available for implementing sequential circuits. These include

PAL 16R4
PAL 16R6
PAL 16R8
PAL 22V10
PAL 23S8

The **R** in devices 16R4, 16R6, and 16R8 indicates that these devices have built-in D flip-flops: four in 16R4, six in 16R6, and eight in 16R8. Each output pin in these devices has eight product terms associated with it. The seven product terms in a group drive an OR gate, while a dedicated product term controls the enable input of an inverting tristate buffer. When this product term is activated, the output is enabled and the sum of the product terms is gated to the output. On the other hand, if the product term is not activated, the tristate buffer remains in the high-impedance state and the output pin can be used as an input.

■ **EXAMPLE 6.12**

Let us implement the following 3-bit random sequence generator using a PAL16R:

$$
\begin{array}{ll}
A & 000 \leftarrow \\
  & \downarrow \\
C & 010 \\
  & \downarrow \\
G & 110 \\
  & \downarrow \\
H & 111 \\
  & \downarrow \\
E & 100 \\
  & \downarrow \\
D & 011 \\
  & \downarrow \\
B & 001 \\
  & \downarrow \\
F & 101
\end{array}
$$

Since there are 3 bits in a pattern, three D flip-flops of PAL16R4 have to be used. The ABEL description of the circuit is shown in Figure 6.33. The excitation equations for the flip-flops are derived as the output of the PLAOPT program module. These are

$$q_2.D = \overline{(rst + \overline{q_2}q_1q_0 + \overline{q_1}\overline{q_0} + q_2\overline{q_1})}$$
$$q_1.D = \overline{(rst + \overline{q_0})}$$
$$q_0.D = \overline{(rst + \overline{q_2}\overline{q_0} + q_2q_0)}$$

The JEDEC file corresponding to these equations is verified by applying the test vectors included in the ABEL description. The simulation results are shown in Figure 6.34. ∎

FIGURE 6.33
ABEL source file

```
module trial2
title 'random sequence generator design using p16r4'
    bjdu     device 'p16r4';

    clock         pin 1;
    rst,enab      pin 2,11;
    q2,q1,q0      pin 16,15,14 istype 'reg,_invert';
    sreg        = [q2,q1,q0];
    A=000;B=001;C=010;D=011;E=100;F=101;G=110;H=111;
    X=.x.;

equations
    sreg.clk  = clock;
    sreg.oe   = !enab;

state_diagram sreg
    state A:
    IF (!rst) THEN C;
    ELSE A;

    state B:
    IF (!rst) THEN F;
    ELSE A;

    state C:
    IF (!rst) THEN G;
    ELSE A;

    state D:
    IF (!rst) THEN B;
    ELSE A;

    state E:
    IF (!rst) THEN D;
    ELSE A;

    state F:
    GOTO A;

    state G:
    IF (!rst) THEN H;
    ELSE A;

    state H:
    IF (!rst) THEN E;
    ELSE A;
```

FIGURE 6.33
(Continued)

```
test_vectors
  ([clock,rst,enab]->[sreg])
   [  .c.,  1,   0]->[   A];
   [  .c.,  0,   0]->[   C];
   [  .c.,  0,   0]->[   G];
   [  .c.,  0,   0]->[   H];
   [  .c.,  0,   0]->[   E];
   [  .c.,  0,   0]->[   D];
   [  .c.,  0,   0]->[   B];
   [  .c.,  0,   0]->[   F];
   [  .c.,  0,   0]->[   A];
end
```

FIGURE 6.34
Simulation results

```
Simulate ABEL 4.03 Date Tue Jul 13 13:34:37 1993
Fuse file: 'trial2.ttl' Vector file: 'trial2.tmv' Part: 'PLA'
random sequence generator design using p16r4
         c
         l      e
         o  r   n
         c  s   a   q q q
         k  t   b   2 1 0
V0001    C  1   0   L L L
V0002    C  0   0   L H L
V0003    C  0   0   H H L
V0004    C  0   0   H H H
V0005    C  0   0   H L L
V0006    C  0   0   L H H
V0007    C  0   0   L L H
V0008    C  0   0   H L H
V0009    C  0   0   L L L
9 out of 9 vectors passed.
```

PAL22V10: This is the most widely used PLD in industry. The block diagram of the device is shown in Figure 6.35. It has 12 dedicated inputs, with one input also acting as a clock input. In addition, it has 10 I/O lines, which can be configured either as inputs or outputs. It employs a variable product term distribution which allocates from 8 to 16 product terms to each output. Each output pin is driven by a **macrocell**. Figure 6.36 shows the logic diagram of a macrocell. As can be seen in the diagram, there is a D flip-flop at the output of the D flip-flop; thus, a registered output can be obtained by setting the selection inputs of multiplexer MUX2 to $s_1 s_2 = 01$. The D flip-flop can be bypassed by setting $s_1 s_2 = 11$ to provide a combinational output. Furthermore, the output polarity for the registered and the combinational output can be made active-low by making $s_1 s_2 = 00$ and 10 respectively. Multiplexer MUX1 allows the feedback of the registered output to the AND array. Also, when the buffer at the output pin is not enabled (i.e., when the pin is used as an input), MUX1 is set ($s_0 = 1$) to transfer the input value to the AND array. All 10 D flip-flops in a 22V10 share an asynchronous reset product term and a synchronous preset product term.

■ **EXAMPLE 6.13**

Let us implement the Mealy-type sequential circuit shown in Figure 6.37 using a PAL22V10. The ABEL specification of the circuit is shown in Figure 6.38.

6.8 SEQUENTIAL LOGIC DESIGN USING PLDs   211

FIGURE 6.35
Structure of the 22V10

FIGURE 6.36
Macrocell of the 22V10

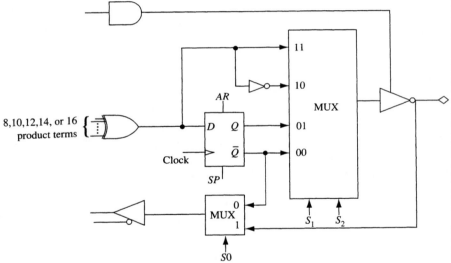

FIGURE 6.37
State table of a sequential circuit

		Input			
Present State		$\bar{a}\bar{b}$	$\bar{a}b$	$a\bar{b}$	ab
	A	E,1	C,0	B,1	E,1
	B	C,0	F,1	E,1	B,0
	C	B,1	A,0	D,1	F,1
	D	G,0	F,1	E,1	B,0
	E	C,0	F,1	D,1	E,0
	F	C,1	F,1	D,0	H,0
	G	D,1	A,0	B,1	F,1
	H	B,1	C,0	E,1	F,1
		Next State, output			

FIGURE 6.38
ABEL source file

```
module trial8
  title 'state machine implementation using PAL22V10'
    bjdu    device 'p22v10';

    clock           pin 1;
    a,b             pin 3,4;
    z               pin 23 istype 'com';
    q2,q1,q0        pin 14,15,22 istype 'reg';
    sreg      = [q2,q1,q0];
    A=000;B=001;C=010;D=011;E=100;F=101;G=110;H=111;
    X=.x.;

  equations
    [sreg].clk   = clock;

  state_diagram sreg
    state A:
    IF  (!a&!b) THEN E WITH z=1;
    ELSE IF (!a&b) THEN C WITH z=0;
    ELSE IF (a&!b) THEN B WITH z=1;
    ELSE E WITH z=1;

    state B:
    IF  (!a&!b) THEN C WITH z=0;
    ELSE IF (!a&b) THEN F WITH z=1;
    ELSE IF (a&!b) THEN E WITH z=1;
    ELSE B WITH z=0;

    state C:
    IF  (!a&!b) THEN B WITH z=1;
    ELSE IF (!a&b) THEN A WITH z=0;
    ELSE IF (a&!b) THEN D WITH z=1;
    ELSE F WITH z=1;

    state D:
    IF  (!a&!b) THEN G WITH z=0;
    ELSE IF (!a&b) THEN F WITH z=1;
    ELSE IF (a&!b) THEN E WITH z=1;
    ELSE B WITH z=0;

    state E:
    IF  (!a&!b) THEN C WITH z=0;
    ELSE IF (!a&b) THEN F WITH z=1;
    ELSE IF (a&!b) THEN D WITH z=1;
    ELSE E WITH z=0;

    state F:
    IF  (!a&!b) THEN C WITH z=1;
    ELSE IF (!a&b) THEN F WITH z=1;
    ELSE IF (a&!b) THEN D WITH z=0;
    ELSE H WITH z=0;
```

FIGURE 6.38
(Continued)

```
      state G:
      IF (!a&!b) THEN D WITH z=1;
      ELSE IF (!a&b) THEN A WITH z=0;
      ELSE IF (a&!b) THEN B WITH z=1;
      ELSE F WITH z=1;

      state H:
      IF (!a&!b) THEN B WITH z=1;
      ELSE IF (!a&b) THEN C WITH z=0;
      ELSE IF (a&!b) THEN E WITH z=1;
      ELSE F WITH z=1;

      test_vectors
        ([clock,a,b]  ->[sreg,z])
        [  .p.,X,X]  ->[   H,X];
        [  .c.,1,0]  ->[   E,1];
        [  .c.,0,0]  ->[   C,0];
        [  .c.,0,1]  ->[   A,0];
        [  .c.,0,1]  ->[   C,0];
        [  .c.,1,0]  ->[   D,1];
        [  .c.,1,1]  ->[   B,0];
        [  .c.,0,1]  ->[   F,1];
        [  .c.,1,1]  ->[   H,0];
        [  .c.,0,1]  ->[   C,0];
        [  .C.,0,1]  ->[   A,0];
      end
```

PAL23S8: This device is used specifically for state machine implementation. It has nine dedicated input lines and eight output lines (Figure 6.39). Four of the output pins are driven by D flip-flops and the remaining four by **macrocells,** which are very similar to those used in 22V10. By blowing appropriate fuses, a macrocell can be configured to provide a direct sum-of-products output (active high or active low) or the output can be propagated via the D flip-flop. In addition, the true and the complement of the output signal can be fed back to the AND array. By disabling the tristate buffer at the macrocell output, the I/O pin can be configured as an additional input to the device.

The four output pins, which are driven by the D flip-flops, can also be used as input pins by disabling the tristate buffers at their outputs. The 23S8 has six **buried** D flip-flops, the output of which are directly fed back to the AND array. Note that like in the 22V10, the OR gates driving the flip-flops in the 23S8 can generate the sum of variable product terms. Two flip-flops have six product terms feeding the OR gate, two have eight, and remaining two have ten product terms. A special product term in the device, known as the **observability product term,** is used to observe the status of the buried flip-flops. When this product term is enabled, the outputs of the two macrocells and four D flip-flops are disabled, and the outputs of the six buried flip-flops are made available on the associated I/O pins.

EXAMPLE 6.14

Let us implement the state machine shown in Figure 6.40 using a 23S8. The state machine has four inputs, two outputs, and nine states. We will use the four available flip-

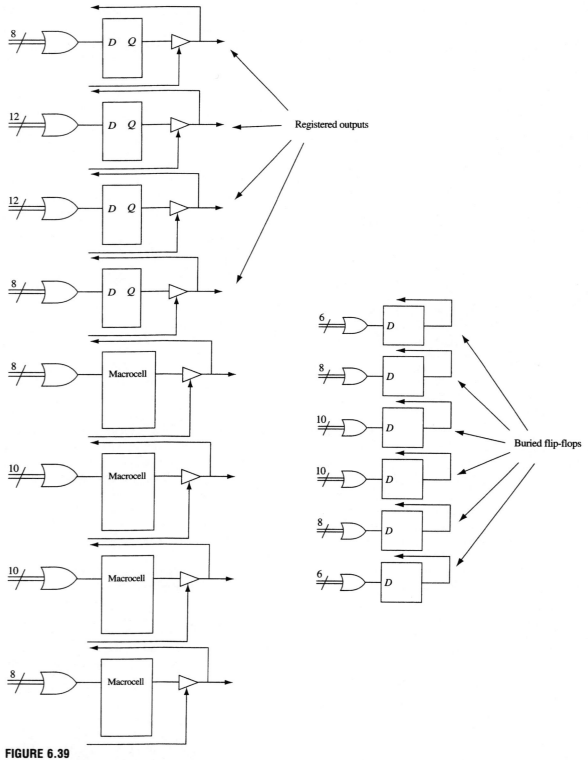

FIGURE 6.39
Block diagram of the 23S8

6.8 SEQUENTIAL LOGIC DESIGN USING PLDs   □ 215

flops to implement the next state logic, and three of the macrocells to implement the output logic of the state machine. The ABEL source file for the circuit is shown in Figure 6.41. The chosen state assignment results in the following next state and output equations, which can be easily programmed into a 23S8.

$q_3.D = bdq_3\bar{q}_2q_1\bar{q}_0 + \bar{a}q_3q_2\bar{q}_1 + bc\bar{d}\,\bar{q}_3\bar{q}_2\bar{q}_1q_0 + \bar{a}\,\bar{b}q_3\bar{q}_2q_1 + \bar{a}q_3\bar{q}_2q_1q_0$
$\qquad + abcd\,\bar{q}_3q_2q_1\bar{q}_0 + \bar{a}\bar{b}\,\bar{d}\,\bar{q}_3\bar{q}_2\bar{q}_1q_0 + bcdq_3\bar{q}_2\bar{q}_1$

$q_2.D = bdq_3\bar{q}_2q_1\bar{q}_0 + d\,\bar{q}_3\bar{q}_2\bar{q}_1q_0 + \bar{a}q_3q_2\bar{q}_1 + ac\bar{d}\,\bar{q}_3\bar{q}_2\bar{q}_1\bar{q}_0 + \bar{a}\,\bar{q}_3q_2\bar{q}_1 + \bar{a}\,\bar{q}_3q_2\bar{q}_0$
$\qquad + abcdq_3\bar{q}_2\bar{q}_1 + bcd\,\bar{q}_3q_2q_1\bar{q}_0 + bc\bar{q}_3\bar{q}_2\bar{q}_1q_0 + bcdq_3\bar{q}_2\bar{q}_1 + bcdq_2\bar{q}_1\bar{q}_0$

$q_1.D = a\bar{b}\,\bar{q}_3\bar{q}_2\bar{q}_1q_0 + d\,\bar{q}_3\bar{q}_2\bar{q}_1q_0 + \bar{a}\,\bar{b}q_3\bar{q}_2q_1 + \bar{a}q_3\bar{q}_2q_1q_0 + abcd\,\bar{q}_3q_2\bar{q}_0$
$\qquad + ac\bar{d}\,\bar{q}_3\bar{q}_2\bar{q}_1\bar{q}_0 + bcd\,\bar{q}_3q_2\bar{q}_1 + \bar{a}\,\bar{q}_3q_2\bar{q}_1$

$q_0.D = bc\bar{d}\,\bar{q}_3\bar{q}_2\bar{q}_1q_0 + \bar{a}\bar{b}\,\bar{d}\,\bar{q}_3\bar{q}_2\bar{q}_1q_0 + \bar{a}q_3\bar{q}_2q_1q_0 + \bar{a}\,\bar{q}_3q_2q_1\bar{q}_0 + abcd\,\bar{q}_3q_2\bar{q}_0$
$\qquad + cd\,\bar{q}_3\bar{q}_2\bar{q}_1\bar{q}_0 + ac\bar{q}_3\bar{q}_2\bar{q}_1\bar{q}_0 + \bar{a}q_3q_2\bar{q}_1q_0$

$z_0 \;\; = a\bar{b}\,\bar{q}_3\bar{q}_2\bar{q}_1q_0 + bdq_3\bar{q}_2q_1\bar{q}_0 + \bar{a}\,\bar{q}_3q_2\bar{q}_1\bar{q}_0 + ad\,\bar{q}_3\bar{q}_2\bar{q}_1\bar{q}_0 + d\,\bar{q}_3\bar{q}_2\bar{q}_1q_0$
$\qquad + \bar{a}q_3q_2\bar{q}_1 + \bar{c}\,\bar{q}_3\bar{q}_2\bar{q}_1\bar{q}_0$

$z_1 \;\; = abcdq_3q_2\bar{q}_1\bar{q}_0 + bc\bar{d}\,\bar{q}_3\bar{q}_2\bar{q}_1q_0 + \bar{a}\,\bar{b}\,\bar{d}\,\bar{q}_3\bar{q}_2\bar{q}_1q_0 + \bar{a}\,\bar{b}q_3\bar{q}_2q_1 + \bar{a}\,\bar{q}_3q_2\bar{q}_1\bar{q}_0$
$\qquad + \bar{a}q_3\bar{q}_2q_1\bar{q}_0 + \bar{a}\,\bar{q}_3q_2q_1\bar{q}_0$ ■

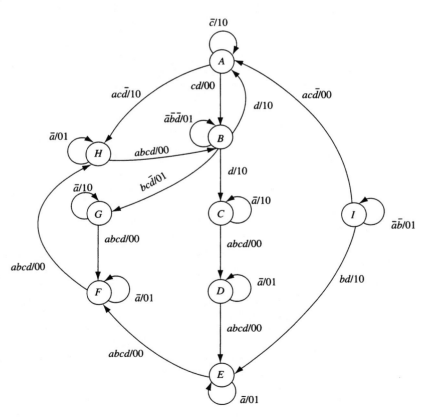

FIGURE 6.40
State diagram

FIGURE 6.41
ABEL source file

```
module examp3
title 'sequential circuit design using PAL23S8'
  bjdu    device 'p23s8';

  clock,reset  pin 1,6;
  a,b,c,d      pin 2,3,4,5;
  z0,z1        pin 18,19 istype 'com';
  q3,q2,q1,q0  pin 14,15,16,17 istype 'reg';
  sreg         = [q3,q2,q1,q0];
  A=0000;B=0001;C=1110;D=0011;E=1100;F=0100;G=1101;H=1111;I=0010;
  X=.x.;


  equations
    sreg.clk = clock;
    sreg.ar  = reset;

  state_diagram sreg
    state A:
    IF (a&c&!d) THEN H WITH z0=1;z1=0;
    ELSE IF (c&d) THEN B WITH z0=0;z1=0;
    ELSE IF (!c) THEN A WITH z0=1;z1=0;

    state B:
    IF (a&!b&!d) THEN I WITH z0=1;z1=0;
    ELSE IF (b&c&!d) THEN G WITH z0=0;z1=1;
    ELSE IF (d) THEN C WITH z0=1;z1=0;
    ELSE IF (!a&!b&!d) THEN B WITH z0=0;z1=1;

    state C:
    IF (a&b&c&d) THEN D WITH z0=0;z1=0;
    ELSE IF (!a) THEN C WITH z0=1;z1=0;

    state D:
    IF (a&b&c&d) THEN E WITH z0=0;z1=0;
    ELSE IF (!a) THEN D WITH z0=0;z1=1;

    state E:
    IF (a&b&c&d) THEN F WITH z0=0;z1=1;
    ELSE IF (!a) THEN E WITH z0=1;z1=0;

    state F:
    IF (a&b&c&d) THEN H WITH z0=0;z1=0;
    ELSE IF (!a) THEN F WITH z0=0;z1=1;

    state G:
    IF (a&b&c&d) THEN F WITH z0=0;z1=0;
    ELSE IF (!a) THEN G WITH z0=1;z1=0;

    state H:
    IF (a&b&c&d) THEN C WITH z0=0;z1=0;
    ELSE IF (!a) THEN H WITH z0=0;z1=1;
```

FIGURE 6.41
(Continued)

```
state I:
IF (a&c&!d) THEN A WITH z0=0;z1=0;
ELSE IF (b&d) THEN E WITH z0=1;z1=0;
ELSE IF (!a&!b) THEN I WITH z0=0;z1=1;

test_vectors
    ([clock,reset,a,b,c,d]  ->[sreg,z0,z1])
    [   .c.,    1,X,X,X,X]  ->[  A,  0,  0];
    [   .c.,    0,1,X,1,0]  ->[  H,  1,  0];
    [   .c.,    0,1,1,1,1]  ->[  C,  0,  0];
    [   .c.,    0,1,1,1,1]  ->[  D,  0,  0];
    [   .c.,    0,1,1,1,1]  ->[  E,  0,  0];
    [   .c.,    0,0,X,X,X]  ->[  E,  0,  0];
    [   .c.,    0,1,1,1,1]  ->[  F,  0,  1];
    [   .c.,    0,1,1,1,1]  ->[  H,  0,  0];
    [   .c.,    0,0,X,X,X]  ->[  H,  0,  1];
    [   .c.,    1,X,X,X,X]  ->[  A,  0,  0];
    [   .c.,    0,X,X,1,1]  ->[  B,  0,  0];
    [   .c.,    0,1,0,X,0]  ->[  I,  0,  0];
    [   .c.,    0,1,X,1,0]  ->[  A,  0,  0];
    [   .c.,    0,X,X,1,1]  ->[  B,  0,  0];
    [   .c.,    0,X,1,1,0]  ->[  G,  0,  1];
end
```

Erasable PLDs

In the past all PLDs were manufactured using fuse-based TTL technology. The advantage of TTL-type PLDs is low propagation delay (as low as 10 ns); in addition, they are capable of driving an I/O bus. Some manufacturers have also used fuse-based ECL technology rather than TTL to manufacture PLDs. Currently available ECL PLDs offer a propagation delay of 6 ns. A drawback of fuse-based programmable logic devices is that they cannot be reprogrammed. Thus, they are not suitable in applications requiring alteration of logic functions or in situations where a new logic design has to be modified several times before a satisfactory design is obtained. This, in addition to high power consumption in bipolar devices, has resulted in a gradual shift to CMOS technology, resulting in PLDs that consume significantly less power than the equivalent bipolar devices.

One significant advantage of CMOS devices is that they use EPROM cells or EEPROM cells instead of fuses as programmable connections. A fuse takes up a large amount of silicon area whereas the EPROM and EEPROM cells are significantly smaller than fuses, thus more functions can be packed into a smaller device. The devices based on EPROM cells are known as **EPLDs** (Erasable Programmable Logic Devices), and those based on EEPROM cells are referred to as **EEPLDs** (Electrically Erasable Programmable Logic Devices).

EPLDs

Altera Corp. introduced the first EPLD in 1984. Since then, many other companies have marketed EPLDs (e.g., VLSI technology, Intel, and Cypress Semiconductors). Altera EPLDs have a programmable AND/fixed OR structure, whose output feeds a macrocell

(Fig. 6.42a). The output multiplexer in Figure 6.42a allows the selection of the output OR gate, the flip-flop, or their complements. The feedback multiplexer allows feedback from the output of the OR gate, the output of the flip-flop, or directly from the I/O pin. If the output of the flip-flop is not made available on the I/O pin (i.e., the flip-flop is buried), then the I/O pin cannot be used. Similarly, if the I/O pin is used as an input pin, the flip-flop cannot be used. In other words, the macrocell of Figure 6.42a allows only a single feedback into the AND array.

Certain EPLD devices allow dual feedback paths into the AND array (Fig. 6.42b). One feedback path comes directly from the I/O pin, and the other comes from the internal logic. Basically, dual feedback is achieved by eliminating the feedback multiplexer of Figure 6.42a. If the tristate buffer in Figure 6.42b is disabled, the internal logic becomes isolated from the I/O pin, thus allowing the internal flip-flop to be buried and the I/O pin to be used as an input line.

Many EPLDs have programmable flip-flops, which can be configured as a *D*, *JK*, *SR* or *T* type. The macrocells in the Altera's EPM5032 (also marketed by Cypress Semiconductors) can be configured as one of the standard flip-flops as well as a conventional latch. In addition, certain EPLDs allow macrocells to be individually programmed for combinational or registered functions. Recently introduced EPLDs, such as the MAX family of chips from Altera, have programmable I/O configurations with variable product term distributions. All chips in the MAX family have eight dedicated input lines. The smallest chip, the EPM5016, has eight additional lines that can be programmed as inputs or outputs. The EPM5128 has 52 programmable I/O lines.

Advanced Micro Devices Inc. (AMD) has developed a generic device, the PALC18U8Q, a 20-pin EPLD that can be configured to emulate the functions of most 20-pin PLDs. It has ten dedicated inputs and eight I/O macrocells. Two versions of the

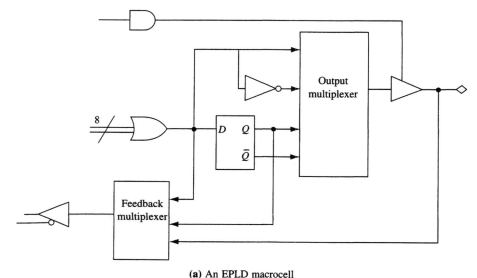

(a) An EPLD macrocell

FIGURE 6.42

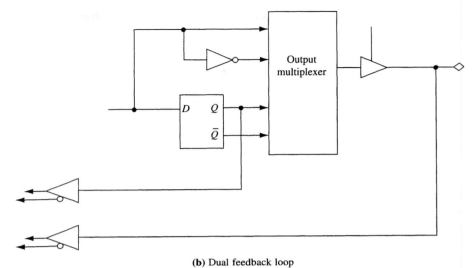

(b) Dual feedback loop

FIGURE 6.42
(Continued)

device with propagation delays of 25 ns and 35 ns are available. Another generic pair of devices for replacing 20-pin and 24-pin PLDs is offered by Cypress Semiconductors. The 20-pin C18G8 has a propagation delay of 12 ns. The 24-pin C20G10 can be programmed to emulate the functions of devices that include but are not limited to the PAL20L10, PAL20L8, PAL20R8, PAL20R6, PAL12R4, PAL12L10, PAL14L8, PAL16L6, PAL18L4, PAL20L2, and PAL20V8.

As mentioned earlier, the bipolar PAL22V10 device, because of its flexibility and ability to replace many earlier PLDs, has been accepted as a standard device by many companies. Texas Instruments has marketed a reprogrammable version of AMD's chip, the TICPAL22V10, which has the additional feature of having a zero-power standby mode. Cypress Semiconductors also offers a reprogrammable version of the PAL22V10, the CY22V10, which has a 15-ns propagation delay. Another erasable and reprogrammable PLD offered by Cypress Semiconductors is the CY7C332. It is similar to the PAL22V10 but has 12 I/O macrocells and 13 input-only macrocells that can be configured to provide registered, latched, or transparent access to the array. Cypress Semiconductors also offers a device, the CY7C330, which is similar to the CY7C332 but contains two pairs of buried flip-flops for state machine implementation. Each pair of flip-flops is driven by 32 product terms. Several versions of the CY7C330 are available with operating frequencies ranging from 25 MHz to 50 MHz.

Atmel Corp. offers a superset version of the PAL22V10, the ATV750. This is also a CMOS EPROM-based device with two flip-flops in each macrocell as well as multiple feedback paths (Fig. 6.43). This simplifies state machine implementation. Furthermore, four to eight product terms can be assigned to each of the 20 sum terms (two for each output) inside the programmable array.

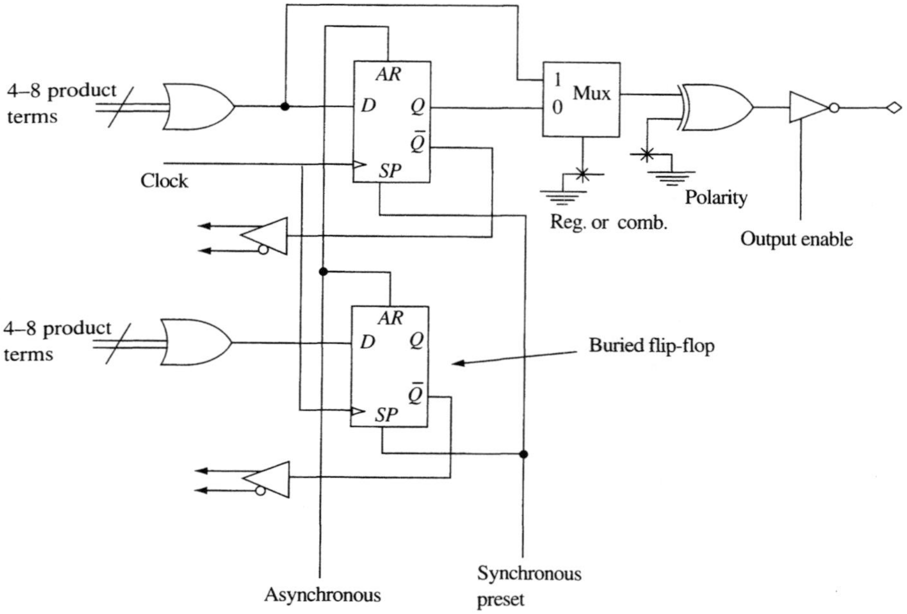

FIGURE 6.43
Macrocell of the ATV750

■ **EXAMPLE 6.15**

Let us implement the sequential circuit of Figure 6.44 using an ATV750. Note that an appropriate state assignment has to be used such that the number of product terms driving the selected flip-flops and the macrocells do not exceed the available product terms. In other words, a formal state assignment technique, which guarantees the minimum number of product terms in the excitation and output equations of a state machine, has to be used in order to check whether the state machine will fit into an ATV750. By using the state assignment technique NOVA (see Sec. 6.5), we obtain such an assignment for the state machine under consideration:

$A = 001, B = 101, C = 000, D = 011, E = 100, F = 110, G = 010, H = 111$

FIGURE 6.44
State table of a sequential circuit

Present State	Input $\bar{a}\bar{b}$	$\bar{a}b$	$a\bar{b}$	ab
A	A,001	D,010	B,001	E,010
B	C,000	D,000	C,010	F,000
C	A,001	A,101	B,001	B,101
D	D,100	E,101	D,010	E,101
E	C,000	D,100	C,010	C,100
F	F,000	H,000	C,010	B,101
G	D,010	A,101	E,010	B,101
H	D,100	E,100	C,010	C,100

The ABEL description of the state machine using this state assignment is shown in Figure 6.45. ∎

FIGURE 6.45
ABEL source file

```
module trial4
title 'state machine implementation using PALP750'
   bjdu  device  'p750';

   clock,preset  pin 1,6;
   a,b           pin 2,3;
   z0,z1,z2      pin 18,19 istype 'com';
   q2,q1,q0      mode 27,28,29 istype 'reg_D';
   sreg        = [q2,q1,q0];
   A=001;B=101;C=000;D=011;E=100;F=110;G=010;H=111;
   X=.x.;
   equations
   [sreg].clk  = clock;
   [sreg].sp = preset;

state_diagram sreg
  state A:
  IF (!a&!b) THEN A WITH z0=0;z1=0;z2=1;
  ELSE IF (!a&b) THEN D WITH z0=0;z1=1;z2=0;
  ELSE IF (a&!b) THEN B WITH z0=0;z1=0;z2=1;
  ELSE E WITH z0=0;z1=1;z2=0;

  state B:
  IF (!a&!b) THEN C WITH z0=0;z1=0;z2=0;
  ELSE IF (!a&b) THEN D WITH z0=0;z1=0;z2=0;
  ELSE IF (a&!b) THEN C WITH z0=0;z1=0;z2=0;
  ELSE F WITH z0=0;z1=0;z2=0;

  state C:
  IF (!a&!b) THEN A WITH z0=0;z1=0;z2=1;
  ELSE IF (!a&b) THEN A WITH z0=1;z1=0;z2=1;
  ELSE IF (a&!b) THEN B WITH z0=0;z1=0;z2=1;
  ELSE B WITH z0=1;z1=0;z2=1;

  state D:
  IF (!a&!b) THEN D WITH z0=1;z1=0;z2=0;
  ELSE IF (!a&b) THEN E WITH z0=1;z1=0;z2=1;
  ELSE IF (a&!b) THEN D WITH z0=0;z1=1;z2=0;
  ELSE E WITH z0=1;z1=0;z2=1;

  state E:
  IF (!a&!b) THEN C WITH z0=0;z1=0;z2=0;
  ELSE IF (!a&b) THEN D WITH z0=1;z1=0;z2=0;
  ELSE IF (a&!b) THEN C WITH z0=0;z1=1;z2=0;
  ELSE C WITH z0=1;z1=0;z2=0;

  state F:
  IF (!a&!b) THEN G WITH z0=0;z1=0;z2=0;
```

FIGURE 6.45
(Continued)

```
         ELSE IF (!a&b) THEN H WITH z0=0;z1=0;z2=0;
         ELSE IF (a&!b) THEN C WITH z0=0;z1=1;z2=0;
         ELSE C WITH z0=1;z1=0;z2=0;

         state G:
         IF (!a&!b) THEN D WITH z0=0;z1=1;z2=0;
         ELSE IF (!a&b) THEN D WITH z0=0;z1=1;z2=0;
         ELSE IF (a&!b) THEN E WITH z0=0;z1=1;z2=0;
         ELSE B WITH z0=1;z1=0;z2=1;

         state H:
         IF (!a&!b) THEN D WITH z0=1;z1=0;z2=0;
         ELSE IF (!a&b) THEN E WITH z0=1;z1=0;z2=0;
         ELSE IF (a&!b) THEN C WITH z0=0;z1=1;z2=0;
         ELSE C WITH z0=1;z1=0;z2=0;

         test_vectors
           ([clock,preset,a,b]->[sreg])
           [  .c.,     1,X,X]->[    H];
           [  .c.,     0,1,1]->[    C];
           [  .c.,     0,0,1]->[    A];
           [  .c.,     0,0,1]->[    D];
           [  .c.,     0,1,1]->[    E];
           [  .c.,     0,1,0]->[    C];
           [  .c.,     0,0,1]->[    A];
           [  .c.,     0,1,0]->[    B];
           [  .c.,     0,1,1]->[    F];
         end
```

EEPLDs

Lattice Semiconductors was the first to develop a family of EEPLDs. Based on a CMOS EEPROM process, the 20-pin GAL16V8 is the first member of this family (Fig. 6.46). It can be substituted for almost all of the 20-pin devices marketed by other companies.

■ **EXAMPLE 6.16**

Let us implement the state machine of Figure 6.47 using a GAL16V8. The ABEL source file for the machine is shown in Figure 6.48, and the simulation results are shown in Figure 6.49. In this example the next state equations have been used to specify the machine rather than the state diagram. The next state equations have been derived using the following state assignment:

	q_3	q_2	q_1	q_0			q_3	q_2	q_1	q_0
A	0	0	0	0		F	0	1	0	0
B	0	0	0	1		G	0	1	1	0
C	1	0	0	0		H	1	0	0	1
D	0	0	1	0		I	1	0	1	0
E	1	1	0	0						

6.8 SEQUENTIAL LOGIC DESIGN USING PLDs □ **223**

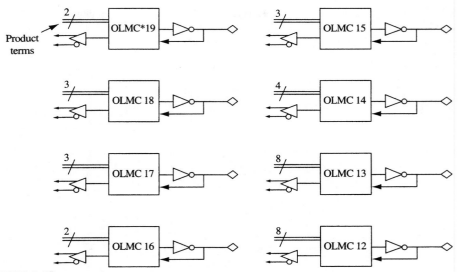

FIGURE 6.46
Block diagram of the GAL16V8
* An OLMC (output logic macrocell) can be programmed to operate in simple (combinational output), complex (combinational outputs with feedback), or registered (outputs obtained via *D* flip-flops) mode.

FIGURE 6.47
State diagram

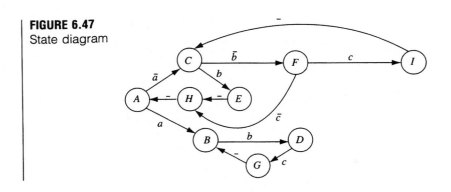

FIGURE 6.48
ABEL source file

```
                   module trial7
         title 'state machine implementation using GALP16V8'
         bjdu      device 'p16v8r';

         clock,enabl   pin 1,11;
         a,b,c         pin 3,4,5;
         q3,q2,q1,q0   pin 16,17,18,19 istype 'reg,invert';
         X=.x.;

         equations
           [q3,q2,q1,q0].clk = clock;
           [q3,q2,q1,q0].oe  = !enabl;
```

FIGURE 6.48
(Continued)

```
q0.D := !(a & !q3 & !q2 & !q1 & !q0
     #  !c & !q3 & !q2 & !q0
     #  q3 & q2 & !q1 & !q0
     #  !q3 & q2 & q1 & !q0);

q1.D := !(b & !q3 & !q2 & !q1 & q0
     #  c & !q3 & q2 & !q1 & !q0
     #  c & !q3 & !q2 & q1 & !q0);

q2.D := !(!b & !q3 & !q2 & !q1 & q0
     #  q3 & !q2 & !q1 & !q0
     #  !q3 & !q2 & q1 & !q0);

q3.D := !(!b & !q3 & !q2 & !q1 & q0
     #  !a & !q3 & !q1 & !q0
     #  b & q3 & !q2 & !q0
     #  !c & !q2 & q1 & !q0
     #  q3 & !q2 & q1 & !q0
     #  q2 & !q1 & !q0);

test_vectors
 ([clock,enabl,a,b,c]->[q3,q2,q1,q0])
  [  .p.,    0,X,X,X]->[ 0, 0, 0, 0];
  [  .c.,    0,0,X,X]->[ 1, 0, 0, 0];
  [  .c.,    0,X,1,X]->[ 1, 1, 0, 0];
  [  .c.,    0,X,X,X]->[ 1, 0, 0, 1];
  [  .c.,    0,X,X,X]->[ 0, 0, 0, 0];
  [  .c.,    0,1,X,X]->[ 0, 0, 0, 1];
  [  .c.,    0,X,1,X]->[ 0, 0, 1, 0];
  [  .c.,    0,X,X,1]->[ 0, 1, 1, 0];
  [  .c.,    0,X,X,X]->[ 0, 0, 0, 1];
  [  .c.,    0,X,0,X]->[ 1, 1, 0, 0];
  [  .c.,    0,X,X,X]->[ 1, 0, 0, 1];
  [  .c.,    0,X,X,X]->[ 0, 0, 0, 0];
  [  .c.,    0,0,X,X]->[ 1, 0, 0, 0];
  [  .c.,    0,X,0,X]->[ 0, 1, 0, 0];
  [  .c.,    0,X,X,1]->[ 1, 0, 1, 0];
end
```

As can be seen from the simulation results of Figure 6.49, the responses to all the test vectors are as expected, which verifies the correct operation of the state machine.

FIGURE 6.49
Simulation results

```
Simulate ABEL 4.03 Date Tue Jul 13 18:15:05 1993
Fuse file: 'trial7.ttl' Vector file: 'trial7.tmv' Part: 'PLA'
state machine implementation using GALP16V8
        ce
        ln
        oa
        cb      qqqq
        klabc   3210
V0001   00000   LLLL
V0002   C0000   HLLL
V0003   C0010   HHLL
V0004   C0000   HLLH
V0005   C0000   LLLL
V0006   C0100   LLLH
V0007   C0010   LLHL
V0008   C0001   LHHL
V0009   C0000   LLLH
V0010   C0000   HHLL
V0011   C0000   HLLH
V0012   C0000   LLLL
V0013   C0000   HLLL
V0014   C0000   LHLL
V0015   C0001   HLHL
15 out of 15 vectors passed.
```

■

Lattice Semiconductors also offers a 24-pin generic circuit, the GAL20V8, which can replace many of the standard 24-pin PLDs. Another generic-array-logic circuit, which can emulate the functions of the currently available 24-pin PAL and FPLA devices, is the GAL39V8. The maximum input-to-output delay of the device is 30 ns. Lattice Semiconductors has also marketed a CMOS EEPROM-based device, the GAL22V10, as a replacement for the bipolar PAL22V10.

As mentioned earlier, the major advantage of erasable PLDs is that they can be reprogrammed. However, it is still necessary to remove such a chip from the system for erasing and reprogramming. Every time a chip is put into or removed from the system there is a risk of damaging the chip. Lattice Semiconductors introduced an in-circuit reprogrammable version of the GAL16V8, the GAL16Z8, which can be reprogrammed with standard 5-V levels while still in the system.

Other PLDs based on CMOS EEPROM technology are the PEEL (Programmable Electrically Erasable Logic) devices from International CMOS Technology. Two of these devices, the PEEL18CV8 (Fig. 6.50) and the PEEL22CV10, can emulate the 20-pin and 24-pin PAL chips respectively. They get their flexibility from very malleable macrocells. Each macrocell can be configured in any of a dozen options—more than any other generic PLD. International CMOS Technology has also developed a superset of the PAL22V10, the PEEL22CV10Z. The primary advantage of the PEEL22CV10Z, apart from reprogrammability, is that its macrocells can be configured into three times as many options as the original 22V10's macrocells.

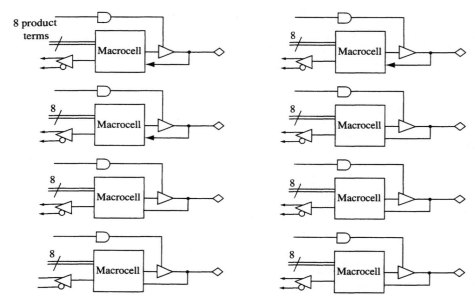

FIGURE 6.50
Block diagram of the PEEL18CV8

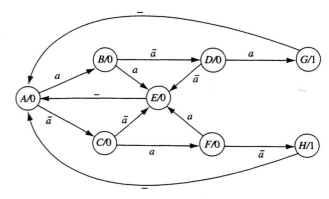

FIGURE 6.51
State diagram

■ **EXAMPLE 6.17**

Let us implement the state machine of Figure 6.51 using a PEEL18CV8. The machine serially receives a 3-bit binary pattern and produces an output of 1 only if the pattern is 101 or 010 is received; otherwise the output is 0. The ABEL description of the machine is shown in Figure 6.52. ■

FIGURE 6.52
ABEL source file

```
module trial10
title 'state machine implementation using PEEL18CV8'
    bjdu     device 'p18cv8';
    clock,a    pin 1,3;
    z          pin 19 istype 'com';
    q2,q1,q0   pin 16,17,18 istype 'reg';
    sreg       = [q2,q1,q0];
    A=0;B=1;C=8;D=2;E=12;F=4;G=6;H=9;
    X=.x.;
equations
    [sreg].clk = clock;
state_diagram sreg
    state A:
    z=0;
    IF (!a) THEN C;
    ELSE B;

    state B:
    z=0;
    IF (a) THEN E;
    ELSE D;

    state C:
    z=0;
    IF (!a) THEN E;
    ELSE F;

    state D:
    z=0;
    IF (!a) THEN E;
    ELSE G;

    state E:
    z=0;
    GOTO A;

    state F:
    IF (!a) THEN H;
    ELSE E;

    state G:
    z=1;
    GOTO A;

    state H:
    z=1;
    GOTO A;

test_vectors
    ([clock,a] ->[sreg])
    [ .p.,X] ->[  0 ];
    [ .c.,X] ->[  8 ];
    [ .c.,X] ->[ 12];
    [ .c.,X] ->[  9 ];
    [ .c.,X] ->[  0 ];
    [ .c.,X] ->[  1 ];
    [ .c.,X] ->[  2 ];
```

FIGURE 6.52
(Continued)

```
            [  .c.,X] ->[   6 ];
            [  .c.,X] ->[   1 ];
            [  .c.,X] ->[  12 ];
            [  .c.,X] ->[   9 ];
            [  .c.,X] ->[   0 ];
            [  .c.,X] ->[   8 ];
            [  .c.,X] ->[   4 ];
            [  .c.,X] ->[  10 ];
         end
```

Currently, AMD's PAL29M16 is one of the most powerful EEPLDs available. It is a 24-pin device and has 29 inputs and 16 programmable I/O macrocells. Eight of these macrocells provide single feedback, and the remaining eight provide dual feedback to the AND array. As in AmPAL22V10, this device uses variable product term distribution, which allocates 8, 12, or 16 product terms to the macrocells. The PAL29M16 can preload the registers/latch in the macrocells to any desired state during testing for full logic verification. As in the PAL23S8, there is a dedicated observability product term in the PAL29M16. When this product term is activated, the contents of the register/latch are available on the I/O pins. All register/latches in the device are cleared on power-up.

■ **EXAMPLE 6.18**

Let us consider the implementation of the sequential circuit specified by Figure 6.54 using a 29M16. The ABEL source file for the machine is shown in Figure 6.53. ■

FIGURE 6.53
ABEL source file

```
module trial5
title 'state machine implementation using PAL29M16'
     bjdu device 'p29m16';

     clock          pin 1;
     a,b,c          pin 11,13,14;
     z0,z1,z2       pin 15,16,17 istype 'com';
     q2,q1,q0       pin 8,9,10 istype 'reg';
     sreg         = [q2,q1,q0];
     A=000;B=100;C=010;D=001;E=110;F=111;G=101;
     X=.x.;

equations
  [sreg].clk   = clock;

state_diagram sreg
  state A:
  IF (a&!b) THEN G WITH z0=1;z1=0;z2=1;
  ELSE IF (!a&c) THEN C WITH z0=0;z1=1;z2=0;
  ELSE IF (!a&b&!c) THEN D WITH z0=1;z1=1;z2=1;
  ELSE A WITH z0=0;z1=0;z2=0;
```

FIGURE 6.53
(Continued)

```
state B:
IF (a&c) THEN E WITH z0=0;z1=1;z2=0;
ELSE IF (!a&c) THEN B WITH z0=0;z1=0;z2=0;
ELSE IF (a&!c) THEN G WITH z0=1;z1=0;z2=0;
ELSE C WITH z0=0;z1=0;z2=1;

state C:
GOTO E WITH z0=0;z1=1;z2=1;

state D:
GOTO A WITH z0=1;z1=1;z2=0

state E:
IF (a&b) THEN F WITH z0=0;z1=0;z2=0;
ELSE IF (!b&c) THEN D WITH z0=1;z1=0;z2=0;
ELSE IF (!a&!b&!c) THEN G WITH z0=0;z1=1;z2=0;
ELSE E WITH z0=1;z1=0;z2=0;

state F:
GOTO A WITH z0=1;z1=0;z2=1

state G:
IF (a&!c) THEN F WITH z0=0;z1=1;z2=0;
ELSE IF (!a&c) THEN G WITH z0=0;z1=1;z2=0;
ELSE IF (!a&!c) THEN B WITH z0=1;z1=0;z2=0;
ELSE D WITH z0=0;z1=0;z2=1;
end
```

FIGURE 6.54
State diagram

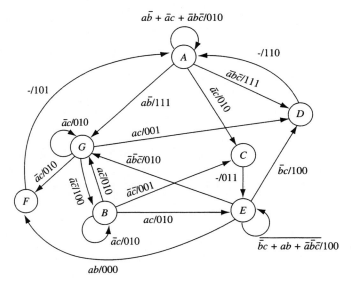

EXERCISES

1. Find a minimal state table for each of the sequential circuits specified:

Present State	Input x = 0	x = 1
A	B,0	F,1
B	D,0	D,0
C	C,0	F,0
D	A,1	A,0
E	D,0	D,0
F	D,0	B,0

(i)

Present State	Input x = 0	x = 1
A	C,0	E,0
B	H,0	G,1
C	B,0	A,0
D	E,1	H,0
E	E,1	C,0
F	H,0	D,1
G	A,1	H,0
H	D,0	F,1

(ii)

2. For the state table shown below, find the output and state sequences corresponding to the input sequence: 01010110 assuming that the circuit starts in state A:

Present State	Input x = 0	x = 1
A	D,0	A,0
B	D,1	A,0
C	C,0	B,0
D	C,0	A,0

3. The state diagram of a sequential circuit is shown below. Implement the circuit using D flip-flop as memory elements. Assume the following state assignment: $A = 000$, $B = 001$, $C = 010$, $D = 011$, $E = 100$, $F = 101$.

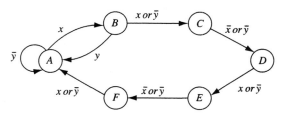

4. Implement the state tables of Exercise 6.3 using JK flip-flops as memory elements.

5. A synchronous sequential circuit with two inputs x_1 and x_2 and an output z is to be designed. The output z is to be 1 whenever both x_1 and x_2 receive identical groups of five input bits. Each input bit is synchronized with a clock pulse applied to the clock input. Show the minimized state table for the circuit and implement it using D flip-flops as memory elements.

6. A synchronous sequential circuit has one input x and one output z. The output is 1 whenever the input sequence is 0110 (e.g., if the input sequence is 000110110, the output sequence would be 000001001). Construct the state diagram for the circuit, and implement it using JK flip-flops as memory elements.

7. A synchronous sequential circuit is to be used for generating the parity of a continuous stream

of binary digits. The output of the circuit produces a logic 1 if the number of 1's received at the input is even; otherwise, the output is at logic 0. Implement the circuit using *JK* flip-flops as memory elements.

8. A synchronous sequential circuit is to be used to detect errors in a message using 2-out-of-4 code. The sequential circuit receives a coded message serially and produces an output of 1 whenever an illegal message is received. Develop a state diagram to meet this specification, and implement the circuit using *D* flip-flops.

9. Derive the state assignment for each of following synchronous sequential circuits represented by their state tables, using the rules stated in Section 6.5.

	Input	
Present State	$x = 0$	$x = 1$
A	A,0	D,0
B	A,0	D,1
C	B,0	C,0
D	A,0	C,0

(i)

	Input	
Present State	$x = 0$	$x = 1$
A	B,0	E,0
B	E,0	D,0
C	D,1	A,0
D	C,1	E,0
E	B,0	D,0

(ii)

10. A sequential circuit having a single input and a single output is to be designed according to the following specification. The output is to be at logic 0 unless an input sequence 0010 is received (e.g., if the input sequence is 0100100100, the output sequence will be 0000001001). Construct a minimum row state table for the circuit.

11. Find a minimum row state table for each of the incompletely specified sequential circuits specified by the following state tables:

	Input	
Present State	$x = 0$	$x = 1$
A	B,1	H,1
B	A,0	G,0
C	-,0	F,-
D	D,0	-,1
E	C,0	D,-
F	A,-	C,0
G	-,-	B,-
H	G,0	E,-

(i)

	Input = $x_1 x_2$			
Present State	00	01	11	10
A	C,1	-,-	G,1	E,1
B	-,-	E,0	-,-	-,-
C	F,1	F,0	-,1	-,-
D	-,-	-,0	-,-	B,1
E	F,0	-,-	D,-	A,0
F	-,-	C,0	C,-	B,0
G	F,0	-,0	-,1	B,-

(ii)

Present State	Input		
	I_1	I_2	I_3
A	A,0	B,1	E,1
B	B,0	A,1	F,1
C	A,1	D,0	E,0
D	F,0	C,1	A,0
E	A,0	D,1	E,1
F	B,0	D,1	F,1

(iii)

12. Find a minimum row state table for each of the sequential circuits whose states are given below by using the partitioning method.

Present State	Input	
	x = 0	x = 1
A	A,0	E,1
B	E,1	C,0
C	A,1	D,1
D	F,0	G,1
E	B,1	C,0
F	F,0	E,1

Next State, Output
(i)

Present State	Input	
	x = 0	x = 1
A	B,0	C,1
B	A,1	E,0
C	F,1	C,0
D	D,0	C,1
E	A,1	B,0
F	B,0	D,1

Next State, Output
(ii)

13. Determine whether or not the state tables shown below can be implemented using a shift register. Derive the necessary excitation equations.

Present State	Input	
	x = 0	x = 1
A	B	F
B	F	B
C	A	A
D	E	C
E	C	E
F	D	D

(i)

Present State	Input	
	x = 0	x = 1
A	E	E
B	E	A
C	B	F
D	F	F
E	G	G
F	G	C
G	D	H
H	D	D

(ii)

14. A one-input, four-state sequential circuit produces the output sequence Z corresponding to an input sequence I as shown:

$$I\ \ 1\ 1\ 0\ 1\ 0\ 0\ 1\ 0\ 1\ 1\ 0\ 0\ 1\ 0$$
$$Z\ \ 1\ 0\ 0\ 0\ 1\ 1\ 1\ 0\ 1\ 0\ 0\ 1\ 0\ 0$$

Determine the state table representation of the circuit, assuming that A is the initial state.

15. A certain sequential circuit has six states (A,B,C,D,E,F), five inputs (s,t,u,v,w), and three outputs (l,m,n). The circuit is specified as shown. Each state of the specification consists of the state of the circuit itself, the outputs associated with the state, and its next state definitions; the transitions are defined conditionally using **IF-THEN-ELSE** statements.

```
State A: l=0, m=0, n=0
   IF s=1 THEN B ELSE A ;

State B: l=0, m=0, n=0;
   IF Z1=1 THEN C                    { Z1=(u+v̄)t̄;  Z2=ūvt̄ }
   ELSE IF Z2=1 THEN B
   ELSE IF t=1 THEN A
   ELSE B;

State C: l=0, m=0, n=0;
   IF Z1=1 THEN C
   ELSE IF Z2=1 THEN D
   ELSE IF t=1 THEN A
   ELSE C;

State D: l=0, m=0, n=0;
   IF Z1=1 THEN E
   ELSE IF Z2=1 THEN D
   ELSE IF t=1 THEN A
   ELSE D;

State E: l=1, m=0, n=1            { Z3=w t̄;  z4=w̄ t̄ }
   IF Z3=1 THEN E
   ELSE IF Z4=1 THEN F
   ELSE IF t=1 THEN A
   ELSE E;

State F: l=0, m=1, n=0;
   IF t=1 THEN A
   ELSE F;
```

Design the circuit using D flip-flops.

16. A sequential circuit is to be designed to monitor the status of a chemical experiment. Every 10s the circuit receives an input pattern between 0001 and 1111. If the input pattern received is 0111, the experiment is assumed to be continuing perfectly; the input patterns 0110 and 1000 are also considered satisfactory. However, if the circuit receives any other binary pattern twice consecutively, the experiment must be stopped. Derive a state diagram of the sequential circuit, and a complete logic diagram using JK flip-flops as memory elements.

17. The shift register implementation of a sequential circuit is shown below. Draw an ASM chart for the circuit.

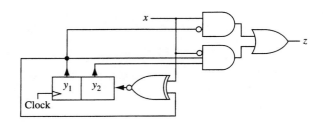

18. Minimize the following incompletely specified sequential circuits if possible; implement the circuits using D flip-flops as memory elements.

Present State	Input	
	$x = 0$	$x = 1$
A	E,0	A,0
B	D,0	B,0
C	E,1	C,–
D	A,1	A,1
E	A,–	B,–

Present State	Input	
	$x = 0$	$x = 1$
A	C,1	–,–
B	B,–	E,–
C	A,1	D,1
D	F,0	G,0
E	G,–	–,–
F	–,–	C,0
G	A,1	B,–

19. A sequential circuit is to be used to identify possible non–code words produced by a 3-out-of-6 code generator circuit. The sequential circuit will examine the output of the code generator circuit one bit at a time. After all six bits have been examined, the circuit produces an output of 1 if it detects a non–code word; otherwise, an output of 0 is produced. Write the ABEL source file for the circuit using the PAL22V10 to implement the circuit.

20. Design a sequential circuit that produces an output of 1 after it has received an input sequence 1101 or 011. Derive the design equations for the circuit assuming D flip-flops as memory elements. Use the principle of partitioning for generating the state assignments.

21. Specify the sequential circuit of Exercise 6.15 in KISS2 format so that the resulting specification can be used as an input file for the SIS system.

22. A sequential circuit produces an output of 1 if and only if it receives an input sequence that contains only one group of 0's. For example, the circuit will produce an output of 1 if the input sequence is 11001111, but will generate an output of 0 if the input sequence is 11001101. Draw an ASM chart for the circuit. Design the circuit using k-out-of-$2k$ state encoding, and using D flip-flops as memory elements.

23. A single input and single output state machine produces an output of 1 and remains at 1 when at least two 0's and two 1's have occurred at the input regardless of the order of occurrence. Write a design specification for the counter in ABEL-HDL.

References

1. J. Hartmanis, "On the state assignment problem for sequential machines," *IRE Trans. on Computers*, December 1972, pp. 593–603.

2. T. Villa and A. L. Sangiovanni-Vincentelli, "NOVA: State assignment for finite state machines for optimal two-level logic implementations," *IEEE Trans. on Computer Aided Design*, September 1990, pp. 1326–1334.

3. B. Lin and A. R. Newton, "Synthesis of multiple level logic from symbolic high-level description languages," *Proc. VLSI'89 Conf.*, Munich, Germany, August 1989.

4. Z. Kohavi, *Switching and Finite Automata Theory*, McGraw-Hill, New York, 1978.

5. B. Holdsworth, *Digital Logic Design*, Butterworths, London, 1987.

Further Reading

1. D. J. Comer, *Digital Logic and State Machine Design,* Saunders, Philadelphia, 1990.
2. J. F. Wakerly, *Digital Design Principles and Practices,* Prentice Hall, Englewood Cliffs, 1990.
3. M. Mano, *Digital Design,* Prentice Hall, Englewood Cliffs, 1984.
4. D. Pellerin and M. Holley, *Digital Design Using ABEL,* Prentice Hall, Englewood Cliffs, 1994.

7
Asynchronous Sequential Circuits

In Chapter 6 we considered clocked (synchronous) sequential circuits; asynchronous circuits do not use clocks. If an input variable is changed in an asynchronous circuit, the circuit goes through a sequence of **unstable** states before settling down to a **stable** state. On the other hand, a synchronous circuit moves from one stable state to another without passing through any unstable states. To illustrate the concept of unstable states, let us consider the SR latch circuit shown in Figure 7.1. When $S = 0$ and $R = 1$, the output Q of the circuit is set to 0, and $\overline{Q} = 1$; this is a stable state because the feedback signals have no effect on the output of the latch. If R is changed to 0, the circuit still remains in the stable condition because there is no change in the output. Now if S is changed to 1, $\overline{Q}$ changes to 0; however, before Q changes to 1, there is a momentary delay during which there is an unstable state $Q = 0$, $\overline{Q} = 0$ with inputs $S = 1$ and $R = 0$. At the end of the delay, the circuit will assume the stable state $Q = 1$, $\overline{Q} = 0$. Thus, an unstable state is said to exist in an asynchronous circuit if the next state at any instant of time after an input has been changed is not equal to the present state. We assume that an asynchronous circuit will eventually assume a stable state, provided the duration of the inputs is longer than the period of time the unstable state or states exist. If an input to an asynchronous circuit is changed only when it is in a stable state, never when it is unstable, then the circuit is said to be operating in **fundamental mode**. The simultaneous change of more than one input variable is avoided in asynchronous circuits because this often leads to serious timing problems.

Figure 7.2 shows the model of asynchronous circuits. Unlike synchronous circuits, these do not require separate memory elements, the propagation delays associated with

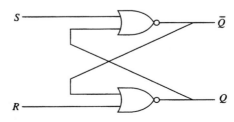

FIGURE 7.1
SR latch constructed from NOR gates

FIGURE 7.2
Model of asynchronous sequential circuit

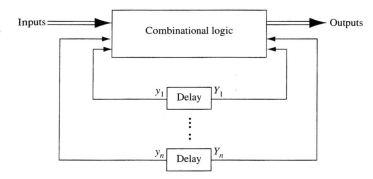

the feedback paths from the outputs to the inputs provide the memory required for sequential operation. Although the delay elements shown in Figure 7.2 can be considered to have the same role as D flip-flops in synchronous circuits, the delays cannot be assumed to be of equal magnitude. The internal state of an asynchronous circuit is represented by the state variable y_i, and the next state variables are denoted by Y_i ($i = 1,n$).

7.1 FLOW TABLE

As in the design of synchronous sequential circuits, the first step in the design of asynchronous circuits is to specify the circuit operation in a formal manner. A conventional way of describing the operation of a fundamental mode circuit is to use a **flow table**. The flow table is very similar to the state table in that it specifies all possible modes of circuit operation. As an example we construct the flow table of a sequential circuit that has two inputs $x_1 x_2$ and one output Z; the circuit produces an output of 1 only if the input sequence $x_1 x_2 = 10, 11, 01$ is received in that order. All other sequences produce an output of 0. Initially the inputs are $x_1 = 0$, $x_2 = 0$, and the circuit is in a state designated as $\widehat{S_0}$ where the circle around S_0 indicates that it is stable. Thus, the first row of the flow table is S_0 and the entry in column $x_1 x_2 = 00$ is $\widehat{S_0}$. The output of an asynchronous circuit is associated with a stable state; hence, the output entry 0 is recorded on the right of $\widehat{S_0}$.

State	$x_1 x_2$ 00	01	11	10
S_0	$\widehat{S_0}$, 0	—	—	—

If the input combination is now changed to $x_1 x_2 = 10$, the circuit enters the stable state $\widehat{S_1}$. To show that the change from $\widehat{S_0}$ to $\widehat{S_1}$ was caused by the input combination $x_1 x_2 = 10$, the uncircled state S_1 is entered in column 10 of the first row. This indicates that the transition from $\widehat{S_0}$ to $\widehat{S_1}$ was via an unstable state. If the input x_2 is changed to 1 while the circuit is in $\widehat{S_0}$, the circuit would enter another stable state $\widehat{S_2}$ via the unstable state S_2. The output of the circuit corresponding to both stable states $\widehat{S_1}$ and $\widehat{S_2}$ is 0. The dash in column 11 indicates that the double input change $x_1 x_2 = 00$ to 11 is not allowed.

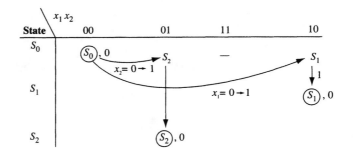

Starting in the second row, an input change from $x_1x_2 = 10$ to 00 will cause the circuit to return to the initial state S_0. On the other hand if the input changes from $x_1x_2 = 10$ to 11, the circuit will enter another stable state S_3.

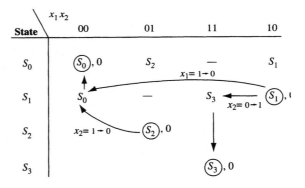

If input x_1 changes from 1 to 0 while the circuit is in S_3, the desired input sequence is complete, and the circuit moves to S_4 with the output $Z = 1$.

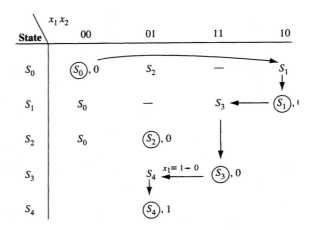

FIGURE 7.3
Flow table for the (10, 11, 01) detector

State	$x_1 x_2$ 00	01	11	10
S_0	(S_0), 0	S_2	—	S_1
S_1	S_0	—	S_3	(S_1), 0
S_2	S_0	(S_2), 0	S_5	—
S_3	—	S_4	(S_3), 0	S_1
S_4	S_0	(S_4), 1	S_5	—
S_5	—	S_2	(S_5), 0	S_1

Starting in (S_2), if x_2 changes from 1 to 0, the circuit is reset to state S_0. If x_1 is changed from 0 to 1, the circuit cannot go to (S_3) because this will indicate that the circuit has received the input sequence (10, 11). So a new state, (S_5), is included; (S_5) has a 0 output associated with it.

The change in x_2 from 1 to 0 when the circuit is in state (S_3) must take it to (S_1), since (S_1) corresponds to the first two symbols of the desired input sequence.

Starting from (S_4), if the inputs change to 00 it is not possible to get $Z = 1$ without resetting the circuit to (S_0). The change from $x_1 x_2 = 01$ to 11 cannot move the circuit to (S_3), however, the circuit can be taken to state (S_5).

Finally, from (S_5), if the inputs change to 01, the circuit cannot move to (S_4) because it is allowed to produce $Z = 1$ only after it resets and receives the proper input sequence. Hence, it must move to (S_2). On the other hand, if $x_1 x_2$ changes to 10, the circuit can go to (S_1) since this starts the desired sequence over again. The complete flow table is shown in Figure 7.3; dashes have been entered wherever input changes are not allowed.

It should be noted that although the primitive flow table of Figure 7.3 looks like that of a Mealy model, it could have been represented in the Moore-model form by adding a column of outputs such that each row has an output same as that associated with the stable next state. The Moore-model version of the primitive table is shown in Figure 7.4.

FIGURE 7.4
Moore-type flow table for the (10, 11, 01) detector

State	$x_1 x_2$ 00	01	11	10	Output Z
S_0	(S_0)	S_2	—	S_1	0
S_1	S_0	—	S_3	(S_1)	0
S_2	S_0	(S_2)	S_5	—	0
S_3	—	S_4	(S_3)	S_1	0
S_4	S_0	(S_4)	S_5	—	1
S_5	—	S_2	(S_5)	S_1	0

Notice that the flow table constructed for the sequence detector has exactly one stable state per row. Such a table is called a **primitive flow table**. In a primitive flow table a change of an input corresponds to a change between columns without any row change. If the state in the new column is uncircled (i.e., unstable), a row change takes place, with the new row corresponding to the next state for the circuit. One important distinction between synchronous and asynchronous circuits is that in an asynchronous circuit the state changes depend on the input changes, whereas in synchronous circuits it is the clock pulse rather than the input changes that triggers the state changes.

7.2 REDUCTION OF PRIMITIVE FLOW TABLES

In general, a primitive flow table constructed from a circuit specification contains redundant states that must be eliminated in order to reduce the hardware required for the circuit implementation. Since each row in a primitive flow table is identified with a unique stable state, the elimination of redundant states results in the reduction of rows in the table. The reduction process is analogous to that of the incompletely specified synchronous circuit; recall that the unspecified entries in a flow table are due to the constraint that only one input variable can change at a time. Thus, two rows in a primitive flow table can be merged if the next state entries in each column corresponding to these rows are the same if both are specified. If two next state entries are the same, with one stable and the other unstable, then after merging the resultant entry is stable. There is no conflict of outputs when two rows are merged, because the outputs are only associated with stable states and two different stable states in the same column are not merged. Since the problem of eliminating redundant states in asynchronous circuits is identical to that encountered in incompletely specified synchronous circuits, the technique of Section 6.6 may be employed.

■ EXAMPLE 7.1

Let us reduce the primitive flow table of Figure 7.5. The implication table corresponding to the flow table is shown in Figure 7.6. Note that a stable state (w) and an unstable state x are compatible if (w) is compatible with (x). An unstable state y is compatible with another unstable state z if (y) is compatible with (z). It is seen from Figure 7.6 that the compatible pairs (i.e., the rows) that can be merged are

$$(S_0,S_5)(S_1,S_6)(S_3,S_4)$$

Row S_2 cannot be merged with any other rows. Thus, the reduced flow table is

	$x_1 x_2$ 00	01	11	10
(S_0,S_5)	(S_0), 0	(S_5), 1	—	S_1
(S_1,S_6)	S_2	(S_6), 1	—	(S_1), 1
(S_3,S_4)	(S_4), 1	S_6	—	(S_3), 1
(S_2)	(S_2), 1	S_5	—	S_3

FIGURE 7.5
A primitive flow table

	$x_1 x_2$			
	00	01	11	10
S_0	ⓢ₀, 0	—	—	S_1
S_1	S_2	—	—	ⓢ₁, 1
S_2	ⓢ₂, 1	S_5	—	S_3
S_3	S_4	—	—	ⓢ₃, 1
S_4	ⓢ₄, 1	S_6	—	—
S_5	S_0	ⓢ₅, 1	—	—
S_6	S_2	ⓢ₆, 1	—	—

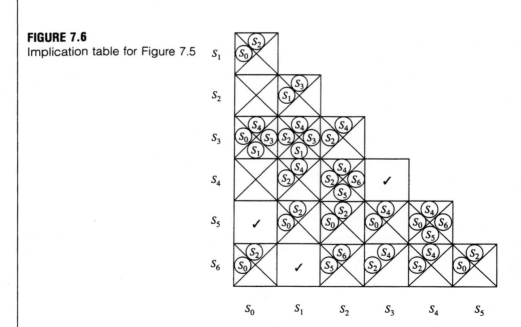

FIGURE 7.6
Implication table for Figure 7.5

By replacing the rows in the reduced flow table as A, B, C, and D the following equivalent flow table can be obtained.

	$x_1 x_2$			
	00	01	11	10
A	Ⓐ, 0	Ⓐ, 1	—	B
B	D	Ⓑ, 1	—	Ⓑ, 1
C	Ⓒ, 1	B	—	Ⓒ, 1
D	Ⓓ, 1	A	—	C

7.3 STATE ASSIGNMENT

This process is also similar to that described for synchronous sequential circuits. However, while the criterion for the selection of unique binary codes for the states in synchronous circuits was that it should result in a minimum hardware implementation, there is a more important requirement for asynchronous circuits. The assignment of secondary variables in asynchronous circuits must be such that only one variable can change during the transition of the circuit from one stable state to another.

Races and Cycles

A race condition results in a fundamental mode circuit if more than one secondary variable is allowed to change during a state transition. The condition arises due to the unequal delays in different feedback paths in the circuit. To illustrate let us consider the flow table shown in Figure 7.7a. By assigning secondary variables $A = 00$, $B = 01$, $C = 11$, and $D = 10$, the excitation table shown in Figure 7.7b results. Assume that the circuit is in state ⑩ and both inputs x_1 and x_2 are equal to 0. Now if input x_2 changes to 1, an unstable condition develops in which the present secondary variables $y_1 y_2 = 00$, but the input combination $x_1 x_2 = 01$ requires that the next stable state should be ⑪. However, due to the unknown delays in the feedback paths, $y_1 y_2$ can change from ⑩ to ⑪ in three different ways:

i. Both y_1 and y_2 change simultaneously, giving a correct transition to ⑪.
ii. y_1 changes before y_2; the circuit will go to state 10 first, followed by transitions to 01 and then to ⑪.
iii. y_2 changes before y_1; the circuit goes to state ⑪.

Thus, if either y_1 or y_2 changes before the other, an incorrect transition to state ⑪ will take place. Such a situation, in which the circuit may reach an incorrect stable state whenever two or more state variables change, is referred to as a **critical race**. As we shall see later, the state assignment in asynchronous circuits can be made in such a way so that critical races are eliminated from the circuit.

Not all races are critical. A race is referred to as **non-critical** if the circuit reaches the correct stable state irrespective of the order in which the state variables change.

FIGURE 7.7

	$x_1 x_2$			
	00	01	11	10
A	Ⓐ	C	Ⓐ	C
B	A	Ⓑ	Ⓑ	D
C	D	Ⓒ	B	Ⓒ
D	Ⓓ	B	A	Ⓓ

(a) Flow table for an asynchronous circuit

$y_1 y_2$	$x_1 x_2$			
	00	01	11	10
00	⓪⓪	11	⓪⓪	11
01	00	⓪1	⓪1	10
11	10	⑪	01	⑪
10	⑩	01	00	⑩

(b) Excitation table

EXAMPLE 7.2

Let us assume that the circuit specified by Figure 7.8 is in state $y_1 y_2 = 00$ and $x_1 = 0$ and $x_2 = 0$. Now if the input x_2 is changed to 1, the circuit must move to ⑪. If both y_1 and y_2 change simultaneously, the desired stable state will be reached. If either y_1 or y_2 changes first, the circuit will move to 10 or 01 respectively. However, irrespective of which variable changes first, the circuit always reaches the correct stable state ⑪.

FIGURE 7.8
Excitation table of an asynchronous sequential circuit

$y_1 y_2$	$x_1 x_2$			
	00	01	11	10
00	⓪⓪	11	10	10
01	00	11	⓪1	00
11	01 ←	⑪	01	01
10	11	11	⑩	11

■

Let us now assume that the circuit of Fig 7.8 is in state (11) with $x_1x_2 = 01$. If the input state changes to 00, the circuit will enter unstable state 01 and then move to the stable state (00), as indicated in the flow table by an arrow leading from one unstable state to another. The absence of an arrow from an unstable state indicates that it is directed to its corresponding stable state. When a circuit goes through a sequence of unstable states before terminating in the desired stable state, it is said to have a **cycle**. It is important that a cycle terminate in a stable state; otherwise, the circuit will sequence through the unstable states indefinitely until the next change of input variable.

■ **EXAMPLE 7.3**

Let us consider the circuit specified by Figure 7.8, and assume that it is in state (00) with input $x_1x_2 = 00$. If input x_1 is now changed to 1, the circuit makes a transition to 10. From 10 the circuit moves to 11, from 11 to 01, and then back to 00, without any further change of input. The circuit will go through the unstable states indefinitely until the next change of an input. ■

Critical Race–Free State Assignment

Critical races can be avoided by assigning binary codes to the states in such a way that when a transition occurs from one state to another, only one secondary variable should change its value. In other words the principle of fundamental mode operation is extended to the state variables as well.

■ **EXAMPLE 7.4**

Let us consider the reduced flow table of Figure 7.9. This is a three-row flow table; hence, it requires at least two secondary variables for unique state assignment. An examination of the flow table reveals that there is a direct transition between rows

A and C (column 10)
B and A (column 01)
B and C (column 10)
C and A (column 00)
C and B (column 11)

This information is summarized in the **transition diagram** shown in Figure 7.10. Each node in the diagram represents a state of the flow table. A directed line connects two nodes if transitions may occur between the corresponding states.

FIGURE 7.9
A three-state flow table

State	x_1x_2 00	01	11	10
A	(A), 0	(A), 0	(A), 0	C
B	(B), 1	A	(B), 1	C
C	A	(C), 1	B	(C), 1

FIGURE 7.10
Transition diagram of Figure 7.9

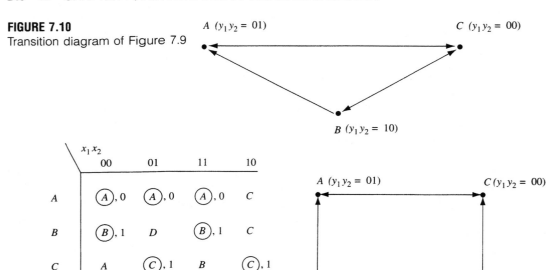

FIGURE 7.11
Critical race–free assignment with an extra state for Figure 7.9

As mentioned previously, critical races can be avoided if only one state variable is allowed to change when a transition is made from one state to another. An arbitrary state assignment is shown in the transition diagram. However, inspection of it reveals that both the state variables y_1 and y_2 must change during the transition from B to A, resulting in a critical race. It is in fact impossible to assign state variables so that only one variable changes during a transition.

However, two state variables can define four states, which implies that there is a fourth or "spare" combination of secondary variables that has not been used. Figure 7.11 shows how this spare combination is utilized to overcome the critical race. If the circuit is in B under the input condition $x_1 x_2 = 00$, and a change to $x_1 x_2 = 01$ occurs, the circuit must go to the desired state A via the "spare" state D. This is achieved by directing the state variable changes to $y_1 y_2 = 10 \rightarrow 11 \rightarrow 01$. ∎

Thus, a critical race-free assignment for a three-row flow table can be achieved by incorporating an additional row in the flow table. This row is employed to generate a path between two stable states, transitions between which would otherwise result in a critical race. The unspecified next state entries in the modified flow table corresponding to the spare state must not be assigned the same state variable combination as the spare state, otherwise an undesired stable state will be created in the fourth row.

A critical race–free state assignment is not always possible for a four-row table using two state variables.

EXAMPLE 7.5

Let us consider the flow table shown in Figure 7.12. The corresponding transition diagram is shown in Figure 7.13. The transitions in the diagonal directions (e.g., D to B and C to A) indicate that no matter how the two state variables are assigned, it is not possible to satisfy the requirement that only a single variable should change during a transition.

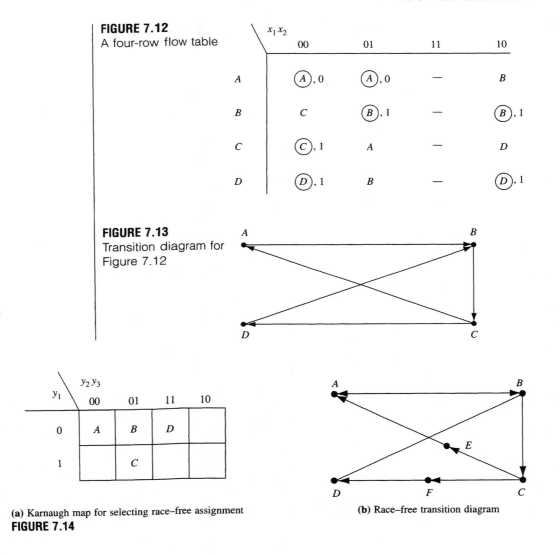

FIGURE 7.12 A four-row flow table

FIGURE 7.13 Transition diagram for Figure 7.12

(a) Karnaugh map for selecting race–free assignment

(b) Race–free transition diagram

FIGURE 7.14

Therefore, a critical race–free assignment can only be achieved by using three state variables. The race-free assignment can be obtained from the Karnaugh map of three variables, as shown in Figure 7.14a. Since in a Karnaugh map adjacent cells differ by a single bit, two states allocated to adjacent cells will have adjacent state assignments. There are $\binom{8}{4} = 70$ different ways of allocating the states to the cells in Figure 7.14a. Analysis shows that these combinations fall into six different patterns [1]. One such pattern is shown in Figure 7.14a. In this case the state assignment is such that A is adjacent to B and B is adjacent to C and D. However, for a critical race–free assignment, A should be adjacent to C as should C to D. Thus, the transition from C to A must be made through the spare state E. Similarly, the transition from C to D must be made via the spare state F. Hence, two additional states must be included in the flow table; the transition diagram of the modified flow table is shown in Figure 7.14b. The modified flow table and the critical race–free assignment are shown in Figures 7.15a and b respectively. ∎

	x_1x_2						y_1	y_2	y_3
	00	01	11	10					
A	Ⓐ,0	Ⓐ,0	—	B	A	0	0	0
B	C	Ⓑ,1	—	Ⓑ,1	B	0	0	1
C	Ⓒ,1	E	—	F	C	1	0	1
D	Ⓓ,1	B	—	Ⓓ,1	D	0	1	1
E	—	A	—	—	E	1	0	0
F	—	—	—	D	F	1	1	1

(a) Modified flow diagram for Figure 7.12 (b) Critical race–free state assignment

FIGURE 7.15

The critical race–free state assignment method considered here requires adding additional rows to the flow table. An alternative method for eliminating critical races assigns two combinations of state variables to each row of the flow table.

■ **EXAMPLE 7.6**

Let us consider the flow table shown in Figure 7.12. Each row of the flow table is replaced with two rows as shown in Figure 7.16. For example, row A in Figure 7.12 is replaced by rows A_1 and A_2, and the stable state in each column of row A is entered in both rows A_1 and A_2. The unstable next states in each row are selected such that they are adjacent to the present state. The assignments can be obtained from the Karnaugh map as shown in Figure 7.17. The two rows corresponding to each row of the original flow table are assigned binary combinations that are complements of each other. Thus, A_1 and A_2 corresponding to row A are assigned $y_1y_2y_3 = 000$ and 111 respectively. In Figure 7.12, a transition from Ⓐ under the input change $x_1x_2 = 00 \rightarrow 10$ is directed to Ⓑ. Since the transition should change one variable only, B_1 is entered in the first row, and B_2 in the second row of column 10 in Figure 7.17; as can be seen in Figure 7.17, A_1 is adjacent to B_1 and A_2 is adjacent to B_2. ■

Note that this method also utilizes the spare states, each state being replaced by a pair, one of which is not used in the original flow table. However the behavior of the circuit remains unchanged irrespective of which state in a pair the circuit is in.

The major drawback of some state assignments is that the transition from one state to another in general requires a sequence of state variable changes. This has a detrimental effect on the circuit as far as speed is concerned. Significant improvement in speed can be achieved if the state assignments are made such that all the required state variable changes during a transition are completed simultaneously. Such assignments are known as **single transition time assignments** or **one-shot assignments**.

7.3 STATE ASSIGNMENT 249

y_1	y_2	y_3		x_1x_2 00	01	11	10
0	0	0	A_1	$(A_1), 0$	$(A_1), 0$	—	B_1
1	1	1	A_2	$(A_2), 0$	$(A_2), 0$	—	B_2
0	0	1	B_1	C_1	$(B_1), 1$	—	$(B_1), 1$
1	1	0	B_2	C_2	$(B_2), 1$	—	$(B_2), 1$
0	1	1	C_1	$(C_1), 1$	A_2	—	D_1
1	0	0	C_2	$(C_2), 1$	A_1	—	D_2
0	1	0	D_2	$(D_1), 1$	B_2	—	$(D_1), 1$
1	0	1	D_2	$(D_2), 1$	B_1	—	$(D_2), 1$

FIGURE 7.16
Modified flow table diagram of Figure 7.12

We consider one such method for critical race–free state assignment [2]. Let us assume a flow table with C columns and let column C_i ($1 \le i \le C$) have x_i stable states. For example, an asynchronous circuit with 2 inputs will have 4 ($= 2^2$) columns. The procedure for the single transition time assignment consists of the following steps:

i. Assign a unique combination to each stable state in a column C_i using $\lceil x_i \rceil$ number of variables.
ii. If the next state in column C_i corresponding to a present state is not the same as the present state itself (i.e., the present state is unstable), it is assigned the same combination of variables as the next state.
iii. If a column covers another column, delete the covered column. Column C_i is said to **cover** another column C_j if C_i has a 1 in every row in which C_j has a 1 whenever C_j is specified. Note that C_i also covers C_j even if the complement of C_i covers C_j.

These steps are repeated for each column of the flow table.

FIGURE 7.17
Multiple state assignment

y_1	y_2y_3 00	01	11	10
0	A_1	B_1	C_1	D_1
1	C_2	D_2	A_2	B_2

■ **EXAMPLE 7.7**

Let us illustrate the above procedure by applying it to the flow table shown in Figure 7.18a. The column $x_1x_2 = 00$ has three stable states—S_0, S_2, and S_4—so 2 ($= \log_2 \lceil 3 \rceil$) state variables y_0 and y_1 are needed to assign a unique combination to each of these states. The next state corresponding to S_1 is S_2, so S_1 is assigned $y_0y_1 = 01$, the combination corresponding to S_2. Similarly, S_3 is assigned $y_0y_1 = 10$.

Column 01 has two stable states and hence requires only a single state variable y_2 to distinguish between them. Unstable states S_0, S_2, and S_4 are assigned $y_2 = 0, 1, 0$ respectively. Columns 11 and 10 have three stable states each, and hence require two variables, y_3/y_4 and y_5/y_6 respectively. The complete state assignment table is shown in Figure 7.18b. The variable y_5 is deleted because it is covered by both y_0 and y_3. Since y_0 is identical to y_3 (i.e., they cover each other), one of them could be deleted; we delete y_3. Similarly y_4 (identical to y_1) and y_6 (identical to y_2) are also deleted. This results in the three-variable assignment shown in Figure 7.18c.

To prove that the state assignment of Figure 7.18c is indeed critical race-free, let us

FIGURE 7.18

State \ x_1x_2	00	01	11	10
S_0	(S_0)	S_1	(S_0)	(S_0)
S_1	S_2	(S_1)	(S_1)	S_0
S_2	(S_2)	S_3	S_1	(S_2)
S_3	S_4	(S_3)	(S_3)	S_2
S_4	(S_4)	S_1	S_3	(S_4)

(a) Flow table

	y_0	y_1	y_2	y_3	y_4	y_5	y_6
S_0	0	0	0	0	0	0	0
S_1	0	1	0	0	1	0	0
S_2	0	1	1	0	1	0	1
S_3	1	0	1	1	0	0	1
S_4	1	0	0	1	0	1	0

(b) State assignment

	y_0	y_1	y_2
S_0	0	0	0
S_1	0	1	0
S_2	0	1	1
S_3	1	0	1
S_4	1	0	0

(c) Reduction of state variables

consider the state transition in each input column of the flow table. In column $x_1 x_2 = 00$ there are two transitions: $S_1 \rightarrow S_2$ and $S_3 \rightarrow S_4$.

For $S_1 \rightarrow S_2$, $y_0 y_1 = 01$; only y_2 changes value during transition. Similarly, for $S_3 \rightarrow S_4$, $y_0 y_1 = 10$ and y_2 changes value. Thus, there will be no critical races during these transitions.

In column $x_1 x_2 = 01$, there are three transitions $S_0 \rightarrow S_1$, $S_2 \rightarrow S_3$, and $S_4 \rightarrow S_1$. Since only y_1 changes during the $S_0 \rightarrow S_1$ transition, no critical races will occur. For the $S_2 \rightarrow S_3$ and $S_4 \rightarrow S_1$ transitions, two state variables (y_0 and y_1) need to change. Since none of the intermediate unstable states (000 or 110 in the case of $S_4 \rightarrow S_1$ and 001 or 111 in the case of $S_2 \rightarrow S_3$) will lead to an erroneous stable state, there is no critical race in the transition.

Similarly, it can be shown that the transitions in columns $x_1 x_2 = 11$ and 10 are also critical race–free. ∎

7.4 EXCITATION AND OUTPUT FUNCTIONS

Once the state assignment to a flow table has been made, the next step is to derive the excitation and output equations to implement the circuit. As mentioned earlier, an asynchronous circuit can be realized as a combinational circuit with feedback.

EXAMPLE 7.8

Let us implement the circuit specified by the flow table of Figure 7.11. By replacing the states in Figure 7.11 with the binary assignment, the excitation table of Figure 7.19a results. The output table, Figure 7.19b, is derived by using the following rules:

i. Assign 0 (1) to the output associated with an unstable state if it is a transient state between two stable states, both of which have outputs 0 (1) associated with them.
ii. Assign a don't care value to the output associated with an unstable state if it is a transient state between two stable states that have different output values.

In order to implement the asynchronous circuit as a combinational circuit with feedback, the logic equations for the state variables and the output variable are derived from the Karnaugh map of functions Y_1, Y_2, and Z. Figure 7.20 shows the Karnaugh maps for Y_1, Y_2, and Z. The logic equations derived in Figure 7.20 are implemented as shown in Figure 7.21. ∎

FIGURE 7.19

y_1y_2 \ x_1x_2	00	01	11	10
01	01	01	01	00
10	10	11	10	00
00	01	00	10	00
11	—	01	—	—

Y_1Y_2

(a) Excitation table

y_1y_2 \ x_1x_2	00	01	11	10
01	0	0	0	—
10	1	—	1	—
00	—	1	1	1
11	—	—	—	—

Z

(b) Output table

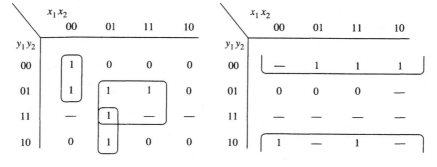

$Y_1 = \bar{x}_1 y_1 \bar{y}_2 + x_1 x_2 \bar{y}_2$

$Y_2 = \bar{x}_1 \bar{x}_2 \bar{y}_1 + x_2 y_2 + \bar{x}_1 x_2 y_1$ $Z = \bar{y}_2$

FIGURE 7.20
Karnaugh maps for the excitation and output functions

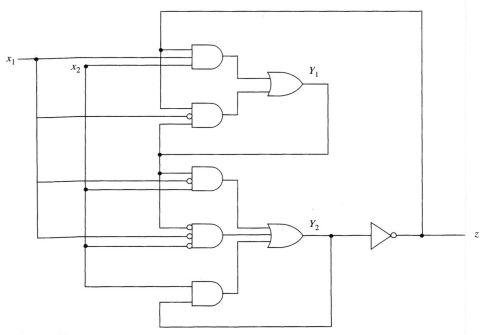

FIGURE 7.21
Realization of the asynchronous circuit

Asynchronous circuits can be implemented by using *SR* latches instead of feedback. The operation of the *SR* NOR latch was discussed in Chapter 5 (Sec. 3); the truth table of the latch is as follows:

S	R	Q	$\bar{Q}$
1	0	1	0
0	0	1	0
0	1	0	1
0	0	0	1

The Karnaugh maps for an *SR* latch realization of the flow table of Figure 7.11 are derived from Figure 7.19; these are shown in Figure 7.22. The implementation of the excitation and output equations is shown in Figure 7.23.

7.5 HAZARDS

As discussed earlier, in asynchronous circuits all input variable and state variable changes are restricted so that only one variable can change at a time. Even then it is still possible to have erroneous outputs at times, because a variable and its complement may not change

FIGURE 7.22
Karnaugh maps for *SR* latch realization

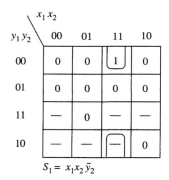

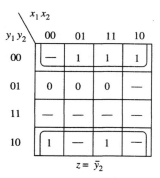

at exactly the same instant. For example, let us consider the circuit shown in Figure 7.24. Assume the inputs x_1 and x_2 are both initially at 1, which will cause the output Z to go to 1. If x_1 changes from 1 to 0, $\bar{x}_1$ also changes value, from 0 to 1, but this change does not occur at exactly the same time because of the propagation delay of the inverter. As a result, the output of the inverter will be 0 for a finite period of time, keeping the output of the AND gate at 0. Moreover, since $x_1 = 0$, the output of the other AND gate will also be 0, thus locking the output of the circuit at 0. Notice that if the propagation delay of the inverter is ignored, the output of the circuit should remain at 1 when the inputs are

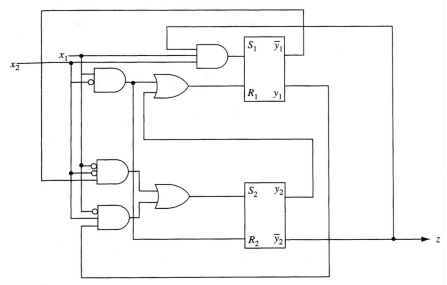

FIGURE 7.23
Implementation of the flow diagram of Figure 7.11

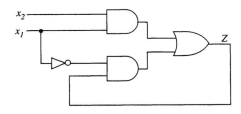

FIGURE 7.24
An asynchronous sequential circuit with a hazard

changed from $x_1x_2 = 11$ to 01. This phenomenon in which the relative differences in delays associated with circuit elements (and interconnections) cause incorrect output is known as a **hazard**.

Hazards may also occur in combinational circuits, although not with such serious consequences as in asynchronous circuits. Figure 7.25 illustrates the occurrence of a hazard. If the circuit receives inputs $x_1x_2x_3 = 100$, the output of the circuit is at 0. If the inputs are changed to $x_1x_2x_3 = 110$, the output should remain at 0; however, since the inverter fed by x_2 cannot change its output instantaneously from 1 to 0, the output of the top OR gate will be at 1. The output of the bottom OR gate will also be at 1 due to input x_2, so the circuit output momentarily goes to 1 while the inverter output is changing from 1 to 0.

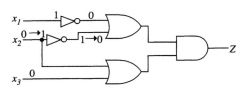

FIGURE 7.25
Combinational circuit with a hazard

There are three types of hazards: **static**, **dynamic**, and **essential**. Static and dynamic hazards may be present in combinational as well as in asynchronous circuits. Essential hazards, however, originate only in fundamental mode asynchronous circuits (see Sec. 7.5).

Static hazards occur when a change in a single input variable may cause a momentary change in the output that is supposed to remain constant during the change. This is often referred to as **glitch**. If the outputs before and after the change of an input variable are both 1, with a transient 0 in between, then the hazard is qualified as a **static-1** hazard. Similarly, if the output before and after the change of an input variable are both 0, with a transient 1 in between, then the hazard is known as a **static-0 hazard**. The type of hazard we came across in Figure 7.25 was a static-0 hazard.

Dynamic hazards occur when the output is supposed to change due to a change in an input variable and it changes three or more times instead of only once before settling down to the proper value. Thus an output required to change from $0 \rightarrow 1$ will go through the sequence $0 \rightarrow 1 \rightarrow 0 \rightarrow 1$ due to a dynamic hazard.

Hazards can be classified into two categories: **logic hazards** and **function hazards**. Function hazards occur when more than one input variable in a circuit changes, whereas logic hazards arise because of how a circuit is realized.

Function Hazards

Function hazards may happen if multiple input variable changes are allowed. To illustrate function hazards let us consider the Karnaugh map for a four-variable function, shown in Figure 7.26. Assume that initially the input combination is $x_1 x_2 x_3 x_4 = 1001$, and hence the corresponding output is 1. If inputs x_3 and x_4 are changed to 1 and 0 respectively, the output will remain at 1, provided both x_3 and x_4 changed simultaneously; notice that the output is also 1 for $x_1 x_2 x_3 x_4 = 1010$. However, if x_3 and x_4 do not change at the same time, the input combination could be either 1011 or 1000, depending on whether x_3 or x_4 changed first. In any case this will result in a transient 0 output. The change of input combination from 1111 to 0011 will also result in a transient 0 output.

Function hazards can also be of either static or dynamic type. The change of two input variables as discussed above resulted in function static-0 hazard. A function dynamic hazard happens when three input variables change (e.g., from $x_1 x_2 x_3 x_4 = 0100$ to 1001). In this case x_1, x_2, and x_4 are changing. Figure 7.27 shows the diagram formed by assuming a single input variable change at a time. For example, x_1 may change before x_2 and x_4; in that case the input combination will correspond to 1100, the leftmost entry in level 1. It is also possible that x_2 or x_4 may change first, as indicated by the middle and

FIGURE 7.26
Karnaugh map with function hazards

$x_3 x_4$ \ $x_1 x_2$	00	01	11	10
00	0	0	1	0
01	0	0	0	1
11	1	0	1	0
10	0	0	0	1

7.5 HAZARDS 257

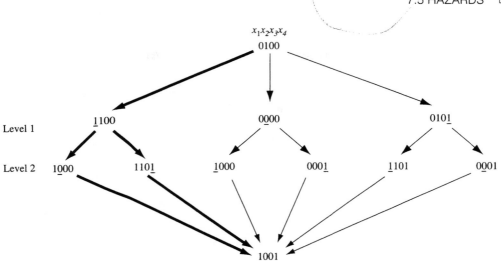

FIGURE 7.27
Transition from $x_1x_2x_3x_4$ = 0100 to 1001 (the underscore indicates that the corresponding input variable has changed)

right entry respectively in level 1. The entries in level 2 correspond to the second input variable change and are derived from level 1. Finally, the input variable change in level 2 completes the transition from the input combination 0100 to 1001.

As can be verified from Figure 7.26, only for the combination 1100 in level 1 is the output 1; for all other combinations the output is 0. The output is 0 for all input combinations in level 2. After the third variable change, the desired input combination $x_1x_2x_3x_4$ = 1001 results, which produces an output of 1. Thus the transition from 0100 to 1001 via 1100 and 1000 (or 1101) results in the output of 0→1→0→1; in other words the output shows dynamic hazard if the input variable x_1 changes before x_2 or x_4. In general, function hazards cannot be completely eliminated by modifying the circuit, hence the constraint of one input variable change at a time. However, because of the unpredictable nature of inputs, this restriction is not always applicable.

Logic Hazards

Logic hazards depend on the realization of a function and can happen even if a single input variable is allowed to change at a time. Such hazards can be either static or dynamic.

Static Logic Hazards

The characteristic of static logic hazards is that during an input variable change, there is a momentary change in the output which is required to stay unchanged. Such hazards can be located from the Karnaugh map of a function. To illustrate let us consider the function

$$f(x_1,x_2,x_3,x_4) = (x_1 + x_4)(x_2 + \bar{x}_4) + \bar{x}_1\bar{x}_2x_3$$

The Karnaugh map for the function is shown in Figure 7.28. Let the input combination at a particular time be $x_1x_2x_3x_4$ = 0111. The change in $\bar{x}_2$ from 1 to 0 makes the input combination 0011. Note that the transition from 0111 to 0011 involves changing the prime

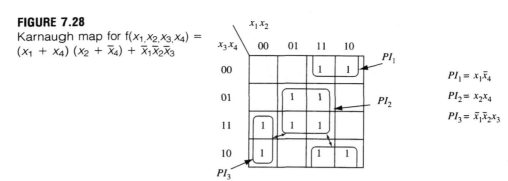

FIGURE 7.28
Karnaugh map for $f(x_1,x_2,x_3,x_4) = (x_1 + x_4)(x_2 + \bar{x}_4) + \bar{x}_1\bar{x}_2\bar{x}_3$

implicant from PI_2 to PI_3. Although the output for both input combinations 0111 and 0011 is 1, during the transition period $\bar{x}_2$ (the only variable that is changing) may not be equal to 1, thus resulting in a static-1 hazard. Similarly, a change in the input combination from 1101 to 1100 will also give rise to a static-1 hazard. Thus, a static hazard can exist when an input variable change causes a movement between two minterms not covered by the same prime implicant, as indicated by the arrows in the Karnaugh map.

Let us now consider how static-0 hazards can be located from the Karnaugh map of a function. Figure 7.29 shows the Karnaugh map for the function

$$f(x_1,x_2,x_3,x_4) = (x_1 + x_3)(x_2 + x_3)(\bar{x}_2 + \bar{x}_3)$$

A static-0 hazard exists if $x_1x_2x_3x_4$ changes from 0101 to 0111; this is because the corresponding cells in that Karnaugh map are not covered by the same prime implicant.

Application of Three-Valued Logic in Hazard Detection

Three-valued (ternary) logic is based on the assumption that when a logic variable changes from 0 to 1 or vice versa, it goes through a transition period where its value may be interpreted as either a 0 or 1. This indeterminate value may be represented by the symbol x, indicating the fact that the value is unknown. The truth tables for the AND, OR, and NOT functions, based on three-valued logic, can be written as follows:

AND			OR			NOT	
a	b	Z	a	b	Z	a	Z
0	0	0	0	0	0	0	1
0	1	0	0	1	1	1	0
0	x	0	0	x	x	x	x
1	0	0	1	0	1		
1	1	1	1	1	1		
1	x	x	1	x	1		
x	0	0	x	0	x		
x	1	x	x	1	1		
x	x	x	x	x	x		

where a and b are inputs and Z is the output.

7.5 HAZARDS   259

FIGURE 7.29
Karnaugh map for $f(x_1x_2x_3x_4) = (x_1 + x_3)(x_2 + x_3)(\bar{x}_2 + \bar{x}_3)$

x_3x_4 \ x_1x_2	00	01	11	10
00	0	0	1	0
01	0	0	1	0
11	1	0	0	1
10	1	0	0	1

It has been shown that a static logic hazard exists in a combinational circuit for the input change

$$I_t = (I_1, ..., I_P, I_{P+1}, ..., I_n)$$

to

$$I_{t+1} = (I_1, ..., I_P, I_{p+1}, ..., I_n)$$

if the output value corresponding to the input combination I_t or I_{t+1} is the same and during the transition from I_t to I_{t+1} the output is indeterminate (i.e., x) [3]. Notice that both single and multiple input changes are allowed.

■ **EXAMPLE 7.9**

Let us analyze the circuit of Figure 7.30a to determine whether or not it contains a static hazard for the input change $x_1x_2x_3 = 010$ to 011. This is accomplished by evaluating the entire circuit using the ternary function of each gate in the circuit. Figure 7.30b shows the output values of each gate in response to the input change. The x at the output of G_5 indicates the existence of a hazard for the given input transition. ■

Static hazards can be eliminated by grouping the minterms that cause the hazard, into one prime implicant. For example, in Figure 7.30 the two minterms in question can be combined to form the prime implicant x_1x_2. An AND gate corresponding to the prime implicant is incorporated into the circuit to make it free of static hazards, as shown in Figure 7.31.

Dynamic Logic Hazards

These hazards give rise to multiple transitions at an output. They result because of the existence of at least three different paths, each with different delay time, along which the

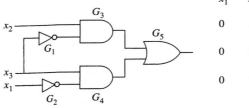

(a) Circuit with a static-1 hazard   (b) Individual gate outputs

FIGURE 7.30

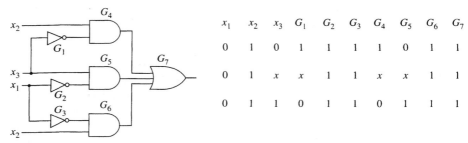

FIGURE 7.31
Hazard-free realization of Figure 7.30a

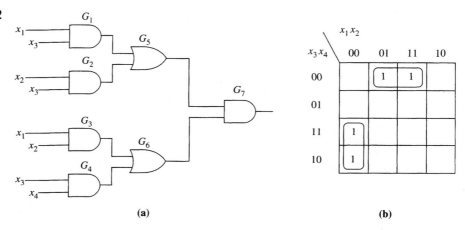

FIGURE 7.32

change in a single input variable can propagate through the circuit. To illustrate let us consider the circuit shown in Figure 7.32a; the corresponding Karnaugh map is shown in Figure 7.32b.

It can be observed from the Karnaugh map that there are no static hazards in the circuit. However, a change in x_3 can cause a dynamic hazard, because there are three different paths through the circuit for the variable x_3. Assume that the initial input combination for the circuit is $x_1x_2x_3x_4 = 0110$, and let x_3 change to 0. The three paths from x_3 to the output are

i. Via gates G_1, G_5, and G_7
ii. Via gates G_2, G_5, and G_7
iii. Via gates G_4, G_6, and G_7

If G_4 changes first from 0 to 1, then the output changes from 0 to 1. Next, if G_1 changes from 1 to 0, the output goes from 1 to 0. Finally, G_2 changes from 0 to 1, which makes the output settle at 1. Thus, as a result of the change in the input from 0110 to 0100, the output has changed $0 \rightarrow 1 \rightarrow 0 \rightarrow 1$, which indicates the existence of a dynamic hazard.

Dynamic hazards cannot be eliminated by adding redundant gates as in the case of static hazards. They can be overcome only by implementing the function in a different way.

Essential Hazards

Essential hazards are peculiar to asynchronous sequential circuits. This type of hazard happens if the change in an input variable propagates through the circuit via two or more paths that have unequal delays. Such a hazard can cause the circuit to terminate in an incorrect stable state.

To illustrate let us consider the circuit shown in Figure 7.33. The excitation table for the circuit is:

$y_1 y_2$ \ $x_1 x_2$	00	01	11	10
00	(00)	(00)	01	(00)
01	—	00	(01)	11
11	(11)	(11)	10	(11)
10	—	11	(10)	00

Let us assume that initially $x_1 = 1$, $x_2 = 0$, $y_1 = 0$, and $y_2 = 0$. Thus, $Y_1 = 0$ and $Y_2 = 0$, and the circuit is in stable state 00. Now let x_2 change from 0 to 1, and assume that the inverter that produces $\bar{x}_2$ has a propagation delay which is larger than the delays of other gates in the circuit including the feedback delay. The sequences of changes resulting from the change of x_2 from 0 to 1 are as follows:

i. x_2 is changed from 0 to 1.
ii. Y_2 changes from 0 to 1, which in turn changes y_2 to 1. Thus the circuit moves to state 01.
iii. Since the output of the inverter generating $\bar{x}_2$ has not changed its value (i.e., its output is still at 1), Y_1 changes from 0 to 1 and Y_2 remains 1. Thus, the circuit comes to state 11 (i.e., $y_1 y_2 = 11$).

FIGURE 7.33
Circuit with essential hazards

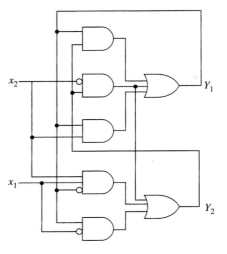

262 □ CHAPTER 7 / ASYNCHRONOUS SEQUENTIAL CIRCUITS

iv. Assuming that the output of the inverter generating $\bar{x}_2$ has now changed from 1 to 0, Y_2 changes from 1 to 0 whereas Y_1 remains 1. The circuit now moves to the stable state 10. Thus, the circuit terminates in the stable state 10 after passing through two unstable states as shown:

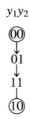

However, if the change of x_1 from 0 to 1 and the change of x_2 from 0 to 1 occurred at the same time, then the circuit would have moved from 00 to 01, as can be seen in the excitation table of the circuit.

The circuit has some additional essential hazards. These happen if x_2 is changed while the circuit is in 01, 11, and 10. The essential hazards cannot be eliminated by circuit modifications, as in the case of static hazards. The only way they can be avoided is to insert delay elements in the feedback paths from the next state variables. This ensures that the change in the state variables is not fed back until the change of the input variable has been completed.

EXERCISES

1. An asynchronous sequential circuit has two inputs, x_1 and x_2, and an output Z. Input x_2 is driven by a noise-free switch. The circuit is to be designed such that a pulse on input x_1 occuring after x_2 has been pressed is transmitted to Z. Derive the excitation and output equations for the circuit.

2. The flow table of an asynchronous sequential circuit is shown below. Identify all the races that will occur if the circuit is implemented from the state table. Determine whether the races will be critical or non-critical. Derive a race-free state assignment for the circuit.

Present State	$\bar{x}_1\bar{x}_2$	$\bar{x}_1 x_2$	$x_1 x_2$	$x_1 \bar{x}_2$
A	Ⓐ	Ⓐ	D	B
B	D	Ⓑ	C	Ⓑ
C	A	A	Ⓒ	Ⓒ
D	Ⓓ	C	Ⓓ	C

3. Derive a race-free state assignment for the flow table shown below. Implement a circuit corresponding to the flow table using this state assignment. Use NAND gates only.

Present State	$\bar{x}_1\bar{x}_2$	$\bar{x}_1 x_2$	$x_1 x_2$	$x_1 \bar{x}_2$
A	Ⓐ,0	Ⓐ,0	B,0	Ⓐ,0
B	-,-	C	Ⓑ,0	A
C	Ⓒ,1	Ⓒ,1	Ⓒ,1	D
D	A	-,-	C	Ⓓ,1

4. Find all the races in the following flow table and indicate if they are critical or not.

y_1y_2	$\bar{x}_1\bar{x}_2$	$\bar{x}_1x_2$	x_1x_2	$x_1\bar{x}_2$
00	⓪⓪	11	⓪⓪	11
01	11	⓪①	11	11
10	00	①⓪	11	11
11	①①	①①	00	①①

Find a state assignment that is critical race free.

5. A circuit with three inputs (x_1, x_2, and x_3) and four outputs (Z_1, Z_2, Z_3, and Z_4) is to be designed. Z_1 takes on the value 1 if input x_1 is changed first, followed by x_2 and then x_3. Z_2 takes on the value 1 if x_2 is changed first, followed by x_1. If the order of change is x_3 first followed by x_1 and then x_2, output Z_3 assumes the value 1. For any other order of input change, output Z_4 takes on the value 1, otherwise it remains at 0.

   a. Find a minimum row flow table and a state assignment.
   b. Implement the circuit using NAND gates only.

6. A circuit for implementing a combinational lock is to be designed. The circuit has two inputs, x_1 and x_2, and an output Z. The lock opens (Z = 1) if there is a sequence of four consecutive changes with x_2 set at 1. Find a minimum row flow table for the circuit.

7. A sequential circuit has two inputs (x_1 and x_2) and one output (z). The output becomes 1 whenever x_1 changes from 0 to 1, and becomes 0 whenever x_2 changes from 1 to 0. Find a minimum row flow table for the circuit.

8. The excitation and the output equations for a two-input (x_1, x_2) and one output (Z) asynchronous sequential circuit are as follows:

$$Y_1 = (x_1 + x_2)y_1 + \bar{x}_1\bar{x}_2 y_2$$
$$Y_2 = (x_1 + x_2)\bar{y}_1 + \bar{x}_1\bar{x}_2 y_2$$
$$Z = y_1 + y_2$$

Derive the flow table for the circuit.

9. Reduce the primitive flow table shown below using the approach discussed in Section 7.2.

Present State	$\bar{x}_1\bar{x}_2$	$\bar{x}_1x_2$	x_1x_2	$x_1\bar{x}_2$
A	Ⓐ,0	-,-	-,-	B
B	C	-,-	-,-	Ⓑ,1
C	Ⓒ,1	F	-,-	D
D	E	-,-	-,-	Ⓓ,1
E	Ⓔ,1	G	-,-	-,-
F	A	Ⓕ,0	-,-	-,-
G	C	Ⓖ,1	-,-	-,-

10. The primitive flow table of an asynchronous sequential circuit is shown below.

Present State	$\bar{x}_1\bar{x}_2$	$\bar{x}_1x_2$	x_1x_2	$x_1\bar{x}_2$
A	Ⓐ,1	B	-,-	C
B	E	Ⓑ,0	D	-,-
C	A	-,-	F	Ⓒ,1
D	-,-	G	Ⓓ,1	H
E	Ⓔ,0	B	-,-	C
F	-,-	G	Ⓕ,0	H
G	E	Ⓖ,1	D	-,-
H	A	-,-	F	Ⓗ,0

264 ◻ CHAPTER 7 / ASYNCHRONOUS SEQUENTIAL CIRCUITS

Reduce the flow table using the approach discussed in Section 7.2. Also, derive the logic diagram of the circuit.

11. Determine a one-shot state assignment for each of the following flow tables assuming fundamental mode operation:

Present State	x_1x_2			
	00	01	11	10
A	B	Ⓐ	Ⓐ	-,-
B	Ⓑ	F	D	Ⓑ
C	B	Ⓒ	Ⓒ	-,-
D	E	Ⓓ	Ⓓ	-,-
E	Ⓔ	C	A	Ⓔ
F	B	Ⓕ	C	-,-

(i)

Present State	x_1x_2			
	00	01	11	10
A	Ⓐ	B	Ⓐ	C
B	A	Ⓑ	C	Ⓑ
C	A	D	Ⓒ	Ⓒ
D	Ⓓ	Ⓓ	A	B

(ii)

12. Determine whether the following circuit has a static hazard or not. If there is, specify the input condition that creates the hazard.

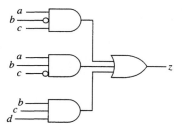

13. Find all the static hazards in the Karnaugh map of a four-variable function. Implement the circuit such that these hazards will be eliminated.

	$\bar{a}\bar{b}$	$\bar{a}b$	ab	$a\bar{b}$
$\bar{c}\bar{d}$			1	
$\bar{c}d$	1		1	
cd	1	1	1	
$c\bar{d}$				

14. Find a static hazard–free realization of the following function.

$$f(a,b,c,d) = \Sigma m(2,3,6,7,10,11,13,15)$$

15. A dynamic hazard occurs in the following circuit when the output changes from 1 to 0. Specify the values of the input variables before and after the occurrence of the hazard. Assume that the inverters in the circuit are much slower than the other gates in the circuit.

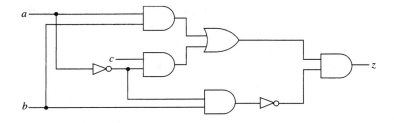

References

1. M. P. Marcus, *Switching Circuits for Engineers,* Prentice Hall, Englewood Cliffs, 1975.

2. C. N. Liu, "A state variable assignment procedure for asynchronous sequential circuits," *Jour. ACM*, April 1963, pp. 209–215.

3. E. B. Eichelberger, "Hazard detection in combinational and sequential switching circuits," *IBM Jour. Res. and Dev.*, March 1965, pp. 90–99.

Further Reading

1. M. Mano, *Digital Design*, Prentice Hall, Englewood Cliffs, 1984.

2. C. Roth, *Fundamentals of Logic Design,* West Publishing, 1985.

8
Registers and Counters

Circuits for counting events are frequently used in computers and other digital systems. Since a counter circuit must remember its past states, it has to possess memory. Thus, the sequential logic design principles discussed in Chapter 5 can be utilized in designing counter circuits. Like all other sequential logic circuits, counter circuits can be classified into two categories:

i. Synchronous
ii. Asynchronous

In synchronous counters all memory elements are simultaneously triggered by a clock, whereas in asynchronous counters the output of each memory element activates the next memory element.

Many types of counters are used in practice. In some cases they count in pure binary; in other cases the count may differ considerably from straight binary (e.g., decade or BCD counters). This chapter examines the construction and operation of the most important types of counters.

8.1 RIPPLE (ASYNCHRONOUS) COUNTERS

In a ripple counter there is no clock or source of synchronizing pulses; however, the state changes still occur due to pulses at clock inputs of the flip-flops. Figure 8.1a shows a 3-bit ripple counter constructed from *JK* flip-flops. It is assumed that the flip-flops change state on the negative-going (falling) edge of the pulses appearing at their clock inputs.

The counter is first cleared by applying a reset pulse; thus, counting begins from 000. A timing diagram representing the sequence of logic states through which flip-flops *A*, *B*, and *C* go in counting from 0 to 7 is shown in Figure 8.1b. As can be seen in Figure 8.1a, the normal output of flip-flop *C* acts as the clock pulse of flip-flop *B*. Similarly, the output of flip-flop *B* is used as the clock pulse source for flip-flop *A*. The input (i.e., the count pulses) is applied only to the clock input of flip-flop *C*. Note that the *J* and *K* inputs of all the flip-flops are tied to logic 1.

The first negative-going pulse at the clock input of flip-flop *C* changes its output to 1. Thus, the counter shows an output of 001. Since the output of the flip-flop *C* is 1, the clock input of flip-flop *B* is also 1. When the second negative-going pulse occurs at the

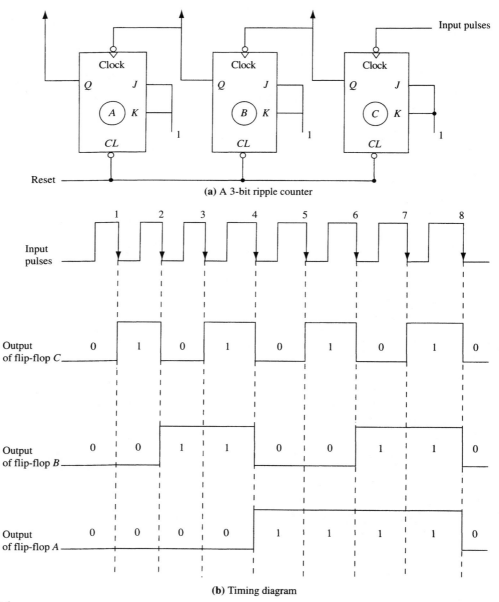

FIGURE 8.1

clock input of flip-flop C, its output makes a transition from 1 to 0. This negative-going transition at the clock input of flip-flop B changes its output from 0 to 1. Hence, the output of the counter is now 010. On the third negative-going pulse at the clock input of flip-flop C, its output changes from 0 to 1. This positive-going transition cannot change the output of the flip-flop B. Therefore, after the third negative-going pulse, the counter output is 011. It continues to operate in this manner until the count value 111 is reached.

The next negative-going input pulse at the clock input of flip-flop C will change its output from 1 to 0. This transition in the output of C will cause the output of flip-flop B to change from 1 to 0, which in turn will change the output of flip-flop A to 0. In other words, the counter is reset to 000 after eight pulses and is ready to begin counting again as subsequent input pulses are applied at the clock input of flip-flop C. Thus, any change in the output of flip-flop C moves through the counter like a ripple on water, hence the name "ripple counter."

An alternative way of implementing a 3-bit ripple counter is shown in Figure 8.2. This implementation uses T flip-flops. The T flip-flops are triggered on the positive-going (rising) edge of the pulses appearing at their clock inputs.

The 3-bit ripple counter circuit has eight ($=2^3$) different states, each one corresponding to a count value. Similarly, a counter with n flip-flops can have 2^n states. The number of states in a counter is known as its **mod (modulo) number**. Thus, a 3-bit counter is a mod-8 counter. Similarly, a 6-bit counter is mod-64 counter (i.e, it has 64 distinct states, 000000 through 111111). A mod-n counter may also be described as a **divide-by-n** counter, in the sense that the most significant flip-flop produces one pulse for every n pulses at the clock input of the least significant flip-flop. Thus, the counter of Figure 8.1a is a divide-by-8 counter.

The ripple counter poses a problem when there is a large number of flip-flops. In such a case, the most significant flip-flop cannot change state until the propagation delay times of all other flip-flops have elapsed. For example, if each flip-flop in a 6-bit ripple counter has a propagation delay of 10 ns, then it will take 60 ns when changing from a count of 31 (011111) to a count of 32 (100000).

Since the states of the flip-flops in a ripple counter do not change simultaneously, undesirable transient states may be produced during the change from one valid state to another. For example, in going from state 011 to 100, the counter of Figure 8.1a generates two transient states 010 and 000:

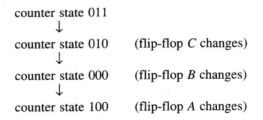

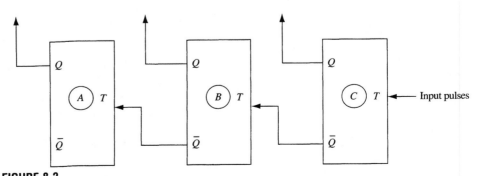

FIGURE 8.2
Three-bit ripple counter implemented with T flip-flops

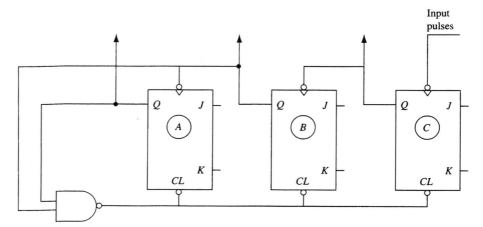

FIGURE 8.3
Mod-6 counter

Such transient states in a counter may have undesirable effects on other parts of a digital system, so they must be guarded against.

It is also possible to design a ripple counter such that it will only count up to a value less than the maximum possible. For example, the 3-bit ripple counter can count up to 7 (111) before resetting to 0 (000). By adding a NAND gate to the counter circuit, it may be made to reset for every fifth pulse— i.e., when the count is 5 (101), as shown in Figure 8.3. The inputs to the NAND gate are the outputs of flip-flops A and B, so the output of the NAND gate will go to 0 whenever the outputs of flip-flops A and B are 1. This happens only when the counter makes a transition from state 101 to state 110. This makes the output of the NAND gate go to 0, which in turn clears the flip-flops within a few nanoseconds. Thus, the counter essentially counts up to 101 and then resets to 000 (i.e., it is a mod-6 counter). In a similar manner, counters with any other modulus can be designed by using a NAND gate to detect the appropriate state for resetting the counter. For example, using a NAND gate with inputs A and C, a 4-bit counter can be reset to 0000 when the count value 1010 is reached (Fig. 8.4). In other words, the normal count will be from 0000 to 1001, resulting in a mod-10 counter, also known as a **decade counter**.

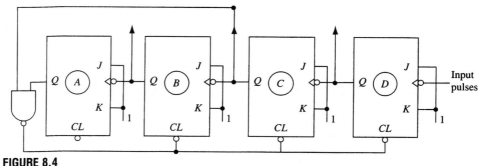

FIGURE 8.4
Decade counter

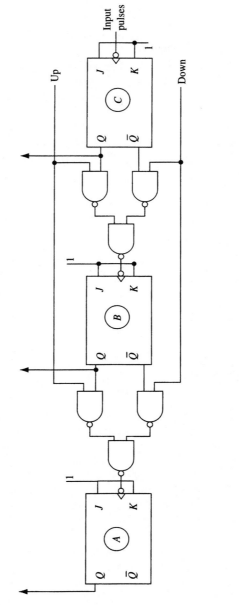

FIGURE 8.5
A 3-bit up-down counter

8.2 ASYNCHRONOUS UP-DOWN COUNTERS

In certain applications a counter must be able to count both up and down. Figure 8.5 shows the circuit for a 3-bit up-down counter. It counts up or down depending on the status of the control signals UP and DOWN. When the UP input is at 1 and the DOWN input is at 0, the NAND network between flip-flops B and C will gate the non-inverted output of flip-flop C into the clock input of flip-flop B. Similarly, the non-inverted output of flip-flop B will be gated through the other NAND network into the clock input of flip-flop A. Thus, the counter will count up. When the control input UP is at 0 and DOWN is at 1, the inverted outputs of flip-flop C and flip-flop B are gated into the clock inputs of flip-flops B and A respectively. If the flip-flops are initially reset to 0's, then the counter will go through the following sequence as input pulses are applied.

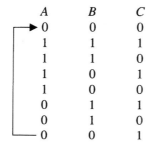

An asynchronous up-down counter is slower than an up counter or a down counter because of the additional propagation delay introduced by the NAND networks.

8.3 SYNCHRONOUS COUNTERS

In asynchronous counters the state of a flip-flop changes only when a transition occurs at the output of its preceding flip-flop. Since each flip-flop has a propagation delay, the time required by an asynchronous counter to complete its response to an input sequence is of the order of the sum of the propagation delays of the flip-flops in the counter. If the total propagation delay of the counter is greater than or equal to the period of the input pulse, then the counter does not function properly. Thus, the maximum frequency at which the input pulse can be applied to a ripple counter has to be lowered if the number of flip-flops in the counter is large.

In synchronous counters, the clock inputs of all the flip-flops are connected together and are triggered by the input pulses. Thus, all the flip-flops change state simultaneously (i.e., in parallel). Figure 8.6a shows the circuit of a 3-bit synchronous counter. The J and K inputs of flip-flop C are connected to logic 1. The B flip-flop has its J and K inputs connected to the output of flip-flop C, and the J and K inputs of flip-flop A are connected to the output of an AND gate that is fed by the outputs of flip-flops B and C. Assuming that the outputs of all the flip-flops are initially reset to 0, the rising edge of the first clock pulse will change the output of flip-flop C to 1. This will result in a 1 at the J and K inputs of flip-flop B. The rising edge of the second clock pulse will cause flip-flop B to change its output from 0 to 1 and flip-flop C to change its output from 1 to 0. On the positive edge of the third clock pulse, the output of flip-flop C will change again, from 0 to 1. Therefore, both the inputs of the AND gate will be at 1 at the end of the third clock

8.3 SYNCHRONOUS COUNTERS 273

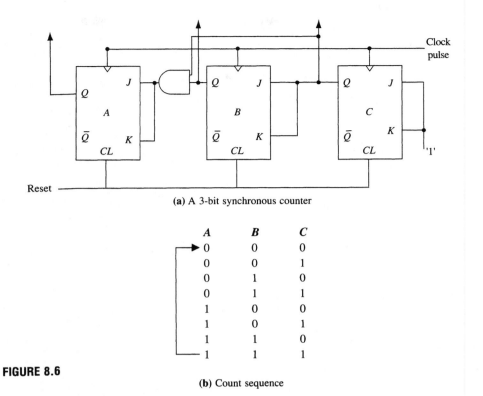

FIGURE 8.6

(a) A 3-bit synchronous counter

A	B	C
0	0	0
0	0	1
0	1	0
0	1	1
1	0	0
1	0	1
1	1	0
1	1	1

(b) Count sequence

pulse. The positive edge of the fourth clock pulse will cause flip-flop C to change its output again; flip-flop B will change at the same time because its J and K inputs are at 1. Flip-flop A will also change state on the positive edge of the fourth clock pulse because its J and K inputs are at 1 due to the AND gate. The count sequence for the 3-bit counter is shown in Figure 8.6b.

The most important advantage of this synchronous counter is that there is no cumulative time delay because all the flip-flops are triggered in parallel. Thus, the maximum operating frequency for this counter will be significantly higher than that for the corresponding ripple counter. For example, if the propagation delay of each flip-flop is 20 ns and that of the AND gate is 10 ns, the minimum clock period for the 3-bit synchronous counter is flip-flop propagation delay + AND-gate propagation delay = 20 + 10 = 30 ns, whereas the minimum clock period for the 3-bit ripple counter (Fig. 8.1) is 3 × flip-flop propagation delay = 3 × 20 = 60 ns. The 3-bit synchronous counter we considered is an up counter; its circuit can be slightly modified so that it can perform as a down counter (Fig. 8.7a). The corresponding count sequence is shown in Figure 8.7b.

Synchronous counters can be designed using the techniques employed for designing synchronous sequential circuits (Chap. 6). The first step in the design is to list the required count sequence in a two-column transition table; the first column represents the present state of the counter, and the second column gives the next state of the counter. The counter moves from a present state to the corresponding next state after a clock pulse has been applied.

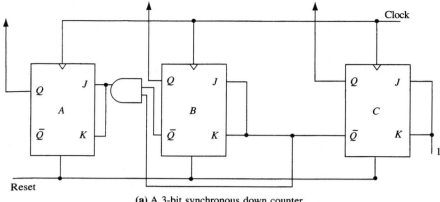

FIGURE 8.7

(a) A 3-bit synchronous down counter

A	B	C
1	1	1
1	1	0
1	0	1
1	0	0
0	1	1
0	1	0
0	0	1
0	0	0

(b) Count sequence

■ **EXAMPLE 8.1**

Let us design a mod-5 counter using *JK* flip-flops. Figure 8.8 shows the transition table for the mod-5 counter circuit. Since there are five unique states, we need 3 ($=\log_2 5$) flip-flops to implement the counter circuit. Next we form the excitation map for the *J* and *K* inputs of each flip-flop in the circuit, as shown in Figure 8.9. The resulting input equations for the flip-flops are:

$$J_A = BC \quad\quad K_A = 1$$
$$J_B = C \quad\quad K_B = C$$
$$J_C = \overline{A} \quad\quad K_C = 1$$

FIGURE 8.8
State table for mod-5 synchronous counter

Present State			Next State		
A	B	C	A	B	C
0	0	0	0	0	1
0	0	1	0	1	0
0	1	0	0	1	1
0	1	1	1	0	0
1	0	0	0	0	0

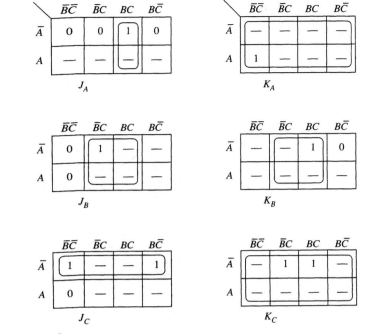

FIGURE 8.9
Excitation maps for the *JK* flip-flops in a mod-5 counter

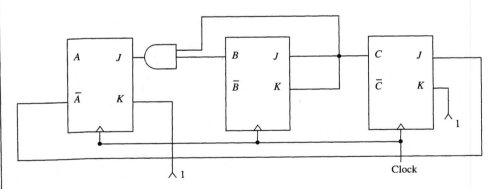

FIGURE 8.10
Mod-5 counter circuit using *JK* flip-flops

Implementation of the counter is shown in Figure 8.10. ∎

In many applications it is necessary to determine whether a counter has reached a particular count value before a certain operation is started. Thus, a decoder circuit has to be used to indicate the presence of the desired count value. For example, we can connect a 4-to-10 decoder at the output of the binary decade counter to convert its output to a decimal representation (Fig. 8.11). The decoder output in turn can be used to drive a display device to indicate the decimal number corresponding to the binary count.

FIGURE 8.11
Conversion of the binary outputs of a counter to decimal form

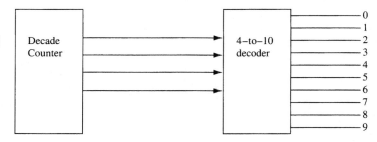

In the counter design technique we considered, all redundant states (i.e., states for which no next states are specified) are used as don't care terms. One very important point in the design of counters (or any sequential circuit) is the starting state of the counter when power is turned on. If the counter circuit starts in a redundant state, it must be ensured that the counter eventually returns to one of the states in the count sequence; a counter having this property is known as a **self-starting counter**.

Unless it is important that a counter go to a particular valid state from a redundant state, it is usually possible to simplify counter logic using the redundant states and to ensure that it is self-starting. For example, in the mod-5 counter circuit of Figure 8.8 there are three redundant states:

A	B	C
1	0	1
1	1	0
1	1	1

By specifying a valid next state for each of the redundant states, the counter can be made self-starting. The next states for the redundant states are selected such that there is a reduction in the number of gates required for implementing the excitation equations for the flip-flops. The complete state diagram of the mod-5 counter is shown in Figure 8.12. If the counter starts in one of the redundant states, the next clock pulse will transfer it to one of the valid states and it will continue to count properly. The excitation maps for the D flip-flop implementation of the self-starting counter are shown in Figure 8.13a, and the corresponding circuit is shown in Figure 8.13b.

8.4 GRAY CODE COUNTERS

In a Gray code, only one bit changes in going from one code combination to another. Thus, a Gray code counter can be used to eliminate the problem of momentary false count

FIGURE 8.12
State diagram for the self-starting mod-5 counter

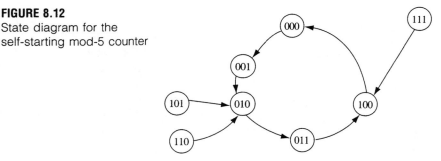

FIGURE 8.13

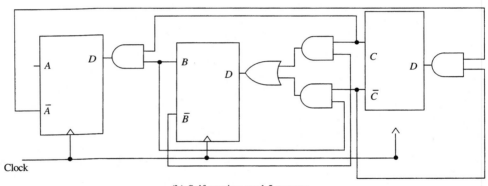

(a) Excitation maps

(b) Self-starting mod-5 counter

values that results from a binary counter, where more than one bit is required to change states during a transition.

■ **EXAMPLE 8.2**

Let us design a 4-bit Gray code counter using *JK* flip-flops. The transition table for the counter is shown in Figure 8.14.

Present State				Next State			
A	B	C	D	A	B	C	D
0	0	0	0	0	0	0	1
0	0	0	1	0	0	1	1
0	0	1	1	0	0	1	0
0	0	1	0	0	1	1	0
0	1	1	0	0	1	1	1

0	1	1	1	0	1	0	1
0	1	0	1	0	1	0	0
0	1	0	0	1	1	0	0
1	1	0	0	1	1	0	1
1	1	0	1	1	1	1	1
1	1	1	1	1	1	1	0
1	1	1	0	1	0	1	0
1	0	1	0	1	0	1	1
1	0	1	1	1	0	0	1
1	0	0	1	1	0	0	0
1	0	0	0	0	0	0	0

FIGURE 8.14
Transition table for 4-bit Gray code counter

The excitation maps for the A, B, C, and D flip-flops are plotted in Figure 8.15a. The simplified functions for the J and K inputs are obtained from these maps; they are

$$J_A = B\overline{C}\overline{D} \qquad K_A = \overline{B}\overline{C}\overline{D}$$
$$J_B = \overline{A}C\overline{D} \qquad K_B = AC\overline{D}$$
$$J_C = \overline{A}\overline{B}D + ABD \qquad K_C = \overline{A}BD + A\overline{B}D$$
$$J_D = \overline{A}\overline{B}\overline{C} + AB\overline{C} + \overline{A}BC + A\overline{B}C \qquad K_D = \overline{A}B\overline{C} + A\overline{B}\overline{C} + \overline{A}BC + ABC$$

The implementation of the counter is shown in Figure 8.15b. ∎

8.5 INTEGRATED-CIRCUIT COUNTERS

Although the techniques just described can be used to design various types of counters, in practice they are rarely constructed by using gates and flip-flops. A wide variety of ready-built counters are available in all of the major integrated circuit logic families—e.g., the SN7490, SN7492, and SN7493 in transistor-transistor logic (TTL) and the CD4017, CD4018, CD4022, CD4026, and CD4029 in complementary MOS (CMOS) logic.

The SN7490 is a decade counter. It consists of an independent mod-2 counting stage followed by a mod-5 ripple counter. A decade counter is produced by connecting the output of the mod-2 counter to the input of the mod-5 counter. Figure 8.16a shows the block diagram of the SN7490. All four outputs A, B, C, and D can be reset to 0 by holding the reset lines $R_{0(1)}$ and $R_{0(2)}$ at logic 1. The outputs can also be reset to a count of 9 ($ABCD = 1001$) by putting logic 1's on the $R_{9(1)}$ and $R_{9(2)}$ lines. Figure 8.16b shows the truth table for the Reset inputs. It can be seen from the truth table that Reset to 9 takes precedence over Reset to 0 lines (i.e., if all reset lines are held at logic 1, the outputs will go to 9). The A output of the counter is connected to the $\overline{\text{Clock}}_{B-D}$ input, in order to make it operate in the decade mode. It can also act as a mod-6 counter if the D output is connected to $\overline{\text{Clock}}_{B-D}$ and the B and C outputs are connected to $R_{0(1)}$ and $R_{0(2)}$ respectively (Fig. 8.17).

Both the SN7492 and the SN7493 have similar structures. In both, the first stage is a mod-2 counter, followed by a mod-6 (SN7492) or a mod-8 (SN7493) counter; these form mod-12 and mod-16 counters respectively.

FIGURE 8.15

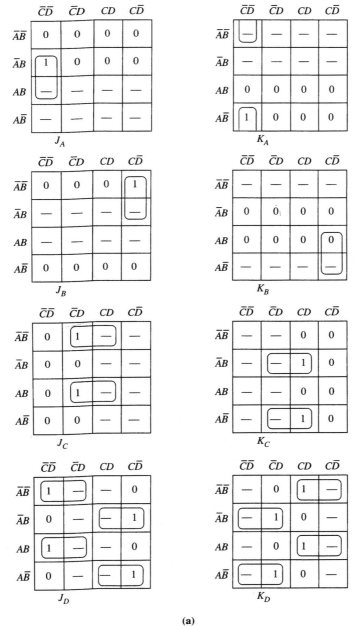

(a)

Figure 8.18 shows the block diagram for the following TTL counters:

    SN74160: Synchronous decade counter
    SN74161: Synchronous mod-16 counter
    SN74162: Synchronous decade counter
    SN74163: Synchronous mod-16 counter

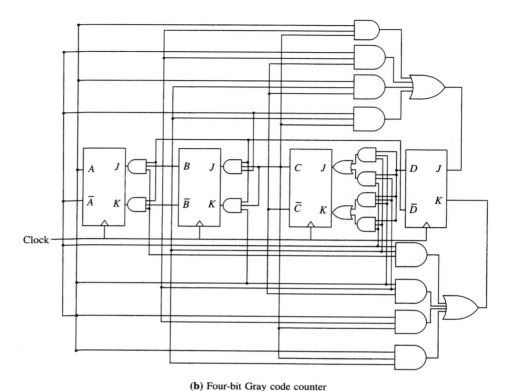

(b) Four-bit Gray code counter

FIGURE 8.15
(Continued)

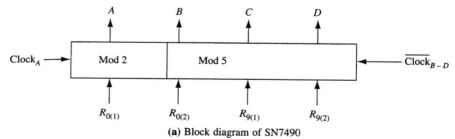

(a) Block diagram of SN7490

FIGURE 8.16

These four counters differ from the three counters discussed previously, in that they can be preset to some count state by loading the corresponding data via the D_3–D_0 inputs on the positive edge of the clock pulse; the /LD (parallel load) input line must be at logic 0 during the clocking in of the data. These devices have two count enable lines: *EP* (enable parallel) and *ET* (enable trickle). For normal counting the *EP* and the *ET* inputs must be held at logic 1. The output lines Q_3–Q_0 are weighted in 8-4-2-1 BCD code for the

Reset Inputs				Output			
$R_{0(1)}$	$R_{0(2)}$	$R_{9(1)}$	$R_{9(2)}$	A	B	C	D
1	1	0	—	0	0	0	0
1	1	—	0	0	0	0	0
—	—	1	1	1	0	0	1
—	0	—	0	Count			
0	—	0	—	Count			
—	0	0	—	Count			
0	—	—	0	Count			

(b) Reset truth table

FIGURE 8.16
(Continued)

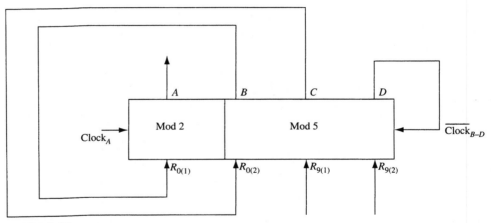

FIGURE 8.17
Mod-6 counter

74160/74162 and in binary for the 74161/74163. Functionally, the 74162 is exactly similar to the 74160. However, in the 74162 all outputs are cleared to 0 only after a positive clock edge, even though the /CLR (master clear) signal is at logic 0 (i.e., the reset function is synchronized), whereas in the 74160 a logic 0 at the /CLR input clears the outputs regardless of the clock pulse, thus providing an asynchronous reset capability. Similarly, the only difference between the 74161 and the 74163 is that the former has asynchronous reset, whereas in the latter the reset function is synchronized with the clock. The *RCO* (ripple carry) output of the counters is at logic 1 when *ET* input is at logic 1 and the 74160/74162 counters have reached their maximum count values (15 for the 74161/74163 and 9 for the 74160/74162). The count sequences for the 74160 (74162) and the 74161 (74163) are shown below (assuming the initial state is 0000).

282 ◻ CHAPTER 8 / REGISTERS AND COUNTERS

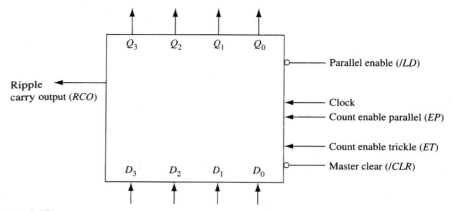

FIGURE 8.18
Block diagram of Signetics 74160/61/62/63 counters

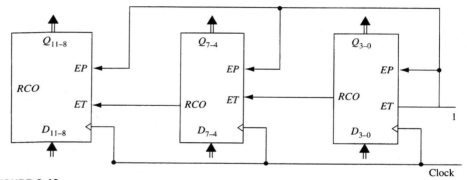

FIGURE 8.19
Three-stage ripple carry counter

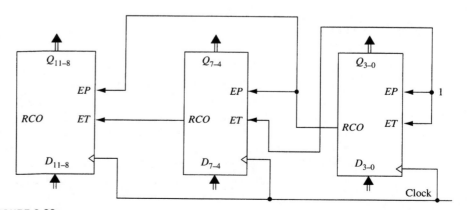

FIGURE 8.20
Three-stage carry-lookahead counter

8.5 INTEGRATED-CIRCUIT COUNTERS □ 283

74160 (74162)				74161 (74163)			
Q_3	Q_2	Q_1	Q_0	Q_3	Q_2	Q_1	Q_0
0	0	0	0	0	0	0	0
0	0	0	1	0	0	0	1
0	0	1	0	0	0	1	0
0	0	1	1	0	0	1	1
0	1	0	0	0	1	0	0
0	1	0	1	0	1	0	1
0	1	1	0	0	1	1	0
0	1	1	1	0	1	1	1
1	0	0	0	1	0	0	0
1	0	0	1	1	0	0	1
0	0	0	0	1	0	1	0
0	0	0	1	1	0	1	1
0	0	1	0	1	1	0	0
0	0	1	1	1	1	0	1
0	1	0	0	1	1	1	0
0	1	0	1	1	1	1	1

These counters can be cascaded to form count sequences of any length.

■ **EXAMPLE 8.3**

Three SN74161s can be cascaded to form a mod-4096 counter. The *RCO* output of a lower state is connected to the *ET* input of the next higher stage. Thus, the $(i + 1)$th stage of the counter can count up only when the ith stage has already reached the top of its count sequence. In other words, this scheme counts in the ripple carry mode. Alternatively, the 74161s can be cascaded to count in the carry-lookahead mode as shown in Figure 8.20. This scheme is significantly faster than the ripple carry scheme when the number of cascaded stages is high. ■

One of the CMOS counters mentioned previously (the CD4029) is a presettable up/down counter. It consists of four stages and can operate in either binary or in BCD mode with lookahead carry capability. Figure 8.21 shows the block diagram of the device. The Binary/decade input selects the mode of operation; BCD mode is selected when this input is low, whereas binary counting is accomplished when the input line is high. The counter counts up or down depending on whether the signal applied to the Up/down input is logic 1 or logic 0 respectively. Four "jam" inputs are provided to permit presetting the counter to any state; the preset enable signal must be at logic 1 in order to accomplish this. A logic 0 on each jam line resets the counter to its zero count if the preset enable signal is at logic 1. The counter increments or decrements its count value by 1 on the positive edge of the clock pulse, provided the carry-in and the preset enable

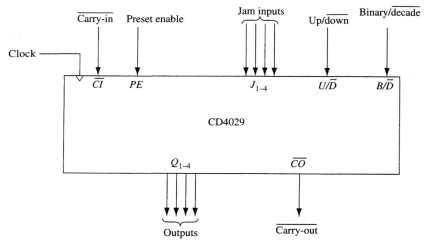

FIGURE 8.21
Block diagram of RCA CD4029 up/down counter

signals are at logic 0. The $\overline{\text{carry-out}}$ is normally at logic 1 and goes to logic 0 when the counter reaches its maximum count in the up mode or the minimum count in the down mode, provided the $\overline{\text{carry-in}}$ input is at logic 0. The CD4029 devices can be cascaded to construct a mod-n counter.

■ | **EXAMPLE 8.4**

Two CD4029s can be cascaded to construct a mod-256 counter as shown in Figure 8.22.

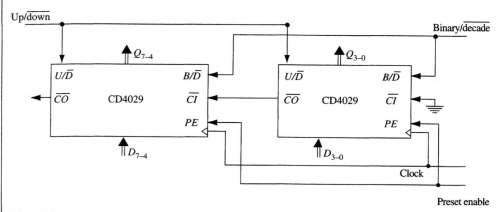

FIGURE 8.22
Two CD4029s cascaded to form a mod-256 counter   ■

8.6 SHIFT REGISTERS

A shift register is a group of flip-flops connected in a chain so that the output from one flip-flop becomes the input of the next flip-flop. All the flip-flops are driven by a com-

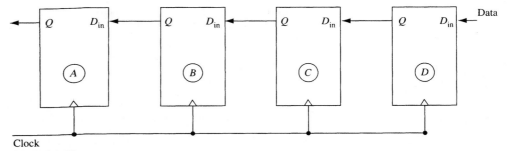

FIGURE 8.23
A four-stage shift register

mon clock, and all are set or reset simultaneously. Figure 8.23 shows a shift register with D flip-flops. As shown, the output of each flip-flop is the input of the next more significant flip-flop. On each positive clock edge, the data input at the D flip-flop is loaded into it, the output of D is loaded into C, the output of C is loaded into B, and the output of B is loaded into A.

To illustrate the operation of the shift register let us assume that all the flip-flops in the 4-bit shift register are initially reset (i.e., the stored bits are all 0's) and an input sequence 1010 is applied serially to the data input of flip-flop D. The first positive clock edge sets flip-flop D to 1, and the contents of the shift register become

A	B	C	D
0	0	0	1

On the next clock pulse, flip-flop C will receive the 1 from the output of flip-flop D and flip-flop D will be loaded with the 0 bit at its data input. Hence,

A	B	C	D
0	0	1	0

The third clock pulse results in

A	B	C	D
0	1	0	1

and on the fourth positive clock edge the content of the shift register becomes

A	B	C	D
1	0	1	0

This completes the operation of entering the input pattern 1010 into the shift register. Note that each clock pulse shifts the stored data one position to the left; hence, this is known as a **shift-left register**.

Figure 8.24a shows a 4-bit **shift-right register**. It works on the same principle as the shift-left register except that on each positive edge of the clock pulse the stored bits move one position to the right. Serial data is now applied to the leftmost flip-flop. The contents of the shift register after each clock pulse are shown in Figure 8.24b for the input 1010.

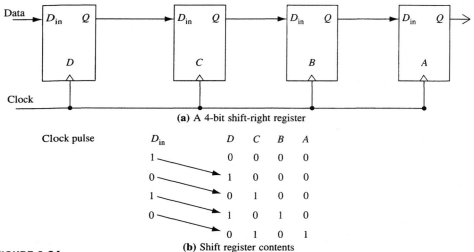

FIGURE 8.24

Shift registers can be classified into four groups:
  i. **Serial in/serial out,** in which data can be shifted in or out of the register one bit at a time (Fig. 8.25a).
 ii. **Serial in/parallel out**, in which each flip-flop output is directly available; thus, if data is loaded serially it is available at the outputs of the flip-flops and can be read simultaneously (Fig. 8.25b).

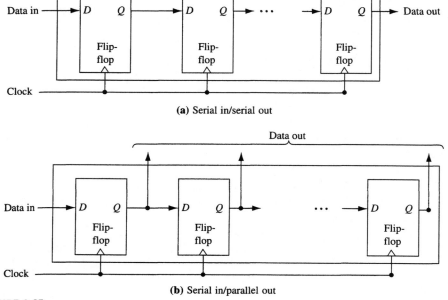

FIGURE 8.25

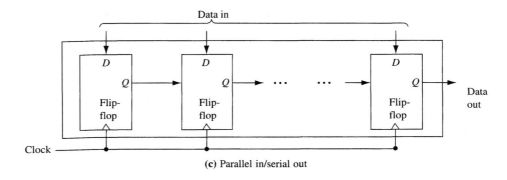

(c) Parallel in/serial out

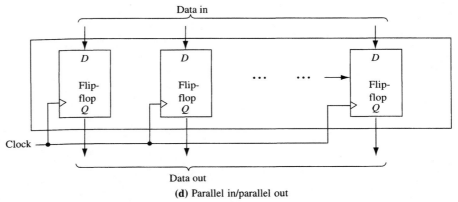

(d) Parallel in/parallel out

FIGURE 8.25
(Continued)

iii. **Parallel in/serial out**, in which data bits are loaded simultaneously into the flip-flops and shifted out one bit at a time (Fig. 8.25c).
iv. **Parallel in/parallel out**, in which data bits can be loaded directly into the flip-flops and their outputs can be read simultaneously (Fig. 8.25d).

A wide variety of completely self-contained IC shift register chips are available in TTL and CMOS. In many chips it is possible to change to serial or parallel mode of operation by applying control signals. Some also have control inputs to determine the direction of the shift operation (left to right or vice versa). Figure 8.26a shows the functional diagram of a commercially available shift register chip (the SN74194). It is a 4-bit **universal shift register** that can be configured to shift data in either direction, to load serially or to load in parallel, and to output data either serially or in parallel. Separate shift-left and shift-right inputs are provided, with right serial output. An asynchronous clear input is also provided, which can be used to reset the flip-flops independent of the clock. The operation of the SN 74194 is controlled by two mode lines, S_0 and S_1, as shown in Figure 8.26b.

*8.7 SHIFT REGISTER COUNTERS

A counter is constructed from a serial in/parallel out shift register by connecting the output of the individual flip-flops to a logic network and connecting the output of that net-

FIGURE 8.26

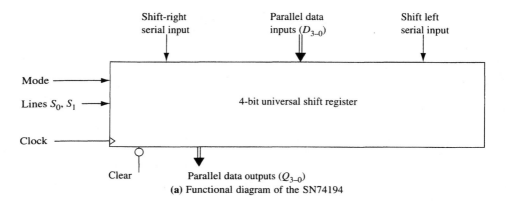

(a) Functional diagram of the SN74194

S_1	S_0	Function
0	0	Hold (i.e., the current state of the register is unchanged)
0	1	Shift right
1	0	Shift left
1	1	Parallel load

(b) Function table

work to the input of the shift register (Fig. 8.27) so that it cycles through some predetermined number of states. For example, in the case of a 3-bit shift register the sequence of states might be

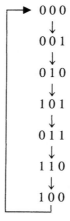

Such a counter may be considered to be a mod-7 counter; it is also known as a **nonbinary counter** because the sequence of states does not form a consecutive sequence of binary coded numbers. The design of shift register counters can be considerably simplified by using a **state sequence tree** [1]. The state sequence tree shows how many possible sequences of different lengths can be obtained from a shift register of a given length. The derivation of the state sequence tree starts with the all 0's state of the shift register at the root of the tree. The effect of a 0 and a 1 on the counter is recorded as shown:

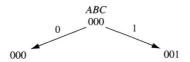

Since the result of shifting in a 0 does not change the state, the left side of the tree need not be considered any further. The effect of shifting in a 0 or a 1 while the shift register is in state 001 is shown next:

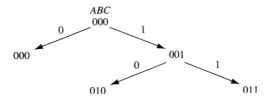

Since there are two new states, the next level of the tree should be

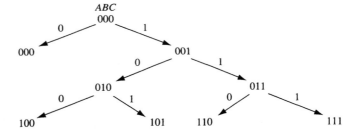

This process is continued for each branch of the tree until a newly derived state already exists in the path from the root to this node of the tree (i.e., a state repeats itself). Figure 8.28 shows the complete state sequence tree for the 3-bit shift register. A shift counter of mod m, where m is less than or equal to the total number of possible states in the shift register, can be constructed by selecting the desired sequence of states from the sequence tree.

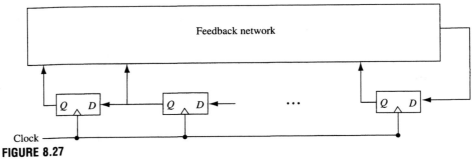

FIGURE 8.27
Shift register counter

EXAMPLE 8.5

Let us design a mod-6 counter using a 3-bit shift register. It can be seen from the state sequence tree for the shift register (Fig. 8.28) that there are three different ways of achieving a state sequence of length 6:

I	II	III
ABC	ABC	ABC
001	011	000
010	110	001
101	101	011
011	010	111
110	100	110
100	001	100
000	000	010
111	111	101

In each of these three state sequences the unused states are underlined. Looking at the first state sequence, we see that if the present content of the shift register is 001 ($=\overline{A}\,\overline{B}C$), it must change to 010 ($\overline{A}B\overline{C}$) after a clock pulse. In other words when the shift register contains 001, the input to it must be a 0. Similarly, when the preset content is 010, the shift input must be 1 so that after the next clock pulse its content is 101. Thus, the shift input must be equal to the lowest significant bit of the next state to which the shift register should move

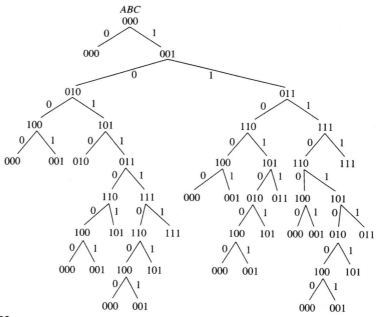

FIGURE 8.28
State sequence tree for a 3-bit shift register

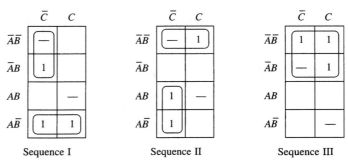

FIGURE 8.29
Karnaugh maps for feedback logic networks

after the clock pulse is applied. Referring to Figure 8.27, it can be seen that the output of the feedback network is fed to the shift input. The appropriate feedback network for the three state sequences can be designed by using Karnaugh maps (Fig. 8.29). The minimal Boolean expressions for implementing the feedback networks are derived as follows:

Sequence I: $D_C = \overline{A}\,\overline{C} + A\overline{B}$

Sequence II: $D_C = A\overline{C} + \overline{A}\,\overline{B}$

Sequence III: $D_C = \overline{A}$

State 111 will produce a "lock up" in sequence I (i.e., it will result in a 1 output from the feedback network, hence locking the shift register in state 111, from which it will not "return" to the main counting sequence). Hence, the don't care term corresponding to state 111 in the Karnaugh map is ignored. For similar reasons, state 111 is not considered while minimizing the feedback expression for sequence II. The implementations of the counters based on state sequences I, II, and III are shown in Figure 8.30.

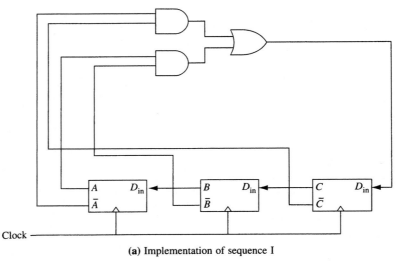

(a) Implementation of sequence I

FIGURE 8.30

FIGURE 8.30
(Continued)

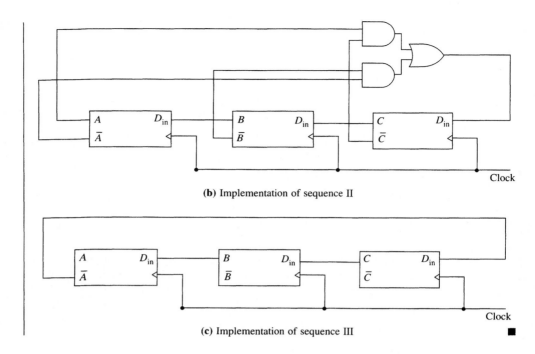

(b) Implementation of sequence II

(c) Implementation of sequence III

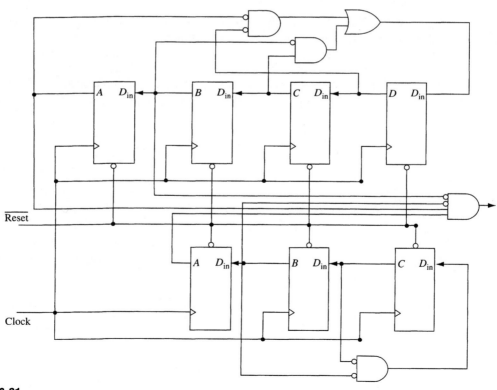

FIGURE 8.31
Mod-40 counter

8.7 SHIFT REGISTER COUNTERS   293

The design of shift counters based on a state sequence tree is not practicable using shift registers with more than four stages. However, a three-stage shift counter may be combined with a four-stage shift counter to form a composite counter having a state sequence length (i.e., mod no.) higher than what is possible using either of the individual counters. The sequence length of the composite counter is equal to the smallest number that is divisible by the sequence length of each of the individual counters. If the sequence lengths have no common factors, then the sequence length of the composite counter is equal to their product. For example, a mod-5 (three-stage) counter can be combined with a mod-8 (four-stage) counter to form a mod-40 counter, but a combination of a mod-5 and a mod-10 counter will give only a sequence length of 10, not 50. Figure 8.31 shows the implementation of a mod-40 counter based on a mod-5 and a mod-8 counter. The count sequence is as follows:

Clock Pulse	Mod 5	Mod 8
"	0	0
"	1	1
"	3	2
"	6	3
"	4	7
"	0	14
"	1	12
"	3	8
"	6	0
"	4	1
"	0	2
"	1	3
"	3	7
"	6	14
"	4	12
"	0	8
"	1	0
"	3	1
"	6	2
"	4	3
"	0	7
"	1	14
"	3	12
"	6	8
"	4	0
"	0	1
"	1	2
"	3	3
"	6	7
"	4	14
"	0	12
"	1	8
"	3	0
"	6	1
"	4	2
"	0	3
"	1	7
"	3	14
"	6	12
"	4	8

It is assumed that each counter produces a logic 1 output when it reaches its maximum count value. Thus, the counter circuit of Figure 8.31 gives a 1 output when the mod-6 and mod-8 counters have settled in states 100 and 1000 respectively, indicating that the count sequence has been completed.

An alternative approach to designing shift counters is based on toggling the least significant flip-flop of the shift register whenever the output of the most significant flip-flop is 0. For an n-bit shift register, this approach results in a state sequence of length $2^n - 1$.

■ **EXAMPLE 8.6**

Let us implement a mod-15 counter using the Signetics SN74195 parallel load shift register device. The block diagram of the device is shown in Figure 8.32a. If the $\overline{PE}$ (Parallel Enable) input is active-low, data on inputs (D_0–D_3) can be transferred in parallel to the outputs (Q_0–Q_3) of the device. Alternatively, if the $\overline{PE}$ input is active-high, data can be shifted in serially via the J and $\overline{K}$ inputs. By tying the J and $\overline{K}$ input pins together, the JK flip-flop can be used to emulate a D flip-flop. The outputs of the SN74195 can be asynchronously reset to 0's by applying an active-low signal at the $\overline{MR}$ (Master reset) input.

Figure 8.32b shows the implementation of the mod-15 counter using an SN74195. The state sequence for the counter is

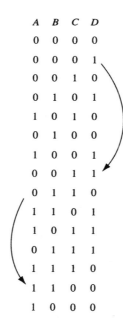

It can be seen from the state sequence that a mod-15 counter can be converted into a mod-6 counter by forcing a jump from state 0001 to state 0011, and from state 0110 to state 1100. This can be achieved by connecting $\overline{A}\,\overline{C}$ to the J and $\overline{K}$ inputs of the SN74195, as shown in Figure 8.32c.

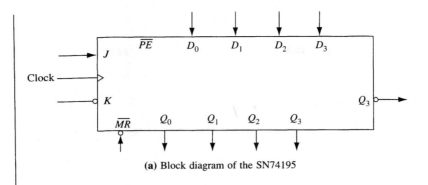

(a) Block diagram of the SN74195

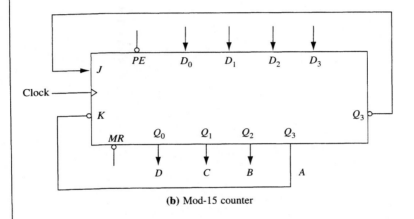

(b) Mod-15 counter

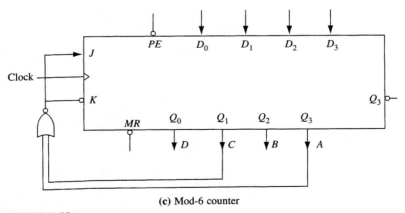

(c) Mod-6 counter

FIGURE 8.32

8.8 RING COUNTERS

A ring counter is basically a **circulating** shift register in which the output of the most significant stage is fed back to the input of the least significant stage. Figure 8.33 is a ring counter constructed from D flip-flops. The output of each stage is shifted into the next stage on the positive edge of a clock pulse. The "reset" signal clears all the flip-flops except the first one. Because flip-flop 1 is preset by a logic 1 on the reset line, the initial content of the counter is

A	B	C	D
0	0	0	1

The first positive clock edge shifts the output of flip-flop 4 into flip-flop 1 and the outputs of flip-flops 1, 2, and 3 into flip-flops 2, 3, and 4 respectively. Therefore, the state of the counter becomes

A	B	C	D
0	0	1	0

The second clock pulse changes the state to

A	B	C	D
0	1	0	0

At the end of the third clock pulse the counter state becomes

A	B	C	D
1	0	0	0

On the fourth clock pulse the 1 output from flip-flop 4 is transferred into flip-flop 1, thus

A	B	C	D
0	0	0	1

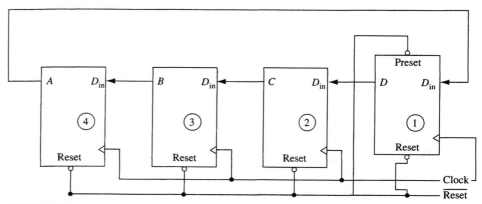

FIGURE 8.33
A 4-bit ring counter

which is, of course, the initial state of the counter. Since the count sequence has 4 distinct states, the counter can be considered as a mod-4 counter. Thus, the 4-bit ring counter includes only 4 of the 16 states that are possible using 4 flip-flops. Since an n-bit ring counter uses only n of its 2^n possible states, leaving $2^n - n$ states unused, it makes very inefficient use of flip-flops. For example, a decimal ring counter has only 10 counting states but requires 10 flip-flops, whereas a binary counter having 10 flip-flops will have $1024\ (=2^{10})$ states and can count up to 1023. The major advantage of a ring counter over a binary counter is that it is **self-decoding** (i.e., no extra decoding circuit is needed to determine what state the counter is in). Each state is uniquely identified by a logic 1 at the output of the corresponding flip-flop. On the other hand, in an n-bit binary counter (except for a few count values) more than one flip-flop is on at a particular count, so additional gates are needed to generate a decoding signal for each state.

The ring counter technique can be effectively utilized to implement synchronous sequential circuits. A major problem in the realization of sequential circuits is the assignment of binary codes to the internal states of the circuit in order to reduce the complexity of circuits required (see Chap. 6). By assigning one flip-flop to one internal state, it is possible to simplify the combinational logic required to realize the complete sequential circuit. When the circuit is in a particular state, the flip-flop corresponding to that state is set to logic 1 and all other flip-flops remain reset.

■ EXAMPLE 8.7

Let us design the sequential circuit described by the following state table:

Present State	Input $x = 0$	$x = 1$	Output
A	F	C	0
B	D	C	1
C	A	D	0
D	A	E	1
E	C	F	0
F	C	G	1
G	B	G	0
	Next State		

Since the sequential circuit has seven states, a 7-bit ring counter is required. Let us assume that states A, B, C, D, E, F, and G correspond to flip-flops 1, 2, 3, 4, 5, 6, and 7 respectively in the ring counter. State A is the next state for both state C and state D when input $x = 0$. Therefore, whenever either flip-flop 3 or flip-flop 4 is set (i.e., the ring counter state is 0000100 or 0001000) and $x = 0$, flip-flop 1 should receive a logic 1 input signal, causing the counter to move to state 0000001. Assuming JK flip-flops are used to implement the ring counter, the excitation equations for flip-flop 1 is

$$J_1 = \bar{x}C + \bar{x}D$$
$$K_1 = 1$$

In other words, flip-flop 1 is set to logic 1 only if the counter is in state C or D; otherwise, it is reset. The excitation equations for the other flip-flops may be derived in a similar manner and are given by

$$J_2 = \bar{x}G \qquad K_2 = 1$$
$$J_3 = \bar{x}E + \bar{x}F + xA + xB \qquad K_3 = 1$$
$$J_4 = \bar{x}B + xC \qquad K_4 = 1$$
$$J_5 = xD \qquad K_5 = 1$$
$$J_6 = \bar{x}A + xE \qquad K_6 = 1$$

The counter should move to state G whenever the present state is F and the input $x = 1$; hence, the excitation equation for the J input of flip-flop 7 is

$$J_7 = xF$$

Furthermore, whenever the counter is in state G and $x = 1$, it should remain in state G; therefore, flip-flop 7 should not be reset for this input/state combination. Thus, the excitation equation for the K input is

$$K_7 = \bar{x}$$

The output signal Z can be written as

$$Z = B + D + F$$

The logic diagram of the complete sequential circuit is shown in Figure 8.34.

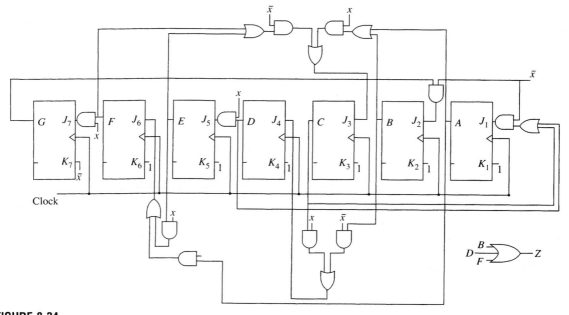

FIGURE 8.34
Implementation of a seven-state sequential circuit

*8.9 JOHNSON COUNTERS

Johnson counters are a variation of standard ring counters, with the inverted output of the last stage fed back to the input of the first stage. They are also known as **twisted ring** counters. An n-stage Johnson counter yields a count sequence of length $2n$, so it may be considered to be a mod-$2n$ counter. Figure 8.35a shows a 4-bit Johnson counter. The state sequence for the counter, assuming that initially all the flip-flops are reset to 0, is given in Figure 8.35b. At any count step only one flip-flop changes state, so in a Johnson counter a state can be decoded using a 2-input AND gate regardless of the number of flip-flops in the counter. For example, the combination $A = 0$ and $D = 0$ occurs only in one state (i.e., the first state of the count sequence shown in Fig. 8.35b). Hence, an AND gate with inputs $\overline{A}$ and $\overline{D}$ can be used to decode this state. The decoding function for each state of the 4-bit Johnson counter is shown in Figure 8.35b. Note that for exactly the same number of flip-flops a Johnson counter has twice the mod number of a ring counter. However, a Johnson counter requires decoding gates, whereas a ring counter does not. In general, an n-bit Johnson counter provides state sequences of even length, $2n$. However, it is possible to achieve count sequences of odd length, $2n - 1$, by using a 2-input gate for the feedback. A 4-bit **pseudo Johnson** counter of this type and its count sequence are shown in Figures 8.36a and b respectively.

Both the ring and the Johnson counter must initially be forced into a valid state in the count sequence because they operate on a subset of the available number of states. Since a master reset signal is usually provided, this is usually not a problem; in the case of the ring counter, one of the flip-flops must be preset while the others are being reset. An alternative way to force these counters to a valid state is to use appropriate feedback logic. In the 3-bit Johnson counter shown in Figure 8.37a, the non-valid states are 010 and 101. If the counter starts in one of these states, it will cycle through them indefinitely. The incorporation of feedback logic, $\overline{A} (\overline{B} + C)$, will make the counter self-correcting within two clock cycles (Fig. 8.37b).

Another apparent disadvantage of Johnson-type counters is that they are not very efficient for high count cycles. A mod-32 Johnson counter, for example, requires 16 flip-flops, whereas only 5 are required for a binary counter. However, by cascading a mod-4 and a mod-8 Johnson counter, it is possible to reduce the number of flip-flops to only 6. This arrangement is shown in Figure 8.38. The clock signal is applied to the mod-8 counter via the AND gate, which allows the clock signal to pass through only when the

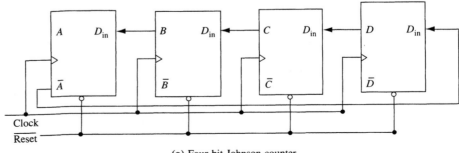

(a) Four-bit Johnson counter

FIGURE 8.35

FIGURE 8.35
(Continued)

State				Decoder
A	B	C	D	
0	0	0	0	$\bar{A}\bar{D}$
0	0	0	1	$\bar{C}D$
0	0	1	1	$\bar{B}C$
0	1	1	1	$\bar{A}B$
1	1	1	1	AD
1	1	1	0	$C\bar{D}$
1	1	0	0	$B\bar{C}$
1	0	0	0	$A\bar{B}$

(b) Count sequence and state decoder for 4-bit Johnson counter

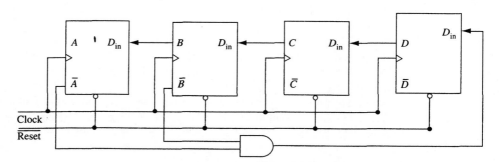

(a) Four-bit pseudo Johnson counter

FIGURE 8.36

State			
A	B	C	D
0	0	0	0
0	0	0	1
0	0	1	1
0	1	1	1
1	1	1	0
1	1	0	0
1	0	0	0

(b) Count sequence

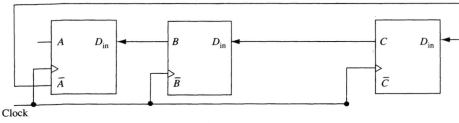

(a) Three-bit Johnson counter

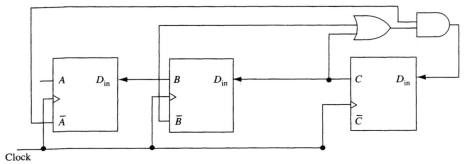

(b) Three-bit self-correcting Johnson counter

FIGURE 8.37

mod-4 counter is in state $EF = 10$ (i.e., is in the final state of its count sequence). The count sequence for the mod-32 counter is

$$
\begin{array}{cccccc}
A & B & C & D & E & F \\
0 & 0 & 0 & 0 & 0 & 0 \\
0 & 0 & 0 & 0 & 0 & 1 \\
0 & 0 & 0 & 0 & 1 & 1 \\
0 & 0 & 0 & 0 & 1 & 0 \\
0 & 0 & 0 & 1 & 0 & 0 \\
0 & 0 & 0 & 1 & 0 & 1 \\
& & \vdots & & & \\
1 & 0 & 0 & 0 & 1 & 0
\end{array}
$$

Several CMOS Johnson counters are commercially available (e.g., the CD4017, CD4018, and CD4022). The CD4017 is a 5-stage Johnson counter having ten decoded outputs (Fig. 8.39). It is a positive-edge-triggered device and has three inputs: **clock, reset,** and **clock inhibit.** A logic 1 on the clock inhibit line disables counting with the counter remaining in its current state. A logic 1 reset signal clears the counter to its zero value. The carry-out signal completes one cycle for every ten input cycles and is used in multi device cascading. The CD4022 is very similar to the CD4017 except that it is a four-stage Johnson counter having eight decoded outputs.

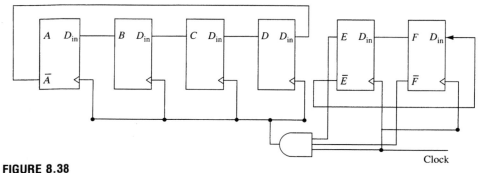

FIGURE 8.38
Mod-32 Johnson counter

The CD4018 contains a five-stage Johnson counter and has the following inputs: **clock, reset, data, preset enable,** and five individual **jam inputs** (Fig. 8.40a). A logic 1 on the preset enable input allows data on the jam inputs to be loaded into the counter. The counter is cleared to all 0's if a logic 1 is applied to the reset input. The CD4018 can provide counter configurations of mod nos. 2, 4, 6, 8, or 10 by feeding $\overline{Q}_0$, $\overline{Q}_1$, $\overline{Q}_2$, $\overline{Q}_3$, and $\overline{Q}_4$ respectively back to the data input. However, in each case an appropriate initial count value must be loaded into the CD4018 by using the jam inputs; the initial count values are

	Q_4	Q_3	Q_2	Q_1	Q_0
Mod 2	0	1	0	1	0
Mod 4	0	0	1	1	0
Mod 6	0	0	0	1	1
Mod 8	0	0	0	0	1
Mod 10	0	0	0	0	0

It is also possible to obtain counter configurations of odd mod nos. (e.g. 3, 5, 7, and 9) by using a two-input AND gate to feed the outputs of two stages in CD4018 to the data input; the inputs to the AND gate for the appropriate mod no. are selected as shown:

FIGURE 8.39
Functional diagram of the CD 4017

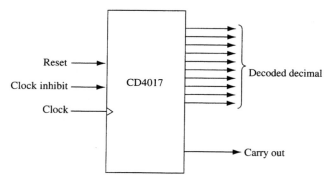

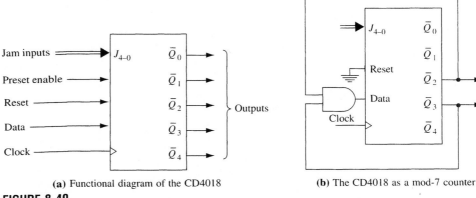

(a) Functional diagram of the CD4018   **(b)** The CD4018 as a mod-7 counter

FIGURE 8.40

Mod 3	$\overline{Q}_1, \overline{Q}_0$
Mod 5	$\overline{Q}_2, \overline{Q}_1$
Mod 7	$\overline{Q}_3, \overline{Q}_2$
Mod 9	$\overline{Q}_4, \overline{Q}_3$

Figure 8.40b shows the implementation of a mod-7 counter using an AND gate and a CD4018; it is assumed that the counter is preset to the initial state $Q_4Q_3Q_2Q_1Q_0 = 00001$ via the jam inputs.

*8.10 PSEUDO-RANDOM SEQUENCE GENERATION USING SHIFT REGISTERS

Pseudo-random sequences are used in many applications e.g. as extremely high quality audio noise, and as test patterns in logic circuit testing. They are sufficiently random in nature to replace truly random sequences and can be easily repeated.

Pseudo-random binary sequences can be generated by connecting the outputs of a selected number of stages in a shift register to its input through an EX-OR network. Since the feedback path involves only EX-OR operation (i.e., Mod-2 addition of the chosen stages of the shift register), it is said to be a **linear feedback shift register** (**LFSR**). The output of any stage in an LFSR is a function of the initial state of the bits in the register and of the outputs of the states that are fed back. Thus, the selection of feedback paths is crucial in constructing an LFSR that performs as required.

An n-bit LFSR may be represented by a polynomial that is **irreducible** and **primitive**. A polynomial that cannot be factored is called irreducible. For example, the polynomials $f(x) = x^3 + x + 1$ and $g(x) = x^3 + 1$ are irreducible polynomials of **degree** 3; the degree of a polynomial is the largest superscript in the polynomial. A polynomial $p(x)$ of degree n is primitive if the remainders generated from the divisions of all polynomials with degree $c(\leq 2^n - 1)$ by $p(x)$ correspond to all possible non-zero polynomials of degree less than n. For example, the division of $x, x^2, x^3, \ldots x^7$ by polynomial $g(x)$ results in the same three remainders 1, x, and x^2 repeatedly. On the other hand, if we divide $x, x^2, x^3, x^4, x^5, x^6, x^7$ by polynomial $f(x)$, the remainders obtained are $x, x^2, x + 1, x^2 + x, x^2 + x + 1, x^2 + 1$, and 1 respectively. In other words, the division of all polynomials of

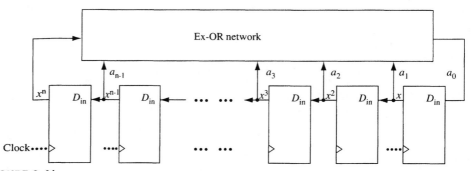

FIGURE 8.41
General representation of an LFSR

degree 1 to 7 by $f(x)$ results in all seven non-zero polynomials of degree less than 3. Thus, although both polynomials $f(x)$ and $g(x)$ are irreducible, only $f(x)$ is primitive.

If an LFSR is implemented using $f(x)$ it will repeatedly generate seven binary patterns in sequence. In general, if the polynomial used to construct an LFSR is of degree n, then the LFSR will generate all possible 2^n-1 non-zero binary patterns in sequence; this sequence is termed the **maximal length sequence** of the LFSR.

Figure 8.41 shows the general representation of an LFSR based on the primitive polynomial

$$a(x) = x^n + a_{n-1}x^{n-1} + \ldots + a_2x^2 + a_1x + a_0 \ldots \tag{1}$$

The feedback connections needed to implement an LFSR can be derived directly from the chosen primitive polynomial. To illustrate, let us consider the following polynomial of degree 4:

$$x^4 + x + 1$$

This can be rewritten in the form of expression (1):

$$a(x) = 1 \cdot x^4 + 0 \cdot x^3 + 0 \cdot x^2 + 1 \cdot x + 1 \cdot x^0$$

If the coefficient of x^i is 1, there is feedback from stage i, whereas if the coefficient is 0, the output of stage i is not connected to the EX-OR network. Figures 8.42a and b show

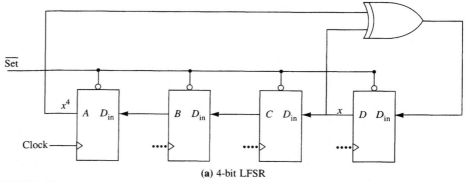

(a) 4-bit LFSR

FIGURE 8.42

FIGURE 8.42
(Continued)

A	B	C	D
1	1	1	1
1	1	1	0
1	1	0	1
1	0	1	0
0	1	0	1
1	0	1	1
0	1	1	0
1	1	0	0
1	0	0	1
0	0	1	0
0	1	0	0
1	0	0	0
0	0	0	1
0	0	1	1
0	1	1	1

(b) maximal length sequence

the four-stage LFSR constructed by using the above polynomial and the corresponding maximal length sequence respectively.

It is possible to construct LFSRs from certain shift registers by using feedback paths from only two stages; a partial list of such registers is given in Table 8.1 [2]. All of these LFSRs produce maximal length sequences.

TABLE 8.1
Maximal length LFSRs with two feedback taps

No. of Stages	Stages From Which Feedback Paths Are Derived (Corresponding Polynomial)	Length of Sequence
7	1, 7 ($x^7 + x + 1$) or 3, 7 ($x^7 + x^3 + 1$)	127
9	4, 9 ($x^9 + x^4 + 1$)	511
10	3, 10 ($x^{10} + x^3 + 1$)	1,023
11	2, 11 ($x^{11} + x^2 + 1$)	2,047
15	1, 15 ($x^{15} + x^4 + 1$) or 4, 15 ($x^{15} + x^4 + 1$) or 7, 15 ($x^{15} + x^7 + 1$)	32,767
17	3, 17 ($x^{17} + x^3 + 1$) or 5, 17 ($x^{17} + x^5 + 1$) or 6, 17 ($x^{17} + x^6 + 1$)	131,071
18	7, 18 ($x^{18} + x^7 + 1$)	262,143
20	3, 20 ($x^{20} + x^3 + 1$)	1,048,575

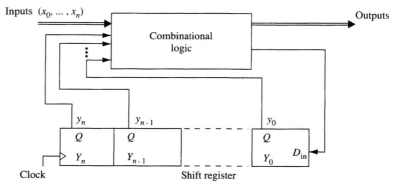

FIGURE 8.43
Sequential circuit mechanization using a shift register

*8.11 SEQUENTIAL CIRCUIT REALIZATION USING SHIFT REGISTERS

The assignment of binary codes to the internal states of a sequential circuit is known as state assignment. Since the complexity of the sequential circuit realization depends heavily upon such an assignment, those assignments that lead to a simpler circuit realization are desirable. One simple form of realization of a sequential circuit is shown in Figure 8.43, in which the memory elements are connected in the form of a shift register. In this configuration the outputs of the serial in/parallel out shift register correspond to the present state of the sequential circuit. Clearly, in an n-bit shift register the function of each next state variable (except for the first one) depends only on its previous stage, whereas the function of the first next state variable Y_0 is

$$Y_0 = f(x_0, \ldots, x_n, y_0, \ldots, y_n)$$

■ EXAMPLE 8.8

Let us implement the sequential circuit specified by Figure 8.44 with a single shift register. The crucial step in the shift register implementation of a sequential circuit is to find an appropriate state assignment. The complete state assignment for the sequential circuit specified by the state table is

	y_2	y_1	y_0
A	0	0	1
	1	0	1
B	1	1	1
	0	1	1
C	0	1	0
D	1	0	0
E	1	1	0
F	0	0	0

Notice that both state A and state B have been assigned two binary codes. The transition table corresponding to this state assignment is:

*8.11 SEQUENTIAL CIRCUIT REALIZATION USING SHIFT REGISTERS   307

FIGURE 8.44
State table of a sequential circuit

Present State	Input $x = 0$	$x = 1$
A	C,0	B,0
B	E,0	B,0
C	D,0	A,0
D	F,0	A,1
E	D,0	D,0
F	F,0	A,1

Next State, Output

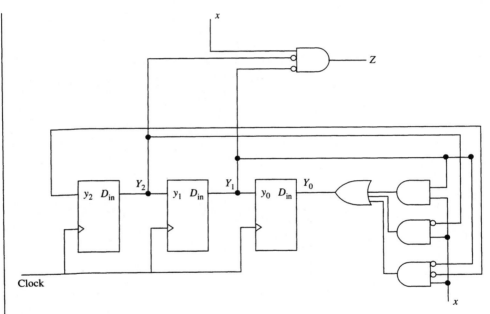

FIGURE 8.45
Single shift register realization of the state table of Figure 8.44

$y_2 y_1 y_0$	$x = 0$ $Y_2 Y_1 Y_0, Z$	$x = 1$ $Y_2 Y_1 Y_0, Z$
001	010,0	011,0
101	010,0	011,0
111	110,0	111,0
011	110,0	111,0
010	100,0	101,0
100	000,0	001,1
110	100,0	100,0
000	000,0	001,1

The next state is specified by the following equations

$$Y_2 = \bar{x}(y_1y_0 + y_1\bar{y}_0) + x(y_1y_0 + y_1\bar{y}_0)$$
$$= \bar{x}y_1 + xy_1 = y_1$$
$$Y_1 = \bar{x}(\bar{y}_1y_0 + y_1y_0) + x(\bar{y}_1y_0 + y_1y_0)$$
$$= \bar{x}y_0 + xy_0 = y_0$$
$$Y_0 = x(\bar{y}_1y_0 + y_1y_0 + \bar{y}_2\bar{y}_0 + \bar{y}_1\bar{y}_0)$$
$$= xy_0 + x\bar{y}_1 + x\bar{y}_2\bar{y}_0$$

The equations for Y_2 and Y_1 clearly indicate the shift register form of the realization. The output equation is

$$Z = x\bar{y}_1y_0$$

The realization of the sequential circuit is shown in Figure 8.45. ∎

In most cases, two shift registers and appropriate combinational logic are needed to realize any arbitrary sequential circuit. For example, the sequential circuit specified by the state table of Figure 8.46 cannot be implemented with a single shift register [3]. However, it is possible to realize the sequential circuit with two shift registers by using the state assignment given in Figure 8.47. The resulting next state equations are

$$Y_0 = x_1x_2\bar{y}_2 + y_2y_1 + \bar{x}_1y_2y_0$$
$$Y_1 = x_2y_1\bar{y}_0 + x_1x_2\bar{y}_0 + x_1y_1$$
$$Y_2 = y_1(x_1 + x_2)$$

FIGURE 8.46
State table of a sequential circuit

Present State	Input		
	$x_1x_2 = 01$	$x_1x_2 = 11$	$x_1x_2 = 10$
A	A	E	A
B	C	F	C
C	—	F	F
D	H	A	A
E	G	F	C
F	D	F	F
G	A	B	A
H	H	H	A
	Next State		

FIGURE 8.47
State assignment for shift realization of the sequential circuit specified in Figure 8.46

State	y_2	y_1	y_0
A	0	0	0
B	0	1	0
C	1	1	0
D	1	0	1
E	0	1	1
F	1	1	1
G	1	0	0
H	0	0	1

Equations Y_1 and Y_2 can be realized with a 2-bit shift register, whereas another 1-bit shift register (i.e., a flip-flop) is required to implement Y_0.

The major advantage of using shift registers in implementing sequential circuits is that it considerably simplifies the testing of such circuits. Although several techniques were proposed in the 1960s for shift register realization of sequential circuits, there is no single efficient procedure for determining whether a sequential circuit is realizable with the minimum number of fixed-size shift registers.

8.12 RANDOM ACCESS MEMORY

A memory unit in a digital system (e.g., a computer) is used to store large amounts of information. In general, a memory unit consists of several locations, each location being identified by a unique address. Thus, if a memory unit has $l(= 2^m)$ locations, then an m-bit address is needed to access each location. The maximum number of information bits that can be stored in a memory is specified as its **capacity**. For example, if a memory has 256 locations, each of which stores 8 information bits, then the memory capacity is 2K (= 2048) bits. The units **Kilo** (= 2^{10} bits), **Mega** (= 2^{20} bits), and **Giga** (= 2^{30} bits) are used to refer to the number of information bits in a memory. Typically, the capacity of a memory unit is expressed in terms of **bytes** (1 byte = 8 bits). For example, a 3.5-inch floppy disk has a capacity of 720 Kbytes or 1.4 Mbytes.

Memories in general can be classified in two categories:

Sequential access memory

Random access memory

In **sequential access memory**, units of data, called **records**, must be accessed in a specific linear sequence until the desired record is found. On the other hand, in a **random access memory (RAM)**, any of the locations may be accessed in any order with the time to access a given location being constant. In other words, data stored in any location of a RAM can be accessed at random. The process of storing information bits into RAM locations is known as a **write** operation. The accessing of information bits from memory locations is known as a **read** operation. The time required for the contents of a location to be available at the output of the memory, after the address of the location has been provided, is known as the **access time**. The time elapsed from the start of one memory operation (read or write) to the time another memory operation can be initiated, is called the **cycle time**. Figure 8.48 shows the block diagram of a RAM unit. An m-bit address is decoded by a 1-out-of-2^m decoder to uniquely select each memory location. If the read operation is selected, the contents of an addressed location are stored in the data register. Alternatively, during a write operation the contents of the data register (loaded externally) are transferred to an addressed location.

Modern digital systems almost exclusively use RAM chips for information storage. RAM chips can be categorized into two types: static and dynamic. In a **static RAM** each bit of information is stored in a cell composed of several transistors. On the other hand, a single transistor is used in a **dynamic RAM** to create an extremely small capacitor. If the capacitor is charged, the transistor cell is interpreted to store 1; the absence of charge in the capacitor is interpreted as a 0. Since the charge in the capacitor tends to leak away very rapidly, it must be refreshed periodically to maintain the original charge. A dynamic

FIGURE 8.48
Block diagram of RAM

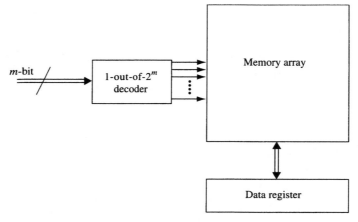

RAM chip uses smaller cells, so it can have a higher density than static RAMs because more cells can be packed into a device. However, static RAMs are in general faster than dynamic RAMs. Both static and dynamic RAMs are volatile (i.e., the stored information is lost in the absence of power to the devices).

One of the popular types of RAM is read only memory (ROM). This is programmed by the manufacturers to the user's specification and cannot be altered after manufacturing. Certain types of ROMs are user programmable; these can be used to implement combinational logic functions (see Chap. 4).

EXERCISES

1. The output of an 8-bit serial-in/serial-out shift register is connected back to its input. Assuming that the initial content of the shift register is 00001111, determine its content after four clock pulses.

2. The initial content of an 8-bit bidirectional shift register is 10101010. The shift-right and shift left serial inputs are connected to logic 0 and logic 1 respectively. Initially the shift register is in shift-right mode. After applying four clock pulses, the shift register is set to shift-left mode and four more clock pulses are applied. Determine the content of the shift register.

3. Design a circuit that will allow the following sequence of operations:
    i. Load a 4-bit pattern into a parallel-in/parallel-out shift register
    ii. Invert the content of the register, and reload
    iii. Go back to (i)

4. Design a random counter to count the sequence 8, 1, 2, 0, 4, 6, 3, 5, 9, 7. Use D flip-flops.

5. Construct an 8-bit binary counter using T flip-flops as memory elements.

6. Design a ripple counter to count in excess-3 code.

7. Design a counter that can count either in mod-8 pure binary or in mod-8 Gray code, depending on a control signal being 1 or 0 respectively.

8. Implement a self-correcting counter that repeatedly generates the sequence 001, 110, 101, 010, 011. Use D flip-flops as memory elements.

9. Design a mod-12 counter using a shift register and appropriate feedback logic.
10. Implement a 4-bit ring counter using *JK* flip-flops, and show its count sequence.
11. Design a circuit that receives serially the output of the 4-bit ring counter in Exercise 8.9, and produces an output of 1 if the counter produces an illegal output combination.
12. Construct a 5-bit ring counter using *D* flip-flops. Combine the counter with a single *D* flip-flop to produce a decade counter.
13. A 4-bit up-down binary counter with a built-in shift register is to be constructed. The counter/shift register is controlled by three control signals, according to the following table:

s	mode	input	operation
1	—	1	shift left
1	—	0	shift right
0	1	—	count up
0	0	—	count down

14. In many applications binary ripple counters are found to be very slow. One possible approach to speed up counting is to use synchronous binary counters with carry-lookahead. Such a counter can be designed by generating a single carry-lookahead signal for each counter stage from the output of the previous stage. Derive the design equation for a 4-bit binary counter with carry-lookahead.
15. Counters are used in many designs to derive lower-frequency clock signals from the original clock signal. Show how a Johnson counter be used to generate decoded signals of mod 2, mod 3, and mod 6 from an incoming clock signal.
16. As discussed in the text, an *n*-bit Johnson counter is a mod-2*n* counter. Show how a 4-bit Johnson counter can be converted into a mod-7 (i.e., odd modulo) counter by adding simple feedback logic.
17. Design an 8-bit ring counter using *D* flip-flops. Also, implement an error detector circuit that will produce an active-high output if the counter generates an erroneous pattern (i.e., all 0's or more than a single 1).
18. Implement the following state table by using a 3-bit serial-in/parallel-out shift register; show the state assignment for the circuit.

Present State	Input $x = 0$	$x = 1$
A	E	B
B	C	F
C	E	B
D	B	D
E	G	—
F	A	D
G	C	F

19. Show how a 4-bit LFSR generating a maximal length sequence (15 states) can be converted into a counter with 16 states (use the minimum number of additional gates).
20. A four-stage shift register is to be used to generate two sequences of length 7 and 15. A sequence of length 7 is generated when a control signal *c* is set to 1; when *c* is set to 0, a sequence of length 15 is generated. Show the appropriate feedback circuit required to generate the sequence assuming the initial content of the shift register is 1111.

21. A RAM unit has 4096 words of 16 bits per word.
   a. What is the capacity of the memory unit?
   b. What size counter is needed to address each memory location?
   c. What is the memory capacity if there are 10 address lines and 32 bits can be stored in each location?

References

1. J. Muth, "Designing shift counters," *Semiconductors*, Vol. 4, No. 1, 1970, pp. 11–13.

2. S. Golomb, *Shift Register Sequences*, Holden Day, 1967.

3. A. J. Nichols, "Minimal shift register realizations of sequential machines," *IEEE Trans. on Computers*, October 1965, pp. 688–700.

Further Reading

1. S. A. Garrod and R. J. Burns, *Digital Logic: Analysis, Application and Design*, Saunders College Publishing, 1991.

2. M. Mano, *Digital Design*, Prentice Hall, 1984.

9
Arithmetic Circuits

Arithmetic operations such as addition, subtraction, multiplication, and division are frequently performed in digital computers and in many other digital systems. Addition and subtraction are usually realized by combinational circuits, whereas multiplication and division are most often realized by sequential circuits. In this chapter we discuss various techniques for implementing arithmetic operations.

9.1 HALF-ADDERS

A half-adder is a combinational logic circuit that accepts two binary digits and generates a sum bit and a carry-out bit. Figure 9.1 shows the truth table of the half-adder circuit. Columns a_i and b_i correspond to the two binary digits to be added; columns s_i and c_i correspond to the sum and the carry bit respectively. The Boolean expressions for the sum and the carry-out can be derived directly from the truth table and are as follows

$$s_i = \overline{a}_i b_i \oplus a_i \overline{b}_i$$
$$= a_i \oplus b_i$$
$$c_i = a_i b_i$$

The NAND-NAND implementation of the sum and the carry-out is shown in Figure 9.2.

FIGURE 9.1
Truth table for half-adder

a_i	b_i	s_i	c_i
0	0	0	0
0	1	1	0
1	0	1	0
1	1	0	1

9.2 FULL ADDERS

The limitation of a half-adder is that it cannot accept a carry-in bit; the carry-in bit represents the carry-out of the previous low-order bit position. Thus, a half-adder can be used only for the two least significant digits when adding two multibit binary numbers, since there can be no possibility of a propagated carry to this stage. In multibit addition, a carry

FIGURE 9.2
Half-adder circuit

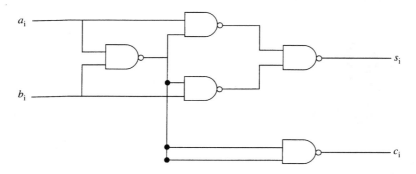

bit from a previous stage must be taken into account, which gives rise to the necessity for designing a full adder. A full adder can accept two operand bits, a_i and b_i, and a carry-in bit c_i from a previous stage; it produces a sum bit s_i and a carry-out bit c_0. Figure 9.3 shows the truth table for a full adder circuit. As can be seen from the truth table, sum bit s_i is 1 if there is an odd number of 1's at the inputs of the full adder, whereas the carry-out c_0 is 1 if there are two or more 1's at the inputs. The sum and carry-out bits will be 0 otherwise. The Boolean expressions for s_i and c_0 obtained from the truth table are as follows:

$$s_i = \bar{a}_i\bar{b}_i c_i + \bar{a}_i b_i \bar{c}_i + a_i \bar{b}_i \bar{c}_i + a_i b_i c_i$$
$$c_0 = \bar{a}_i b_i c_i + a_i \bar{b}_i c_i + a_i b_i \bar{c}_i + a_i b_i c_i$$

These expressions are plotted on the Karnaugh maps shown in Figure 9.4. The expression for s_i cannot be reduced. The expression for c_0 reduces to

$$c_0 = a_i b_i + b_i c_i + a_i c_i$$

The expressions for s_i and c_0 can be rewritten as follows:

$$s_i = \overline{\overline{\bar{a}_i\bar{b}_i\bar{c}_i} + \overline{\bar{a}_i b_i c_i} + \overline{a_i b_i \bar{c}_i} + \overline{a_i \bar{b}_i c_i}}$$
$$= \overline{\overline{(a_i + b_i + c_i)} + \overline{(a_i + \bar{b}_i + \bar{c}_i)} + \overline{(\bar{a}_i + \bar{b}_i + c_i)} + \overline{(\bar{a}_i + b_i + \bar{c}_i)}}$$
$$c_0 = \overline{\overline{\bar{a}_i\bar{c}_i} + \overline{\bar{b}_i\bar{c}_i} + \overline{\bar{a}_i\bar{b}_i}}$$
$$= \overline{\overline{(a_i + c_i)} + \overline{(b_i + c_i)} + \overline{(a_i + b_i)}}$$

FIGURE 9.3
Truth table for a full adder

a_i	b_i	c_i	s_i	c_0
0	0	0	0	0
0	0	1	1	0
0	1	0	1	0
0	1	1	0	1
1	0	0	1	0
1	0	1	0	1
1	1	0	0	1
1	1	1	1	1

The implementation of the expressions for s_i and c_0 using NOR gates is shown in Figure 9.5. It is also possible to implement the full adder by combining two half-adders with some NAND gates as shown in Figure 9.6.

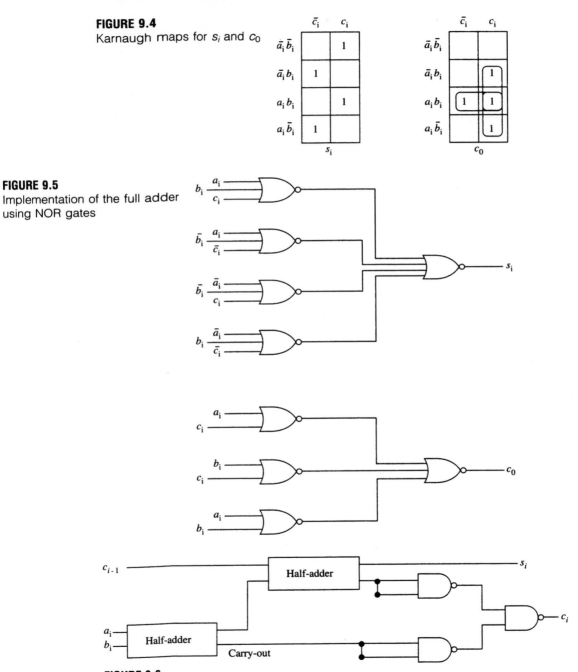

FIGURE 9.4
Karnaugh maps for s_i and c_0

FIGURE 9.5
Implementation of the full adder using NOR gates

FIGURE 9.6
Full adder constructed from half-adders

9.3 SERIAL AND PARALLEL ADDERS

Although a full adder can produce only a single sum bit and a single carry-out bit, it can be used to compute the sum of two binary numbers of any bit length. Figure 9.7 shows a circuit for adding two multibit binary numbers. The bits to be added are applied pair by pair to the a_i and b_i inputs of the adder. The carry-out bit produced by the adder is stored in a D flip-flop, and is used as the carry-in bit for the next pair of bits to be added. Thus, the circuit of Figure 9.7 performs serial addition and is consequently very slow.

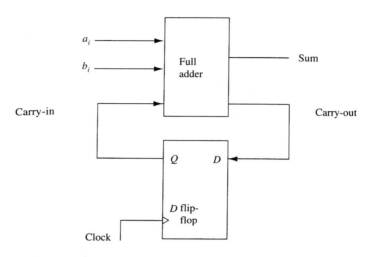

FIGURE 9.7
Multi-bit serial adder

■ EXAMPLE 9.1

Let us add the binary numbers $a_3a_2a_1a_0 = 1001$ and $b_3b_2b_1b_0 = 0101$. Assuming the D flip-flop has been initially cleared to the 0 state, the carry-in input of the full adder is 0. Thus, the input combination to the adder at the time t_0 is $a_0 = 1$, $b_0 = 1$, and carry-in $= 0$; this results in the sum output $s_0 = 0$ and carry-out $= 1$. A clock pulse is applied next, which changes the D flip-flop state to 1. Hence, the input combination to the adder at time t_1 is $a_1 = 0$, $b_i = 0$, and carry-in $= 1$. Proceeding in a similar manner the output of the serial adder corresponding to each input pair a_ib_i can be derived as follows:

Time	a_i	b_i	Carry-in (state of D flip-flop)	Sum	Carry-out
	LSB				
t_0	1	1	0	0	1
t_1	0	0	1	1	0
t_2	0	1	0	1	0
t_3	1	0	0	1	0
	MSB				
	9_{10}	5_{10}		14_{10}	

■

In applications where faster addition is required, an adder that implements parallel addition may be used. A parallel adder is constructed by cascading full adders so that the carry-out from the ith full adder is connected to the carry-in of the $(i + 1)$th adder. The number of adders required is equal to the bit length of the binary numbers to be added. Figure 9.8 shows a 4-bit parallel adder. Since the least significant adder FA_0 cannot have a carry-in, it can be replaced by a half-adder if desired, although in practice a full adder with the carry-in connected to the ground is used. It can be seen from Figure 9.8 that the sum bit S_3 can have a steady value only when its carry-in signal C_2 has a steady value; similarly, S_2 has to wait for C_1 and S_1 has to wait for C_0. In other words, the carry signals must ripple through all the full adders before the outputs stabilize to the correct values; hence, such an adder is often called a **ripple** adder. For example, if the following addition is to be performed

$$\begin{array}{cccc} a_3 & a_2 & a_1 & a_0 \\ b_3 & b_2 & b_1 & b_0 \end{array} \quad \begin{array}{cccc} 0 & 1 & 0 & 1 \\ 0 & 0 & 1 & 1 \\ & 1 & 1 & 1 \end{array} \leftarrow \text{Carry-in}$$
$$\overline{ 1 \ \ 0 \ \ 0 \ \ 0}$$

the carry-out generated from the least significant stage of the adder propagates through the successive stages and produces a carry-in into the most significant stage of the adder. The time required to perform addition in a ripple adder depends on the time needed for the propagation of carry signals through the individual stages of a parallel adder. Thus, parallel addition is not instantaneous. The larger the number of stages in a parallel adder the longer will be the carry propagation time, and consequently the slower the adder.

9.4 CARRY-LOOKAHEAD ADDERS

The long carry propagation time of the ripple adder can be overcome by an alternative implementation of the carry generation circuit known as **carry-lookahead** or **carry-anticipation.**

■ **EXAMPLE 9.2**

Let us consider the 4-bit parallel adder of Figure 9.8 to understand the principle of carry-lookahead. We can write the following equations for the carry-outs:

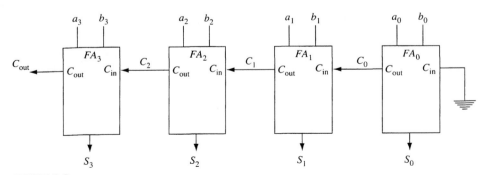

FIGURE 9.8
A 4-bit parallel adder

$$c_0 = a_0 b_0 \qquad (9.1)$$
$$c_1 = a_1 b_1 + (a_1 \oplus b_1) c_0 \qquad (9.2)$$

Substituting Eq. (9.1) into Eq. (9.2) we get

$$c_1 = a_1 b_1 + (a_1 \oplus b_1) a_0 b_0 \qquad (9.3)$$

In a similar manner we can write

$$c_2 = a_2 b_2 + (a_2 \oplus b_2) c_1 \qquad (9.4)$$

By utilizing Eq. (9.3), Eq. (9.4) becomes

$$c_2 = a_2 b_2 + (a_2 \oplus b_2)[a_1 b_1 + (a_1 \oplus b_1) a_0 b_0]$$

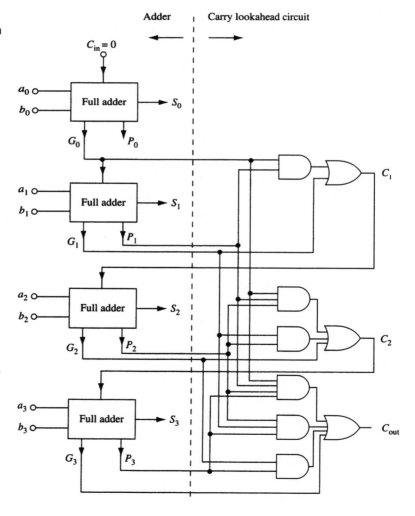

FIGURE 9.9
A 4-bit adder circuit with carry-lookahead

$$= a_2b_2 + (a_2 \oplus b_2)a_1b_1 + (a_2 \oplus b_2)(a_1 \oplus b_1)a_0b_0 \qquad (9.5)$$

Finally we can write

$$c_{\text{out}} = a_3b_3 + (a_3 \oplus b_3)c_2 \qquad (9.6)$$

which with Eq. (9.5) becomes

$$\begin{aligned} c_{\text{out}} &= a_3b_3 + (a_3 \oplus b_3)[a_2b_2 + (a_2 \oplus b_2)a_1b_1 + (a_2 \oplus b_2)(a_1 \oplus b_1)a_0b_0] \\ &= a_3b_3 + (a_3 \oplus b_3)a_2b_2 + (a_3 \oplus b_3)(a_2 \oplus b_2)a_1b_1 \\ &\quad + (a_3 \oplus b_3)(a_2 \oplus b_2)(a_1 \oplus b_1)a_0b_0 \end{aligned} \qquad (9.7)$$

Next we define P_i and G_i as the **carry-propagate** and **carry-generate** signals for the ith stage of the adder, where

$$P_i = a_i \oplus b_i \qquad (9.8)$$
$$G_i = a_i b_i \qquad (9.9)$$

P_i indicates that if $a_i = 0$, $b_i = 1$ or $a_i = 1$, $b_i = 0$, then the carry-in to the ith stage will be propagated to the next stage. G_i indicates that a carry-out will be generated from the ith stage when both a_i and b_i are 1, regardless of the carry-input to this stage.

Substituting Eqs. (9.8) and (9.9) in Eqs. (9.1), (9.3), (9.5), and (9.7), we get

$$\begin{aligned} c_0 &= G_0 \\ c_1 &= G_1 + G_0P_1 \\ c_2 &= G_2 + G_1P_2 + G_0P_2P_1 \\ c_{\text{out}} &= G_3 + G_2P_3 + G_1P_3P_2 + G_0P_3P_2P_1 \end{aligned}$$

Figure 9.9 shows the implementation of a 4-bit carry-lookahead adder. ∎

The propagation delay of the carry in the circuit of Figure 9.9 is independent of the number of bit pairs to be added and equal to the propagation delay of the two-level carry-lookahead circuit. In principle the circuit of Figure 9.9 can be extended to a large number of bit pairs; however, the complexity of the carry-generation equations for large number of stages makes it impractical.

9.5 CARRY-SAVE ADDITION

In a ripple carry or carry-lookahead adder, $m - 1$ additions are required to add m numbers; each addition except the first uses the accumulated sum and a new number. Thus, the total time for addition is $(m - 1)t_d$, where t_d is the time required to do each addition. In a carry-save addition the carry-out of a full adder is not connected to the carry-in of the more significant adder; instead, the sum outputs and the carry-outs of the full adders are stored as sum tuple S and carry tuple C respectively. The final sum is obtained by adding S and C using a carry-propagate adder. Let us illustrate the carry-save addition by adding three 4-bit numbers:

```
    A        1  1  0  0      = 12
    B        0  1  0  1      =  5
    C        1  0  1  1      = 11
           ─────────────
    S        0  0  1  0
    C     1  1  0  1
         ──────────────         ──
   Sum    1  1  1  0  0        28
```

Notice that the carry bits are shifted left to correspond to normal carry-propagation in conventional adders. Figure 9.10a shows the circuit for generating S and C for the above addition. Thus, a carry-save adder reduces three input tuples into two tuples, which constitute a set of sum bits and a set of carry bits. Figure 9.10b shows the symbol for a carry-save full adder.

A series of numbers can be added together by cascading an appropriate number of carry-save full adders. Figure 9.11 illustrates the addition of six 4-bit numbers using the carry-save approach. In general $m - 2$ carry-save additions are required to reduce the m numbers to be added to two numbers, and one ripple carry addition to generate the final sum. Therefore the total time for addition is $(m - 2)t_{cs} + t_d$, where t_{cs} and t_d are the time required to perform a carry-save addition and carry-propagate addition respectively.

*9.6 IMPLEMENTATION OF SYMMETRIC FUNCTIONS USING FULL ADDERS

Symmetric functions are a special class of combinational functions. These are functions that remain unchanged by any permutation of their variables. For example,

$$f(w,x,y,z) = wxy + wxz + wyz + xyz$$

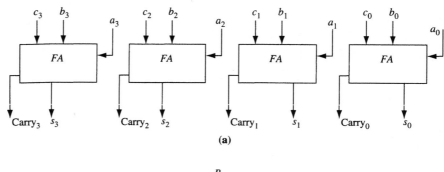

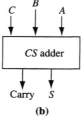

FIGURE 9.10

*9.6 IMPLEMENTATION OF SYMMETRIC FUNCTIONS USING FULL ADDERS □ 321

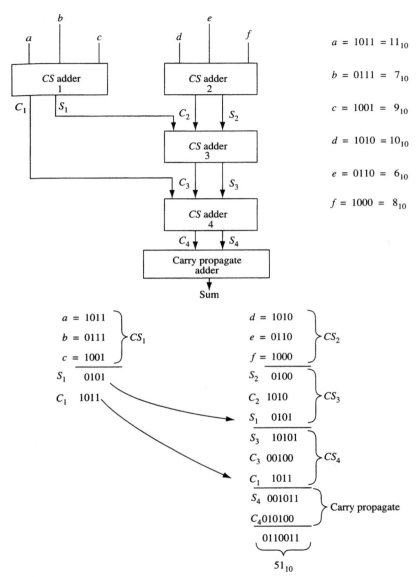

FIGURE 9.11
Carry-save addition of six 4-bit numbers

is a symmetric function, symmetric in w, x, y, and z. This can be verified from the Karnaugh map of the function by interchanging variables. If we interchange w and x in the Karnaugh map for f, an identical map results (Fig. 9.12a and b). Similarly, the interchanges of w and y, w and z, x and y, x and z, and y and z result in identical maps (as can be seen from Figs. 9.12c, d, e, f, and g respectively). A function that is symmetric for every pair of its variables is known as a **totally symmetric function**. Thus the function $f(w,x,y,z) = wxy + wxz + wyz + xyz$ is totally symmetric. It is also possible for a func-

FIGURE 9.12
Karnaugh maps obtained by interchanging variables

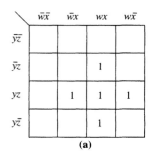

*9.6 IMPLEMENTATION OF SYMMETRIC FUNCTIONS USING FULL ADDERS

tion to be symmetric in only some of its variables. Such a function is called a **partially symmetric function**. For instance,

$$f(w,x,y,z) = w\bar{y} + x\bar{y} + wxz$$

is a partially symmetric function because it is symmetric only in variables w and x. It should be noted that the variables in which a function is symmetric are not necessarily in the uncomplemented form. For example

$$f(w,x,y,z) = w\bar{x} + w\bar{y}z + \bar{y}\bar{x} + xz$$

is symmetric in variables w and $\bar{y}$.

Many functions used in digital system design are totally symmetric (e.g., the sum and the carry function of a full adder, the majority function, and the parity function). A necessary and sufficient condition that a function $f(x_1, x_2, ..., x_n)$ be totally symmetric is that if and only if it can be specified by a set of integers $A = \{a_1, a_2, ..., a_k\}$ where $0 \leq a_j \leq n, j = 1, 2, ..., k$ such that the function has a value 1 if exactly a_j of the variables of symmetry are 1. The numbers in the set A are called ***a*-numbers**. Thus, the sum function and the carry function of a full adder

$$s(w,x,y) = \bar{w}\bar{x}y + \bar{w}x\bar{y} + w\bar{x}\bar{y} + wxy$$
$$c(w,x,y) = \bar{w}xy + w\bar{x}y + wx\bar{y} + wxy$$

are totally symmetric because $A = (1,3)$ for the sum function and $A = (2,3)$ for the carry function. This can be verified from the above expressions. The sum function has a value 1 if and only if exactly one or three variables are 1, and the carry function has a value 1 if exactly two or three variables are 1.

The **majority function** produces an output of 1 if more than half of the variables of the function are 1. Thus, a three-variable majority function

$$f(w,x,y) = wx + xy + wy + wxy$$
$$= wx + xy + wy$$

is totally symmetric because $A = (2,3)$

The even **parity function** $f(x_1, x_2, ..., x_n)$ has the value 1 if and only if an even number of input variables is 1; similarly, an odd parity function is 1 only if an odd number of input variables is 1. Thus, a three-variable (even) parity function

$$f(w,x,y) = \bar{w}xy + w\bar{x}y + wx\bar{y} + \bar{w}\bar{x}\bar{y}$$

is totally symmetric because $A = (0,2)$.

In general, a totally symmetric function $f(x_1, x_2, ..., x_n)$ is denoted as

$$S_{(a_1,a_2, ..., a_k)}(x_1,x_2, ..., x_n)$$

where $a_1, a_2, ..., a_k$ are the ***a*-numbers**. Thus,

$$S_{(2,3)}(w,x,y) = wx + xy + wy$$
$$S_{(0,2)}(w,x,y) = \bar{w}xy + w\bar{x}y + wx\bar{y} + \bar{w}\bar{x}\bar{y}$$
$$S_{(1,4)}(w,x,y,z) = w\bar{x}\bar{y}\bar{z} + \bar{w}x\bar{y}\bar{z} + \bar{w}\bar{x}y\bar{z} + \bar{w}\bar{x}\bar{y}z + wxyz$$

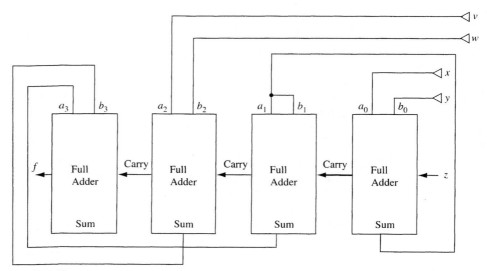

FIGURE 9.13
Implementation of the five-variable majority function using a 4-bit adder

A fully symmetric function can be implemented using full adders available as MSI chips. Figure 9.13 shows the implementation of the five-variable majority function

$$f(v,w,x,y,z) = \bar{v}\bar{w}xyz + \bar{v}w\bar{x}yz + \bar{v}wx\bar{y}z + \bar{v}wxy\bar{z} + v\bar{w}\bar{x}yz + v\bar{w}x\bar{y}z + v\bar{w}xy\bar{z} + vw\bar{x}\bar{y}z$$
$$+ vw\bar{x}y\bar{z} + vwx\bar{y}\bar{z} + \bar{v}wxyz + v\bar{w}xyz + vw\bar{x}yz + vwx\bar{y}z + vwxy\bar{z} + vwxyz$$

using a 4-bit adder chip.

Full adders can also be used to implement fully symmetric functions of a large number of variables. The adders are arranged to count the number of inputs that are 1 and generate the sum in binary form. Additional decoding circuitry is then used to produce the symmetric function output from this binary number.

EXAMPLE 9.3

Let us design a circuit corresponding to the totally symmetric function

$$f(v,w,x,y,z) = S_{(3,4,5)}(v,w,x,y,z)$$

The 5-bit input patterns are applied to the adders as shown in Figure 9.14. The values of the bits in a pattern are added to yield the binary sum $K_4K_2K_1$. The output f of the circuit is 1 when an input pattern generates a binary sum identical to the a-numbers of the totally symmetric function (i.e., $K_4K_2K_1 = 011$, 100, and 101 for the function we are considering). Thus,

$$f = \bar{K}_4K_2K_1 + K_4\bar{K}_2\bar{K}_1 + K_4\bar{K}_2K_1$$
$$f = \bar{K}_4K_2K_1 + K_4\bar{K}_2$$

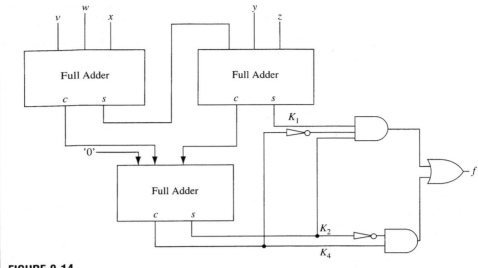

FIGURE 9.14
Implementation of $S_{(3,4,5)}(v,w,x,y,z)$

9.7 BCD ADDERS

It is often more convenient to perform arithmetic operations directly with decimal numbers, especially if the results of the operations are to be displayed directly in decimal form. Each decimal digit is usually represented by 4-bit 8-4-2-1 BCD code; thus, six combinations of the 4-bit code are not valid. When two BCD digits are added, the sum has a value in the range of 0 to 18. If the sum exceeds 9, an adjustment has to be made to the resulting invalid combination. This adjustment is made by adding decimal 6 (i.e., 110_2) to the result, which generates a valid sum as well as a carry-in to the next-higher-order digit.

Figure 9.15 shows the 20 possible sum digits that may result from the addition of two BCD digits and a carry-in. Whenever the sum digit is greater than 9 or the carry bit b_c is 1 for the unadjusted sum, the sum digit is adjusted by adding 6 to it. Consequently, a logic circuit that detects the condition for the adjustment and produces a carry-out C must be used. Such a circuit can be expressed by the following Boolean function:

$$C = b_c + b_8 b_4 + b_8 b_2$$

When $C = 1$, it is necessary to add the correction 0110 to the sum bits $b_8 b_4 b_2 b_1$ and to generate a carry for the next stage. The implementation of one stage of a BCD adder is shown in Figure 9.16. This requires two SN7483 4-bit adder chips, three NAND gates, and an inverter.

9.8 HALF-SUBTRACTORS

The half-subtractor circuit is used to implement a 1-bit binary subtraction. Figure 9.17(a) shows the truth table of a half-subtractor used to subtract Y (subtrahend) from X (minu-

	BCD Sum (without adjustment)					BCD Sum (with adjustment)				
Decimal	b_c	b_8	b_4	b_2	b_1	C	S	S	S	S
0	0	0	0	0	0					
1	0	0	0	0	1					
2	0	0	0	1	0					
3	0	0	0	1	1					
4	0	0	1	0	0					
5	0	0	1	0	1	adjustment not necessary				
6	0	0	1	1	0					
7	0	0	1	1	1					
8	0	1	0	0	0					
9	0	1	0	0	1					
10	0	1	0	1	0	1	0	0	0	0
11	0	1	0	1	1	1	0	0	0	1
12	0	1	1	0	0	1	0	0	1	0
13	0	1	1	0	1	1	0	0	1	1
14	0	1	1	1	0	1	0	1	0	0
15	0	1	1	1	1	1	0	1	0	1
16	1	0	0	0	0	1	0	1	1	0
17	1	0	0	0	1	1	0	1	1	1
18	1	0	0	1	0	1	1	0	0	0
19	1	0	0	1	1	1	1	0	0	1

FIGURE 9.15
Derivation of BCD sum digit

end) and generate the difference bit D and the borrow bit B. The Boolean expressions for the D and B outputs are derived from the truth table and are given by

$$D = \overline{X}Y + X\overline{Y}$$
$$= X \oplus Y$$
$$B = \overline{X}Y$$

9.9 FULL SUBTRACTORS

A full subtractor has three inputs X (minuend), Y (subtrahend), and Z (the previous borrow). The outputs of the full subtractor are the difference bit D and the output borrow B. The truth table of a full subtractor is shown in Figure 9.18a. The output bit D is obtained from the subtraction, $a_i - (b_i + c_i)$. The output bit B is 0 if $a_i \geq b_i$ provided $c_i = 0$. If

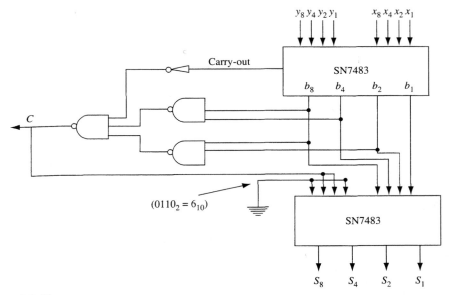

FIGURE 9.16
A single-stage 4-bit BCD adder

$c_i = 1$, output bit B is 1 if and only if $a_i \leq b_i$. The simplified Boolean expressions for outputs B and D are derived from their Karnaugh map plots as shown in Figure 9.18b.

FIGURE 9.17

X	Y	D	B
0	0	0	0
0	1	1	1
1	0	1	0
1	1	0	0

(a) Truth table for half-subtractor

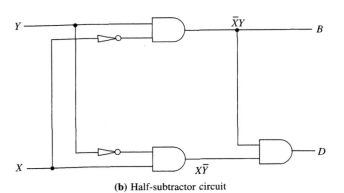

(b) Half-subtractor circuit

a_i	b_i	c_i	D_i	B_i
0	0	0	0	0
0	0	1	1	1
0	1	0	1	1
0	1	1	0	1
1	0	0	1	0
1	0	1	0	0
1	1	0	0	0
1	1	1	1	1

(a) Truth table for a full subtractor

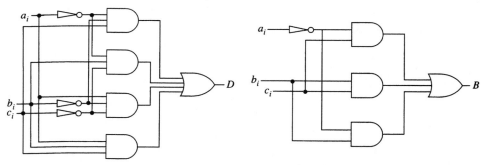

$D = \bar{a}_i\bar{b}_ic_i + \bar{a}_ib_i\bar{c}_i + a_i\bar{b}_i\bar{c}_i + a_ib_ic_i$

$B = \bar{a}_ic_i + \bar{a}_ib_i + b_ic_i$

(b) Karnaugh maps for outputs B and D

(c) Logic implementation for the full subtractor

FIGURE 9.18

The simplified expressions are

$$D_i = \bar{a}_i\bar{b}_ic_i + \bar{a}_ib_i\bar{c}_i + a_i\bar{b}_i\bar{c}_i + a_ib_ic_i$$
$$B_i = \bar{a}_ic_i + \bar{a}_ib_i + b_ic_i$$

The implementations of the expressions for D and B are shown in Figure 9.18c. Notice that the expression for the output D_i is identical to the expression for S_i in the full adder circuit (see Sec. 9.2). Furthermore the expression for B_i is similar to the carry-out expression C_0 in the full adder, except that the input variable a_i is complemented. Thus, a full adder can be used to perform the function of subtraction as well, by applying the complement of input a_i to the circuit which generates the carry output.

9.10 TWO'S COMPLEMENT SUBTRACTORS

All modern digital systems use 2's complement number systems. Subtraction in 2's complement is performed by 2's complementing the subtrahend and adding it to the minuend; any carry-out is ignored. If the sign bit of the resulting number is 0, the numerical part of the number is expressed in magnitude form. However, if the sign bit of the resulting number is 1, the numerical part of the number must be changed to 2's complement in order to get the correct magnitude.

A circuit of a 4-bit 2's complement subtractor is shown in Figure 9.19. The inverters are used to produce the 1's complement of the subtrahend, which is then added to the minuend. The carry-input of the low-order full adder is held at logic 1, which adds 1 in order to implement 2's complementation.

EXAMPLE 9.4

Let us use the circuit shown in Figure 9.19 to subtract $Y = +9$ (01001_2) from $X = +7$ (00111_2). The outputs of the inverter will be 10110. Then 00111 is added to 01001 along with carry-in = 1:

$$\begin{array}{r} 00111 \\ 10110 \\ 1 \\ \hline 11110 \end{array}$$

The 2's complement of the magnitude part gives the difference

$$0001 + 1 = 0010\ (2_{10})$$

Thus, the correct result is -2.

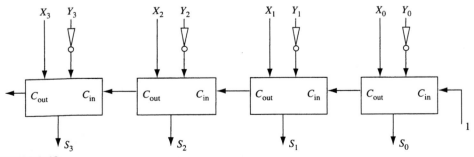

FIGURE 9.19
Two's complement subtractor circuit

■ **EXAMPLE 9.5**

Let us subtract $Y = -6(11010_2)$ from $X = +3(00011_2)$ by using Figure 9.19. The output of the inverter (i.e., 00101) is added to 00011 assuming the carry-in input is at 1:

$$\begin{array}{r} 00011 \\ 00101 \\ 1 \\ \hline 01001 \end{array}$$

Since the result is positive, it is not necessary to take the 2's complement of the magnitude part. Thus, the result of the subtraction is +9. ■

9.11 BCD SUBTRACTORS

Subtraction of two decimal digits can be carried out by using the BCD adder as described in Section 9.7. The 9's complement of the subtrahend is added to the minuend to find the difference. Thus, in addition to the BCD adder a small amount of circuitry is required in BCD subtraction. The 9's complement of a BCD digit can be obtained by subtracting the digit from 9. Figure 9.20 shows the 9's complement representation for the BCD digits. It can be seen from Figure 9.20 that a combinational circuit defined by the following Boolean expressions is required to derive the 9's complement of a BCD digit:

$$f_1 = \overline{b}_1$$
$$f_2 = b_2$$
$$f_4 = b_2 \oplus b_4$$
$$f_8 = \overline{b}_2 \overline{b}_4 \overline{b}_8$$

The logic diagram of the 9's complement circuit is shown in Figure 9.21. In practice the

Decimal Number	BCD				9's complement			
	b_8	b_4	b_2	b_1	f_8	f_4	f_2	f_1
0	0	0	0	0	1	0	0	1
1	0	0	0	1	1	0	0	0
2	0	0	1	0	0	1	1	1
3	0	0	1	1	0	1	1	0
4	0	1	0	0	0	1	0	1
5	0	1	0	1	0	1	0	0
6	0	1	1	0	0	0	1	1
7	0	1	1	1	0	0	1	0
8	1	0	0	0	0	0	0	1
9	1	0	0	1	0	0	0	0

FIGURE 9.20
The 9's complement of BCD digits

FIGURE 9.21
Circuit for a 9's complementer

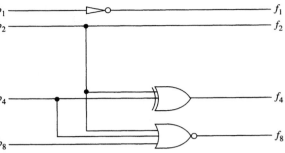

BCD subtractor can also be used as a BCD adder by incorporating a mode control input M to the 9's complement circuit such that

$$M = 0 \quad \text{Add operation}$$
$$M = 1 \quad \text{Subtract operation}$$

The expressions for the 9's complementer circuit are then modified as follows:

$$f_1 = M \oplus b_1$$
$$f_2 = b_2$$
$$f_4 = \overline{M}b_4 + M(b_2 \oplus b_4)$$
$$f_8 = \overline{M}b_8 + M \cdot \overline{b_2}\overline{b_4}\overline{b_8}$$

Figure 9.22 shows one stage of the BCD adder/subtractor circuit.

9.12 MULTIPLICATION

Multiplication schemes used in digital systems are quite similar to pencil-and-paper multiplication. An array of partial products is found first, and these are then added to gener-

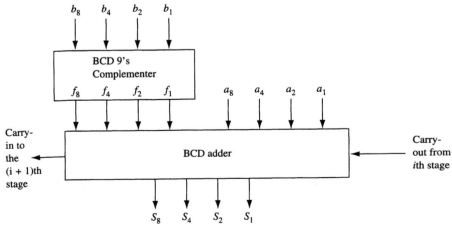

FIGURE 9.22
One stage of a BCD adder/subtractor unit

FIGURE 9.23
Binary multiplication example

```
      Multiplicand =     1 1 1 0
        Multiplier =     1 0 1 0
                         0 0 0 0
Partial product bits     1 1 1 0
                       0 0 0 0
                     1 1 1 0
          Product p = 1 0 0 0 1 1 0 0
```

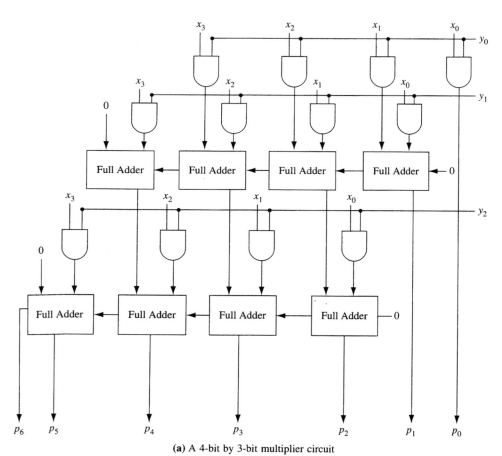

(a) A 4-bit by 3-bit multiplier circuit

FIGURE 9.24

ate the product. Figure 9.23 shows a simple numerical example of the multiplication. As shown in the diagram, the first partial product is formed by multiplying 1110 by 0, the second partial product is formed by multiplying 1110 by 1 and so on. The multiplication of two bits produces a 1 if both bits are 1; otherwise it produces a 0. The summation of the partial products is accomplished by using full adders. In general the multiplication of an m-bit multiplicand X ($=x_{m-1} \cdots x_1 x_0$) by an n-bit multiplier Y ($=y_{n-1} \cdots y_1 y_0$) results in an $(m + n)$-bit product. Each of the mn 1-bit products $x_i y_j$ may be generated by a 2-

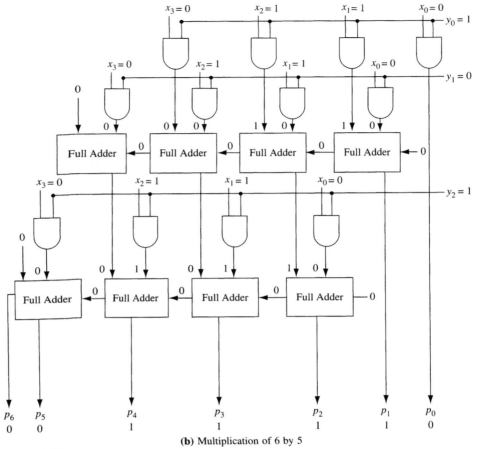

(b) Multiplication of 6 by 5

FIGURE 9.24
(continued)

input AND gate; these products are then added by an array of full adders. Figure 9.24a shows a 4-bit by 3-bit multiplier circuit.

EXAMPLE 9.6

Let us multiply $x_3x_2x_1x_0 = 0110(6_{10})$ by $y_2y_1y_0 = 101(5_{10})$ using the multiplier circuit of Figure 9.24a. The outputs of the AND gates and the full adders in the multiplier circuit corresponding to the applied input values are recorded in Figure 9.24b. Note that the outputs of the AND gates form the partial products and the outputs of the full adders form the partial sum during the multiplication process. The final output pattern is $p_6p_5p_4p_3p_2p_1p_0 = 0011110$ (i.e., 30_{10}), which is the expected result. One of the problems in performing multiplication using this scheme is that when many partial products are to be added, it becomes difficult to handle the carries generated during the summation of partial products. ∎

Multiplication can also be accomplished by means of an adder, a register, and two shift registers. The multiplicand, consisting of n bits, is placed in an n-bit register A. The multi-

334 ☐ CHAPTER 9 / ARITHMETIC CIRCUITS

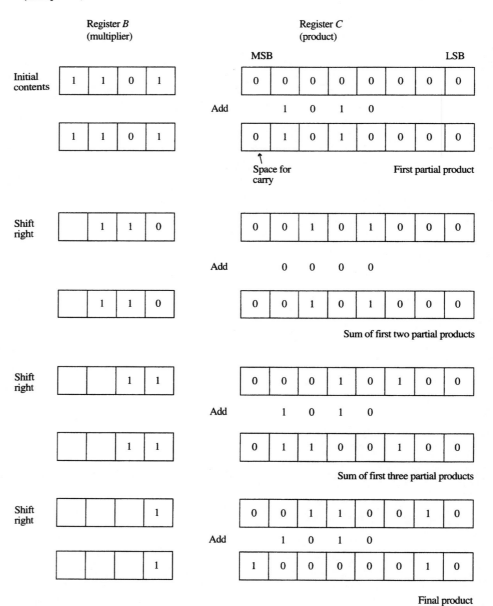

FIGURE 9.25
Multiplication of 10 by 13

plier, consisting of m bits, is loaded into an m-bit shift register B. Each step in multiplication consists of multiplying by either 1 or 0. The resulting partial product is added to the accumulated sum of the previously generated partial products stored in a $2n$-bit shift register C. The multiplication process begins by clearing the shift register C. Then the least significant

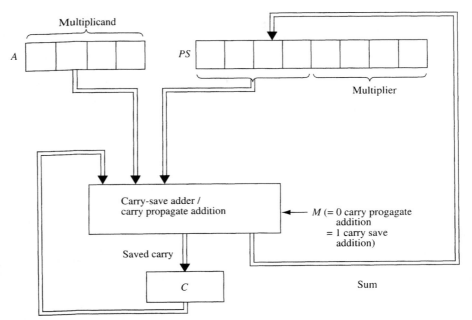

FIGURE 9.26
Carry-save multiplication for $n = 4$

bit of register B is tested. If it is 1, the content of register A and that of register C are added and the sum is placed in register C. Then register C is shifted right one bit position, which has the same effect as shifting the next partial product to the left. The process is repeated n times to get the final product stored in register C. Figure 9.25 illustrates the multiplication of a 4-bit multiplicand 1010 ($=10_{10}$) by a 4-bit multiplier 1101 ($=13_{10}$) using this procedure.

In applications where it is necessary to implement fast multipliers, alternative schemes have to be used. One such scheme is **carry-save multiplication**, which is based on the add-shift multiplication procedure described earlier. In this scheme the multiplicand is stored in an n-bit register A. The multiplier and the product share a $2n$-bit shift register PS, with the multiplier occupying the lower half of the shift register and the product, initially all 0's, occupying the upper half. An additional n-bit register C is used to store the saved carry bits. The carry-save multiplication scheme for $n = 4$ is shown in Figure 9.26. If the rightmost bit of the multiplier is 1 at any time, the contents of registers A, C, and the first n-significant bits of the shift registers are added together. The resulting sum bits are stored in the PS register, and the saved carry bits are stored in register C. The PS register is then shifted right. When the least significant bit in the multiplier is 0, there is no need to add all 0's to the previous partial product because this does not alter its value; nevertheless, register PS is shifted right. After n shift and addition operations, the contents of the carry-save register C and the first n significant bits of the shift register PS are passed through a carry propagate adder to produce the product. An example of carry-save multiplication is illustrated in Figure 9.27.

It should be noted that a separate carry-propagate adder is not needed to obtain the final result. The carry-save adder circuit of Figure 9.26 can be modified by incorporating multipliers to operate either in carry-save or in carry-propagate mode. Figure 9.28 shows the implementation of a 4-bit carry-save/carry-propagate adder.

FIGURE 9.27
Multiplication of 13 by 11

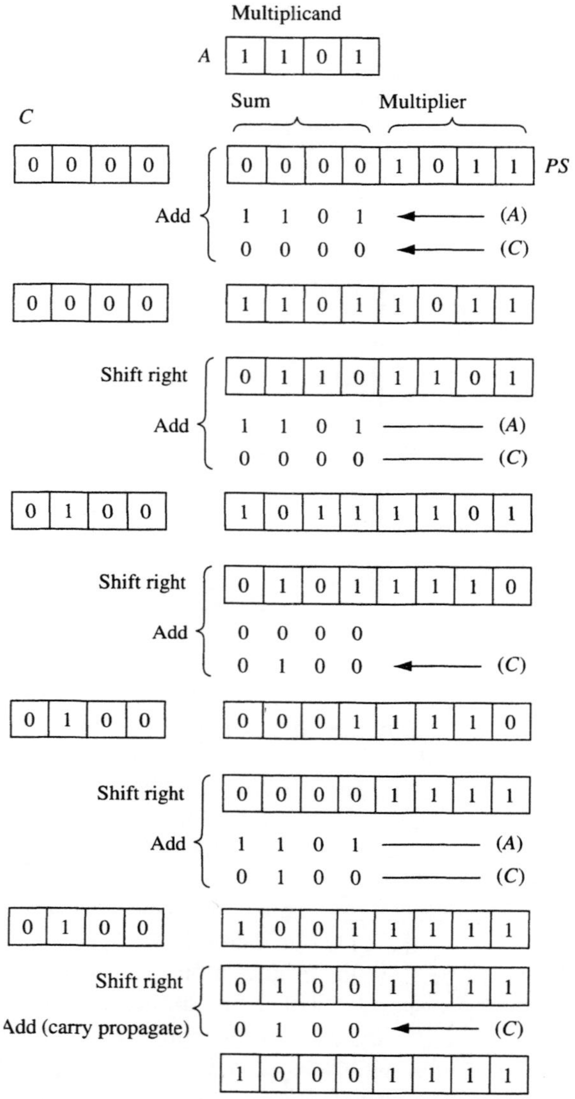

		Multiplicand	
A	1	1	0	1

C

						Sum			Multiplier		
0	0	0	0		0	0	0	0	1	0	1	1	PS

Add
    1 1 0 1 ← (A)
    0 0 0 0 ← (C)

| 0 | 0 | 0 | 0 | | 1 | 1 | 0 | 1 | 1 | 0 | 1 | 1 |

Shift right
| 0 | 1 | 1 | 0 | 1 | 1 | 0 | 1 |

Add
    1 1 0 1 ——— (A)
    0 0 0 0 ——— (C)

| 0 | 1 | 0 | 0 | | 1 | 0 | 1 | 1 | 1 | 1 | 0 | 1 |

Shift right
| 0 | 1 | 0 | 1 | 1 | 1 | 1 | 0 |

Add
    0 0 0 0
    0 1 0 0 ← (C)

| 0 | 1 | 0 | 0 | | 0 | 0 | 0 | 1 | 1 | 1 | 1 | 0 |

Shift right
| 0 | 0 | 0 | 0 | 1 | 1 | 1 | 1 |

Add
    1 1 0 1 ——— (A)
    0 1 0 0 ——— (C)

| 0 | 1 | 0 | 0 | | 1 | 0 | 0 | 1 | 1 | 1 | 1 | 1 |

Shift right
| 0 | 1 | 0 | 0 | 1 | 1 | 1 | 1 |

Add (carry propagate)
    0 1 0 0 ← (C)

| 1 | 0 | 0 | 0 | 1 | 1 | 1 | 1 |

9.13 DIVISION

Division is somewhat more complex than multiplication and involves repetitive shifting and subtraction. Figure 9.29 shows the hardware organization for implementing the division process for a positive dividend and positive divisor. The divisor is placed in register B. The n-bit dividend is loaded into an n-bit shift register Q, which is cascaded to another n-bit shift register A that initially contains all 0's. The output bits of A are connected as one input to the adder, and the other input of the adder is the 1's complement of the divisor (the EX-OR gates act as inverters). The division process can be summarized as follows:

9.13 DIVISION □ 337

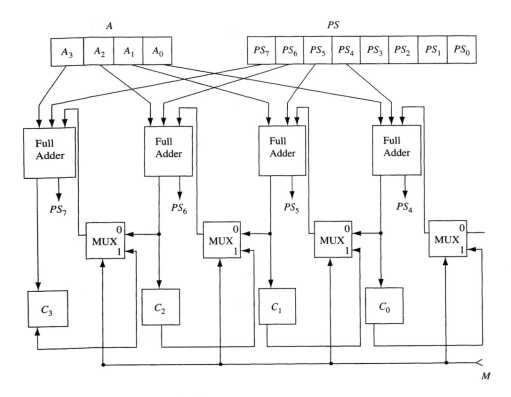

$M = 0$ carry-propagate mode
    $= 1$ carry-save mode

FIGURE 9.28
Implementation of a 4-bit carry-save/carry-propagate adder

Step 1: Shift the registers A and Q left 1 bit position

Step 2: Add A to the 1's complement of B (notice that $C_{in} = 1$; hence the adder output in fact corresponds to $A - B$).

Step 3: If the carry-out $C_{out} = 1$, the sum output produced by the adder is selected by the multiplexer and is loaded into A while C_{out} is loaded into the least significant bit of Q.
    Alternatively, if $C_{out} = 0$, the previous content of A is **restored** (i.e., loaded back into A) and at the same time C_{out} is loaded into the least significant bit of Q.

Step 4: Repeat steps 1 through 3 as many times as there are unused dividend bits in Q.

Step 5: The remainder is in register A and the quotient is in register Q.

■ **EXAMPLE 9.7**

Let us illustrate the division process with a numerical example. Suppose we want to divide $+19$ by $+3$. Then,

	A	Q	$B = 00011, \overline{B} = 11100$
	00000	10011	Initial values
	00001	00110	Shift
	11101		Subtract
$C_{out} = 0$	11110	00110	
	00001	00110	Restore; set $Q_0 = 0$
	00010	01100	Shift
	11101		Subtract
$C_{out} = 0$	11111	01100	
	00010	01100	Restore; set $Q_0 = 0$
	00100	11000	Shift
	11101		Subtract
$C_{out} = 1$	00001	11000	
	00001	11001	Set $Q_0 = 1$
	00011	10010	Shift
	11101		Subtract
$C_{out} = 1$	00000	10010	
	00000	10011	Set $Q_0 = 1$
	00001	00110	Shift
	11101		Subtract
$C_{out} = 0$	11110	00110	
	00001	00110	Restore; set $Q_0 = 0$

Remainder = +1, Quotient = +6

9.14 COMPARATORS

A comparator is used to determine the relative magnitudes of two binary numbers. It compares two n-bit binary numbers and produces three possible magnitude results at the outputs. Figure 9.30a shows the result of comparing two 1-bit numbers A and B. It is seen from the truth table that the various magnitude conditions are satisfied by the following Boolean expressions,

$$f_1 = A\overline{B}$$
$$f_2 = \overline{A}B$$
$$f_3 = \overline{A}\,\overline{B} + AB$$

These expressions can be realized using NAND gates as shown in Figure 9.30b. The Boolean expressions for the 1-bit comparator can be expanded to n-bit operands. In practice, however, it is usually simpler to use an MSI device such as the Signetics 7485 (a 4-bit TTL comparator). This compares two 4-bit numbers and gives $>$, $=$, and $<$ outputs as described earlier. Figure 9.31a shows the logic symbol for the 7485. The expansion inputs $I_{A<B}$ and $I_{A>B}$ must be connected to ground and $I_{A=B}$ to the supply voltage in order to compare two 4-bit numbers A and B.

Two 7485s can be cascaded as shown in Figure 9.31b to compare two 8-bit numbers. Notice that the outputs of the 7485 that handle the low-order bits are connected to

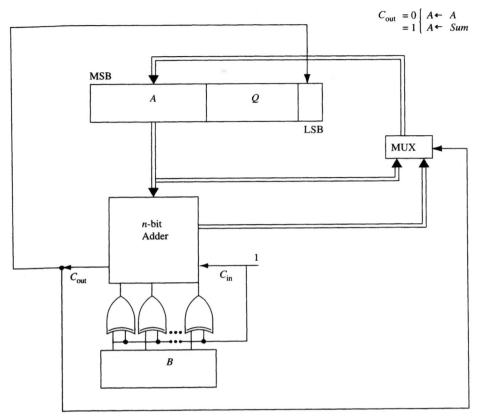

FIGURE 9.29
Division circuit

the expansion inputs of the 7485 that handle the high-order bits, with outputs being taken from the high-order comparator.

The CMOS 4-bit comparator (e.g., the National Semiconductor HC 85) is pin-for-pin compatible with the 7485. Both devices also operate with BCD codes.

FIGURE 9.30

A	B	f_1 (A > B)	f_2 (A < B)	f_3 (A = B)
0	0	0	0	1
1	0	0	1	0
0	1	1	0	0
1	1	0	0	1

(a) Comparison of two 1-bit numbers

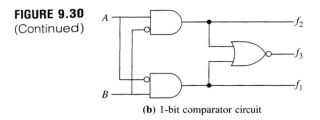

(b) 1-bit comparator circuit

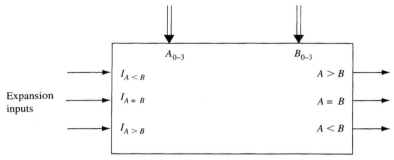

(a) Logic symbol for the Signetics 7485

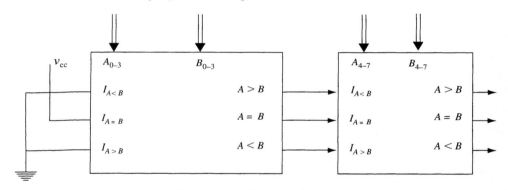

(b) Comparison of two 8-bit numbers

FIGURE 9.31

EXERCISES

1. Design a full adder using two 4-to-1 multiplexers only.
2. Determine the variables of symmetry and the a-numbers for any of the following functions that is symmetric.
   a. $f(w, x, y) = \overline{w} \cdot \overline{z} + xz + w\overline{x}$
   b. $f(w,x,y,z) = x\overline{y}\overline{z} + \overline{w}\overline{y}z + w\overline{x}\overline{y} + \overline{w}\overline{x}y + w\overline{x}\overline{z} + \overline{w}y\overline{z}$
   c. $f(w,x,y,z) = \overline{w}\overline{x}\,\overline{y}z + w\overline{x}\,\overline{y}z + w\overline{x}yz + \overline{w}x\overline{z} + \overline{w}xy + xy\overline{z}$
3. Implement the following symmetric function using full adders only
$$S_{2,4,6}(t,u,v,w,x,y,z)$$

4. Design an 8-bit ripple carry adder using the 1's complement form to represent negative numbers.
5. Design a circuit capable of adding two 4-bit numbers such that its output is the mod-5 sum of the inputs.
6. Design a 4-bit adder/subtractor circuit such that when a select input x is set at logic 0 the circuit adds, and when x is set at logic 1 the circuit subtracts.
7. Using the principle of carry-save addition, multiply the following pairs:
   a. 11011 and 10010 **b.** 1010111 and 1100110
8. A circuit to compare two 4-bit numbers (A_{3-0} and B_{3-0}) is to be designed. The status of the comparison is available on the outputs EQ, NE, LT, and GT as shown:

   If $A = B$, then $EQ = 1$, $NE = 0$, $LT = 0$, $GT = 0$
   If $A \neq B$, then $EQ = 0$, $NE = 1$, $LT = $ —, $GT = $ —
   If $A < B$, then $EQ = 0$, $NE = 1$, $LT = 1$, $GT = 0$
   If $A > B$, then $EQ = 0$, $NE = 1$, $LT = 0$, $GT = 1$

   Sketch the logic diagram of the circuit.
9. Design a 24-bit adder that uses six 4-bit adder circuits, carry-lookahead circuits, and additional logic to generate C_8 and C_{16} from carry-in, carry-generate, and carry-propagate variables.
10. Prove that if two 2's complement numbers are added, the overflow bit is the EX-OR of the carry-in and carry-out of the most significant bit.
11. Show how a 4-bit adder can be used to convert 5-bit BCD representation of decimal numbers 0 to 19 to 5-bit binary numbers.
12. Design a circuit, using full adders only, to multiply a 4-bit number by decimal 10.
13. Show how an 8-bit number ($x_7 \ldots x_0$) can be multiplied by a 4-bit number ($x_3 \ldots x_0$) using four 2-bit multipliers and 6-bit adders.
14. Using the division technique discussed in the text, perform a/b for the following:
    a. $a = 19, b = 5$ **b.** $a = 26, b = 11$

    Show each iteration during the division process.

Further Reading

1. J. F. Cavanaugh, *Digital Computer Arithmetic*, McGraw-Hill, 1984.
2. K. Hwang, *Computer Arithmetic*, Wiley, 1979.

10
Digital Integrated Circuits

The progress in semiconductor technology has contributed to the rapid growth in the density of transistors on a chip. Two basic technologies are used in fabricating chips: **bipolar** and **MOS** (metal oxide semiconductor). The bipolar technology in general provides higher speed, whereas the MOS technology provides higher levels of integration. Digital circuits constructed using MOS technology contain only MOS field-effect transistors (MOSFETs); similarly, bipolar digital circuits contain only bipolar transistors. Both bipolar and MOS transistors are operated as **switches** that generally have two states: ON or OFF, conducting or non-conducting. In this chapter we limit our discussion to the important logic families based on MOS and bipolar transistors.

10.1   NMOS TRANSISTORS

The structure and circuit symbol of an NMOS transistor are shown in Figure 10.1. It consists of two separate regions, the **source** and the **drain**, which are diffused into a p-type substrate; the region between the source and the drain under the oxide layer is called the **channel**. A thin layer of silicon oxide is formed on the surface, and holes are cut into it

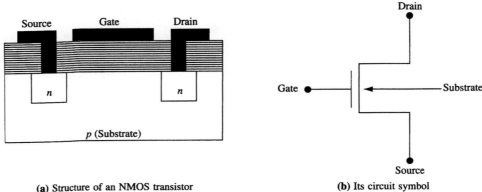

(a) Structure of an NMOS transistor   (b) Its circuit symbol

FIGURE 10.1

to permit access to the source and the drain. A conducting material made of poly-silicon, called the **gate**, is then deposited on the oxide layer covering the channel region.

The substrate and the source of an NMOS transistor are usually connected to ground. When the gate-to-source voltage is less than or equal to 0, the channel between the source and the drain is not established; hence, the transistor remains in a nonconducting condition. The drain-to-source current can flow only if the gate-to-source voltage reaches a certain level, called the **threshold voltage**; for NMOS transistors the threshold voltage is 1 V (for a drain voltage of 5 V).

The NMOS transistor that we have just described conducts only in the presence of the gate voltage. This type of NMOS transistor is called an n-channel **enhancement-mode** transistor and is frequently used in logic circuits. Another type of n-channel MOS transistor, called the **depletion-mode** transistor, has a permanent channel of n-type silicon between the source and the drain. Thus, the transistor is normally conducting, and requires the application of a negative voltage (-4 V or less) to the gate in order to prevent the drain-to-source current flow. Figure 10.2 shows the structure and the symbol of an n-channel depletion-mode transistor.

It is clear from this discussion that the NMOS transistor is essentially an on/off switch. The switch can be turned off by setting the gate-to-source voltage to 0 V (assuming an enhancement-mode transistor) and turned on by setting the gate-to-source voltage equal to the drain voltage (5 V). The switch representation of the NMOS transistor is shown in Figure 10.3. The source and drain are connected if the input voltage (i.e., the gate-to-source voltage) is equal to the drain voltage, and the source and drain are disconnected if the input voltage is 0 V.

10.2 NMOS INVERTERS

The basic circuit element used throughout NMOS integrated circuit design is the inverter. Figure 10.4 shows an inverter circuit constructed by connecting a single NMOS transistor in series with a resistor. The output appears between the resistor and the transistor. The transistor acts as a switch either to make or break the connection between the output and the ground (0 V), depending on the voltage at the input of the transistor. The supply voltage V_{DD} (=5 V) is applied via the resistor R.

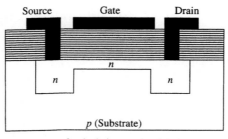

(a) Structure of a depletion-mode NMOS transistor

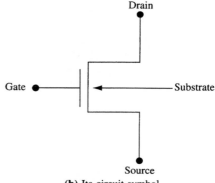

(b) Its circuit symbol

FIGURE 10.2

10.2 NMOS INVERTERS 345

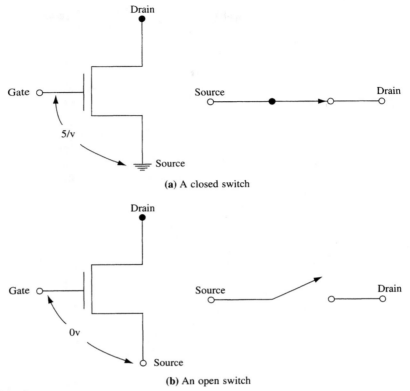

FIGURE 10.3
NMOS transistor

If the voltage at the gate of the transistor is 0 V, the transistor is off (i.e., the connection between the output and the ground is broken). Thus, there is no current flow from V_{DD} through the load resistor R, and consequently there is no voltage drop across the resistor; the output is essentially connected to the supply voltage V_{DD}. In other words when the input is low, the output is high.

FIGURE 10.4
NMOS inverter circuit

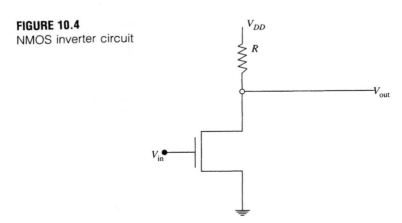

If the input voltage to the transistor is equal to V_{DD}, the transistor is turned ON, and a large current flows through the resistor R, resulting in a substantial voltage drop across it. The output of the circuit is now connected to ground through the low resistance of the transistor, and thus the output falls to 0 V. Hence, a high input voltage produces a low output voltage, behaving precisely as an inverter.

In actual NMOS circuits the load resistor R is replaced by a depletion-mode MOS transistor as shown in Figure 10.5. This is because a resistor occupies a very large area on a chip. The depletion-mode MOS transistor is referred to as the **load** transistor, whereas the other transistors are termed as **drivers**. As mentioned before, a depletion-mode NMOS transistor is normally ON and requires a voltage of about -4 V to turn it OFF. Since in Figure 10.5 the gate of the depletion mode transistor T_L is connected to the source, it is always ON. The driver transistor, T_1, is either ON or OFF depending on the input voltage. In order to make sure that the output voltage of the inverter remains close to either V_{DD} (1) or ground (0) and not halfway between V_{DD} and ground, the physical dimensions of the load and the driver transistors have to be chosen such that

i. the ON resistance of the driver transistor is significantly smaller than the ON resistance of the load transistor.
ii. the OFF resistance of the driver transistor is significantly higher than the ON resistance of the load transistor.

In the first case the output voltage of the inverter will be close to 0 V, whereas in the second case the output voltage will be close to 5 V. An appropriate choice of the channel width (W) and the channel length (L) of the drivers and load transistors, will satisfy the conditions required for proper circuit operation. The **L/W ratio** of the load transistor is usually 4 to 6 times that of the driver transistor [1]. Since the inverter operation depends on the ratio of the transistor geometries, this version of the inverter circuit is known as a **ratioed inverter.**

The speed of the NMOS inverter is heavily dependent upon the load capacitance associated with its output, as illustrated in Figure 10.6. A high input voltage V_{in} to the driver transistor will turn it ON. Consequently, the output voltage V_{out} will almost be 0 V.

FIGURE 10.5
Inverter circuit with NMOS load device

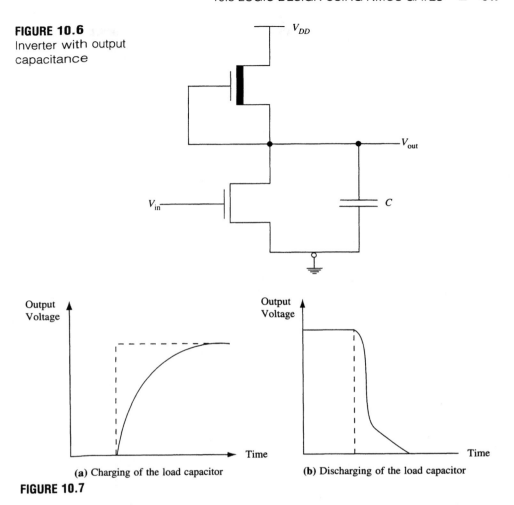

FIGURE 10.6
Inverter with output capacitance

(a) Charging of the load capacitor

(b) Discharging of the load capacitor

FIGURE 10.7

When the input voltage changes to a low, the connection between the output and the ground is broken, since the OFF resistance of the driver transistor is extremely large. The output at this point is 0 V. The ON resistance of the load transistor pulls the output to a high state by charging the capacitor C from V_{DD} through the resistance. Thus, the rate at which the output voltage changes is dependent upon the ON resistance of the load transistor, and reaches high gradually as shown in Figure 10.7a. Next, when the input voltage changes from low to a high, the driver transistor conducts and the load capacitor C discharges through the ON resistance of the transistor. Since the ON resistance of the driver transistor is smaller than the ON resistance of the load transistor, the change in the output voltage from the high to a low state is much faster (Fig. 10.7b) than that in Figure 10.7a.

10.3 LOGIC DESIGN USING NMOS GATES

Basic Gates

Basic logic gates such as NAND and NOR can be constructed by replacing the driver transistor of the inverter with parallel and serial connection of transistors respectively. Fig-

FIGURE 10.8
A two-input NMOS NAND gate and its truth table

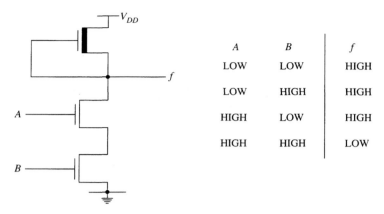

ure 10.8 shows a two-input NAND gate. When inputs A and B are both high the path between the output and the ground is closed, and thus the output is pulled low. Since both the driver transistors are in series, their effective ON resistance is doubled. This means that the resistance of the load transistor must be doubled in order to maintain the same voltage level as that in an inverter. Alternatively, the resistance of each of the driver transistors can be halved by doubling the width of the channel, so that the resistor ratio of the inverter is satisfied.

Figure 10.9 shows the implementation of a two-input NOR gate with NMOS transistors. If either of the inputs A and B is high the corresponding transistor is turned ON, and hence there will be a path from the output to ground, resulting in a low voltage at the output. Since the driver transistors are in parallel, the total ON resistance (i.e., when both the driver transistors are ON), is equal to the resistance of a single driver transistor. In

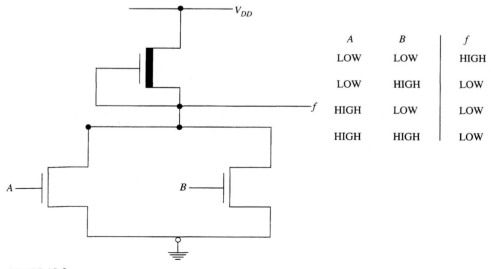

FIGURE 10.9
A two-input NMOS NOR gate and its truth table

10.3 LOGIC DESIGN USING NMOS INVERTERS   349

consequence, a NOR gate is as fast as the corresponding inverter. Moreover, a NOR gate is more efficient than a NAND gate in terms of the area occupied. In general, the area occupied by an n-input NAND gate is roughly of the order of n^2, whereas this is of the order of n for an n-input NOR gate [2]. For these reasons the NOR gate is preferred to the NAND gate in NMOS logic design.

A powerful feature of NMOS design is that with the combinations of series and parallel transistors it is possible to implement arbitrary Boolean expressions directly. For example, the NMOS implementation of $f(A,B,C) = C + (\bar{A} + B)(A + \bar{B})$ is shown in Figure 10.10.

It must be remembered, however, that the series combinations of transistors add their ON resistances; hence, these transistors must be increased in size in order to lower their individual ON resistances.

Dynamic Logic

In the NMOS logic circuits we have considered thus far, the output is low or high depending on whether or not a current flows from the V_{DD} supply to ground through driver

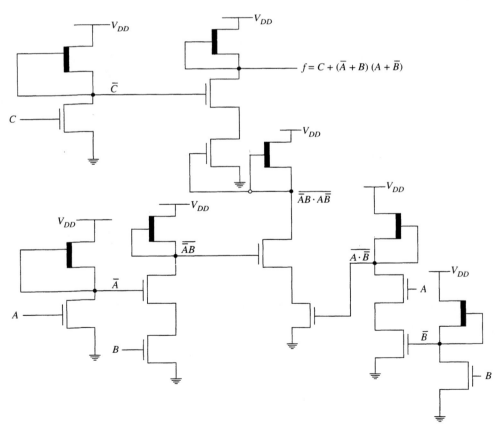

FIGURE 10.10
NMOS transistor implementation of the Boolean expression $C + (\bar{A} + B)(A + \bar{B})$

transistors. Due to circuit load capacitance, the voltages representing the logic high and low values will take a finite amount of time to reach their final values. Thus, the load capacitance only contributes to the speed of operation of a circuit and has no significance as far as the representation of logic values is concerned. Such a circuit is called a **static NMOS logic** circuit. In **dynamic NMOS** logic circuits the load transistor is turned ON or OFF in synchronization with a clock pulse. During the clocking period, the load capacitance is used to retain information in the form of the presence or absence of electric charge. Since the current is drawn only when the load transistor is switched ON, power dissipation is low. Furthermore, the area taken up by the dynamic NMOS circuits is smaller in comparison to the static NMOS circuits, so more complex circuits can be built in a given area.

Two-Phase Ratioed Logic

This is the simplest form of NMOS dynamic logic. The basic characteristic of such circuits is that when both the load and the driver transistor(s) are ON, a potential divider is formed with the voltage level at the output depending on the ratio of load transistor ON resistance to that of the driver transistor(s). Figure 10.11 represents a two-phase ratioed NMOS shift register. It consists of two cascaded inverters connected through switches called **pass transistors** (T_3 and T_6). Notice that the load transistors T_2 and T_4 are enhancement type rather than depletion type. Clk 1 and Clk 2 are two nonoverlapping clocks. When Clk 1 is high, load transistor T_2 conducts, and if V_{in} is greater than the threshold voltage of T_1, current will flow in T_1, resulting in a low-voltage output from the first inverter. Pass transistor T_3 is ON when the Clk 1 is high; thus, the output of inverter 1 (i.e., a logic 0) will be transferred to the input of inverter 2. On the other hand, if V_{in} is less than the threshold voltage of T_1, current does not flow in T_1; hence it is OFF. Consequently, the output of inverter 1 will be high. This charges up capacitance C_1, which temporarily remains charged even after Clk 1 returns to low. Then Clk 2 becomes high, turning T_4 and T_6 ON. Since the gate of T_5 is positive with respect to its source, it too is turned ON. This pulls the output of inverter 2 low, and hence a logic 0 is transferred through T_6 to V_{out}.

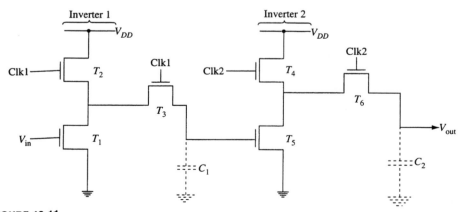

FIGURE 10.11
Two-phase ratioed inverters connected in series

10.3 LOGIC DESIGN USING NMOS INVERTERS

The major advantage of this circuit is that the current flows from V_{DD} to ground only when the load and the drive transistors of the inverters are ON. In the corresponding circuit consisting of static inverters, current flows whenever the drive transistors are ON. Thus, power consumption in the dynamic circuit is much smaller than its static counterpart.

Two-Phase Ratioless Logic

The main drawback of ratioed NMOS designs is that it is necessary to keep a proper ratio between the sizes of load and driver transistors for adequate operation. On the other hand, ratioless circuits depend on the charge stored in a capacitor rather than relying on the voltage divider of the ratioed circuit. Consequently, all transistors can be of equal size, thus permitting more logic within a given area.

There are a number of techniques for realizing ratioless circuits; these techniques differ in the arrangements of clocking and power supplies [3]. Figure 10.12 represents two ratioless inverters connected in cascade. During the time interval when Clk 1 is positive, capacitor C_1 is charged to the level of the input voltage V_{in}. If V_{in} is at logic 1, T_3 will be turned ON, discharging any charge on C_2. When Clk 2 is positive, T_4 is turned ON, discharging C_3 through T_3 and turning T_6 OFF. During the same interval, C_4 is charged to logic 1 through T_5. At the next phase of Clk 1, because the input to T_6 is low, C_4 stays at logic 1 and passes it on to the next stage through T_7. Notice that the capacitor charging time in this configuration is determined by the length-to-width ratio of the load transistor alone, and the discharging time is dependent only on the length-to-width ratio of the driver transistor. Thus, this configuration is ratioless.

The concept of two-phase clocking can be extended to circuit configurations using any number of clocks; in practice, however, the number of clocks is restricted to four. Four-phase logic can operate at higher speed than two-phase logic, while maintaining low power consumption. However, four-phase logic has the disadvantage of being more sensitive to noise.

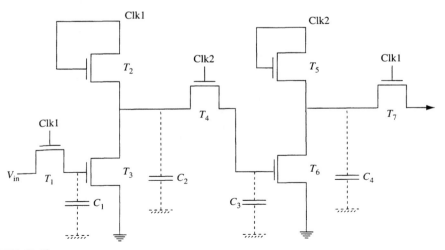

FIGURE 10.12
Two-phase ratioless inverters connected in series

10.4 CMOS LOGIC

CMOS technology is increasingly being used to design high-density chips. The notable advantages of CMOS are low power consumption and high noise immunity (noise immunity is the maximum noise voltage that can be tolerated). There are several ways to process CMOS circuits, each offering its own special advantage. The basic problem is to create both n-channel and p-channel transistors on a single silicon substrate. Figure 10.13 shows the structure of a CMOS inverter. A p-channel transistor is created in an n-type substrate in the normal way; however, an n-channel transistor requires an island of p-type material. The source and the substrate of the p-channel transistor are connected to V_{DD}, whereas the source and the substrate of the n-channel transistor are connected to ground. V_{in} is applied to the gates of both transistors simultaneously. Notice that both the transistors are enhancement-mode transistors and the p-channel transistor is employed as the load element. The schematic circuit representation of the inverter is shown in Figure 10.14.

When V_{in} is high, the n-channel transistor conducts and the p-channel transistor becomes nonconductive. When V_{in} is low, the p-channel transistor becomes conductive and the n-channel transistor does not conduct. Because only one transistor conducts at any given time, there is little current flow from V_{DD} to ground through the transistors and power consumption is low. It should be noted that both the transistors are partially ON during the switching operation itself; however, this happens for only a fraction of the operating interval, so the current flow is in the microampere range. The transfer curve (i.e., the plot of V_{out} against V_{in}) of the inverter is illustrated in Figure 10.15.

The basic CMOS NAND and NOR gates are shown in Figures 10.16a and b. A low input voltage on input A in the NAND circuit turns transistor T_1 ON and turns transistor T_3 OFF. Because no current can flow through T_3, the output voltage approaches V_{DD}. Similarly, a low input voltage on input B results in a high output voltage. Only when there

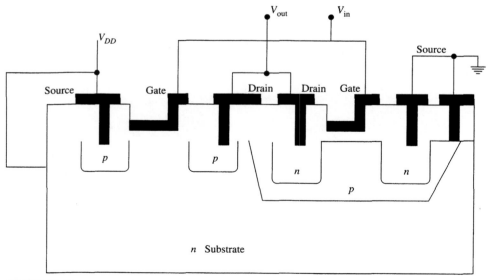

FIGURE 10.13
Physical structure of a CMOS inverter

FIGURE 10.14
Circuit symbol for CMOS inverter

are high input voltages on both inputs A and B does the current flow from V_{DD} to ground and the output go low.

The CMOS NOR circuit (Fig. 10.16b) is implemented by configuring the p-channel transistors in series and the n-channel transistors in parallel. A high input voltage on either input causes one of the p-channel transistors to be OFF and one of the n-channel transistors to be ON, resulting in a low output voltage. Only if both the input voltages are low do both the p-channel transistors turn ON, with both n-channel transistors OFF, to provide a high output.

Transmission Gates

One of the important advantages of CMOS circuits is that it enables the construction of a nearly perfect switch—the **transmission gate**. It consists of a p-channel transistor connected in parallel with an n-channel transistor as shown in Figure 10.17a. The transistor sources are connected to the input and their drains are connected to the output. A control voltage c is applied to the gate of the n-channel transistor, and its inverted value $\bar{c}$ is applied to the gate of the p-channel transistor.

FIGURE 10.15
CMOS transfer curve

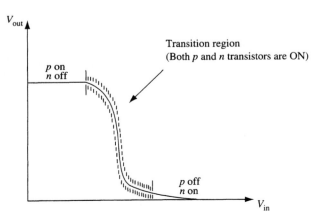

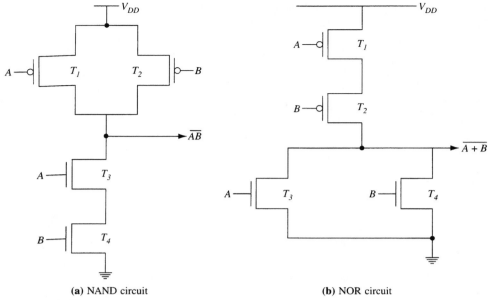

(a) NAND circuit   **(b)** NOR circuit

FIGURE 10.16

When c is at logic 0, both transistors are nonconducting; thus, the output is disconnected from the input. On the other hand, if c is at logic 1, the p-channel transistor transfers a high input voltage to the output, whereas the n-channel transistor transfers a low input voltage to the output. Thus, as long as the control voltage c is high, the input is transmitted to the output. A symbol for the transmission gate is shown in Figure 10.17b. It should be understood that the transmission gate can transfer signals in both directions, although one end is arbitrarily labeled **input** and the other **output**. The behavior of the transmission gate is summarized in Figure 10.18. It can be seen from Figure 10.18 that the transmission gate is a **tristate device**. In other words, it has three possible outputs—open circuit, low, and high. However, it only has two logic levels, since open circuit really means high impedance and is not a logic level.

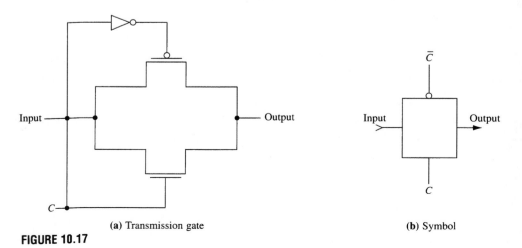

(a) Transmission gate   **(b)** Symbol

FIGURE 10.17

FIGURE 10.18
Input-output relationship of the transmission gate

Control Input c	Input	Output
LOW	LOW	Open circuit
LOW	HIGH	Open circuit
HIGH	LOW	LOW
HIGH	HIGH	HIGH

Figure 10.19 illustrates a typical application of transmission gates. The outputs of four transmission gates are tied together to a common line, Z. It is desired to transmit the signals A, B, C, and D one at a time to line Z. This can be accomplished by making the control input of only one transmission gate at a time high while keeping the other three low. The particular transmission gate to be used can be selected by applying the appropriate input to the 2-to-4 decoder. Several catalog devices consisting of four independent transmission gates are available (e.g., RCA's CD 4066). Each transmission gate in this device has its own active high control line that turns the gate ON when high, OFF when low.

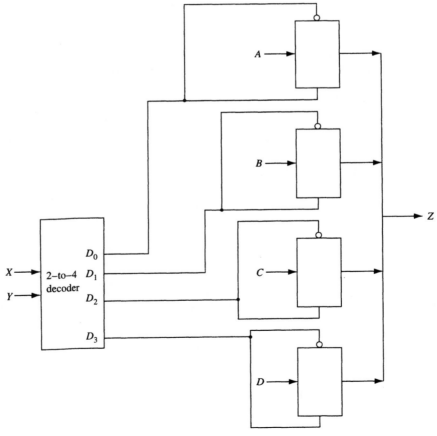

FIGURE 10.19
Four signals connected to a common line using transmission gates

Figure 10.20 shows the implementation of tristate buffers in National Semiconductor's MM54HC240/MM74H device. When the control input is high, the input to the p-transistor is high, which makes it nonconductive. The output of the NOR gate is low, which forces the n-channel transistor to also be nonconductive. Since both transistors are nonconducting, the output of the buffer will be in a high impedance state. When the control input is low, a high on the input will turn the n-channel transistor ON, causing the output to be low, whereas a low on the input will turn the p-channel transistor ON, causing the output to be high.

Clocked CMOS Circuits

The discussion of CMOS has so far concentrated on fully complementary circuits in which each gate consists of a pair of nMOS and pMOS transistors. The problem with the fully complementary approach is that for complex circuits a significant amount of chip area is wasted. This area penalty can be avoided by using clocked CMOS logic. Figure 10.21 illustrates such a circuit. The basic feature of all clocked CMOS circuits is that the output node is **precharged** to V_{DD} when the clock is 0. The inputs of the circuit can be changed only during the precharge phase. When the clock goes to 1, the path to V_{DD} is opened and the path to ground is closed. Therefore, depending upon the input conditions, the output will either remain high or will be pulled down during this phase, which is known as the **evaluate** phase. For example, in Figure 10.21, the output Z is precharged to 1 during the time when the clock = 0. During the evaluate phase, the output Z will be pulled to ground if the function $[(A + C)B + D(E + F)] = 1$; otherwise, it will remain at 1.

The advantage of a clocked CMOS circuit is that it uses only an n-network, together with a p-transistor and an n-transistor. This results in the reduction of the load capacitance, with a consequent increase in speed. However, there are several disadvantages associated with dynamic CMOS circuits (e.g., the inputs must be changed during the precharge phase, and multiple stages cannot be cascaded to realize a function) [4, 5].

CMOS Domino Logic

The CMOS domino circuits and the clocked CMOS circuits have some common characteristics. Figure 10.22 illustrates a domino circuit. When the clock signal is 0, transistor T_1 is switched ON and transistor T_7 is switched OFF. Alternatively, T_1 is OFF and T_7 is ON when the clock signal is 1. Thus, as in clocked CMOS logic, the output is precharged

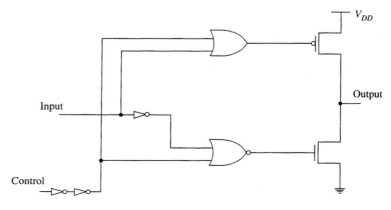

FIGURE 10.20
Tristate output of an inverting buffer

10.4 CMOS LOGIC 357

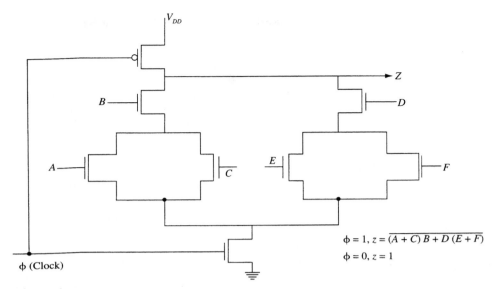

FIGURE 10.21
Clocked CMOS logic

high if the path to ground is open and the precharge is stopped if the path to ground is closed. The output of the clocked CMOS stage is connected to a static CMOS buffer, which feeds all subsequent logic stages. During the precharge phase the dynamic stage

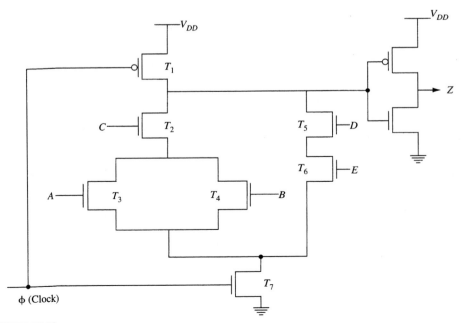

FIGURE 10.22
A domino logic network

has a high output, so the output of the buffer will be 0. This means that all transistors in the subsequent logic stages will be turned OFF during the precharge phase. In addition, the clocked part of the circuit can only make a high-to-low transition during the evaluate phase; therefore, the buffer output can only change from low to high. As a result, the output of the circuit will be hazard-free and cannot change again until the next precharge phase. Many circuits, such as the one in Figure 10.22, may be cascaded to realize a function in which data is transferred from one stage to another like a series of falling dominos; hence the name **domino** logic. Figure 10.23 represents a two-stage domino CMOS circuit. During the precharge phase nodes 1 and 3 are high, and nodes 2 and 4 (output node) are low. Let us assume that the inputs are $A = 1$, $B = 1$, $C = 0$, and $D = 1$. During the evaluate phase, node 1 goes low, which makes node 2 go high. Since one of the inputs to the second stage of the circuit (input D) is high, node 3 is pulled low, causing node 4 to go high.

Domino CMOS circuits provide significant speed enhancement and savings in chip area. One limitation of this structure is that each stage of the circuit must be buffered.

10.5 BIPOLAR LOGIC FAMILIES

We now consider the general features and characteristics of important logic families based on bipolar technology. Common to all such families is the bipolar transistor. There are two types of bipolar transistors: *npn* and *pnp*. In the *npn* transistor a *p*-type region is sandwiched between two *n* regions, whereas a *pnp* transistor is made by inserting an *n*-type region between two *p* regions. The operation of the transistor depends upon the relative voltages applied to the three regions. In general, present-day bipolar technology uses *npn* transistors rather than their *pnp* counterparts, so we shall confine ourselves to *npn* de-

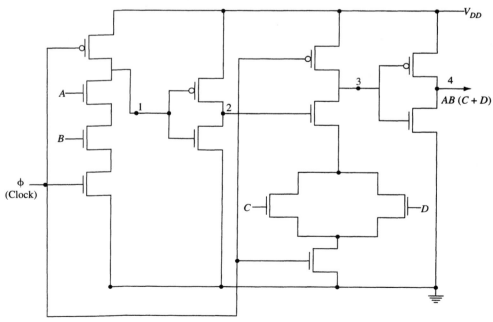

FIGURE 10.23
Two-stage CMOS domino circuit

10.5 BIPOLAR LOGIC FAMILIES □ **359**

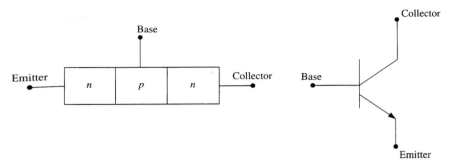

FIGURE 10.24
The structure and the symbol of an *npn* transistor

Figure 10.24 represents the simplified *npn* transistor structure and its symbol. The three adjoining regions within the transistor structure are known as **emitter, base,** and **collector**. Transistor operation depends upon the relative voltages applied to those three regions. The emitter–base junction is usually **forward-biased** (i.e., the base is positive with respect to the emitter) and the collector–base junction is **reverse-biased** (i.e., the collector is positive with respect to the base). The forward-biased emitter–base junction causes electrons to flow from the emitter to the collector, through the base. Since the collector supply voltage is positive, the electrons that pass through the base are attracted by the collector and travel through the external circuit back to the emitter again. Thus, the current flow across the emitter–base junction results in a current flow in the collector–base junction (conventional current flow is opposite to electron flow). The flow of conventional currents in the transistor are indicated in Figure 10.25.

Using Kirchhoff's current law we can write

$$I_E = I_B + I_C$$

In other words, all the current entering the transistor must equal the currents leaving the transistor. Two basic parameters are used to describe the currents through the transistor: α and β. Alpha (α) is the ratio of the collector current to the emitter current,

$$\alpha = \frac{I_C}{I_E}$$

and is also known as the **current transfer ratio**.

FIGURE 10.25
Current flow in transistor

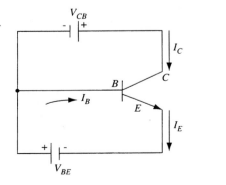

I_E = Emitter current

I_C = Collector current

I_B = Base current

Beta (β) is the ratio of the collector current to the emitter current,

$$\beta = \frac{I_C}{I_B}$$

Since $I_B = I_E - I_C$,

$$\beta = \frac{I_C}{I_E - I_C}$$

Because $I_C = \alpha I_E$,

$$\beta = \frac{\alpha I_E}{I_E - \alpha I_E}$$

and the I_E's factor and cancel out, resulting in

$$\beta = \frac{\alpha}{1 - \alpha}$$

Conversely, α can be expressed in terms of β:

$$\alpha = \frac{\beta}{1 + \beta}$$

Beta (β) is also known as the **d.c. current gain**.

Transistor Switch

A bipolar transistor can be made to approximate a switch. For operation as a switch the transistor must be either in **saturation** or at **cutoff**. When the transistor is at cutoff, the collector current is minimal, which normally implies that the emitter–base voltage as well as the collector–base voltage is reverse-biased. The transistor is saturated when both emitter–base and collector–base junctions are forward-biased.

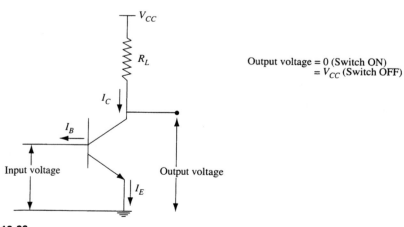

FIGURE 10.26
Transistor switch

10.5 BIPOLAR LOGIC FAMILIES

Figure 10.26 shows a transistor circuit arranged to operate as a switch. The emitter–base voltage (V_{BE}) is the input, (i.e., the controlling voltage) for the transistor switch. The output and the collector–emitter voltage (V_{CE}) for the switch is

$$V_{CE} = V_{CC} - I_C \cdot R_L$$

When the input voltage is such that the emitter–base junction is forward-biased, a base current I_B flows. The collector current I_C is equal to I_B. If the input voltage is high, I_B becomes large enough to make $I_C \cdot R_L$ equal to the supply voltage V_{CC}. Hence,

$$\begin{aligned} V_{CE} &= V_{CC} - V_{CC} \\ &= 0 \end{aligned}$$

Thus, the output voltage is 0 volts, and the transistor is ON. In practice, V_{CE} is about 0.3 volts instead of 0 volts because the saturation resistance of the transistor and R_L form a voltage divider and the output voltage is the voltage across the saturation resistance.

When the input voltage is such that the emitter-base is reverse-biased, the base current I_B is zero. As a result, the transistor is cut off and the collector current I_C is zero. Therefore, $I_C \cdot R_L = 0$ and

$$V_{CE} = V_{CC} - 0 = V_{CC}$$

Thus, the switch is OFF and the output voltage is equal to the supply voltage.

It is clear from the preceding discussion that the circuit of Figure 10.26 **inverts** an input signal. It produces a low output voltage when the input voltage is high enough to saturate the transistor, and it produces a high output voltage if the transistor is cut off (i.e., the input voltage is low).

So far, we have considered the implementation of a transistor switch which can only be switched from saturation to cutoff and vice versa; such a switch is referred to as a **saturated switch.** It is also possible to implement a **non-saturated switch,** in which the ON state occurs with the transistor in the active region. Bipolar logic families, based on both saturated and nonsaturated switching, are used in practice. Transistor-transistor logic (TTL) and integrated-injection logic (I^2L) are examples of the former, whereas emitter-coupled logic (ECL) is an example of the latter. The disadvantage of saturation switching is that extra delay has to be included in the time required to switch a transistor from the ON state to the OFF state. As a result, saturated logic families are slower than nonsaturated logic families.

Transistor-Transistor Logic (TTL)

TTL is probably the most widely used of all logic families. It was introduced in the 1960s and accounted for a variety of integrated-circuit chips, with the levels of integration ranging from small scale (SSI) to medium scale (MSI). There are several variations in the TTL family, such as standard TTL, low-power TTL, high-power TTL, Schottky TTL, low-power Schottky TTL, advanced Schottky TTL, advanced-low-power-Schottky TTL, and Fairchild-advanced-Schottky TTL.

Figure 10.27 shows a basic two-input TTL gate. It consists of a multi-emitter transistor T_1, a conventional transistor T_2, and two resistors R_1 and R_2. The conduction state of T_2 is controlled by the voltage at the collector of T_1.

When one or more of the emitters of T_1 is connected to a low voltage (0.2 V), the emitter–base junction is forward-biased. Current flows from V_{CC} through R_1 into the base of T_1 to keep it in saturation. As a result, its collector–emitter voltage is about 0.2 V (a typical value for a saturated transistor), which is not high enough to turn T_2 ON. Consequently the output of the gate is high. When all the emitters of T_1 are connected to high voltages (typically 5 V), the emitter–base junction of T_1 is reverse-biased and there is no current flow in T_1. However, this results in the base–collector junction of T_1 being forward-biased, which permits current flow from the supply through R_1 to the base of T_2. Hence T_2 is saturated and the output voltage of the gate is 0.2 V (i.e., low). Thus, the output of the TTL gate is low only when all the inputs are high. Otherwise, it produces a low output voltage—in other words the gate implements the NAND function.

It should be noted that a finite time is required for the output voltage of the gate to change between low and high. The time required to change from low to high is called the **rise time**, and the time to change from high to low is called the **fall time**. The rise time is affected by the capacitive load associated with the output of the gate (C in Figure 10.27). The capacitor is charged through R_2 when the output changes from low to high. On the other hand, the capacitor is discharged quickly through the low resistance of a saturated T_2 when the output voltage changes from high to low. Thus, the rise and the fall times of the gate's output voltage are not equal, and the long rise time decreases the switching speed of the gate. One way to reduce the rise time is to lower the value of R_2. However, this increases the power dissipation in the circuit when the output voltage is low, as can be seen from the following

$$\text{Power dissipation} = I^2 R_2$$
$$= \frac{(V_{CC} - V_{CE(\text{sat})})^2}{R_2}$$
$$= \frac{(V_{CC} - 0.2)^2}{R_2}$$

A better solution is to replace R_2 with an active resistance that is large when the output is low and is small when the capacitor C is being charged. This is known as active pull

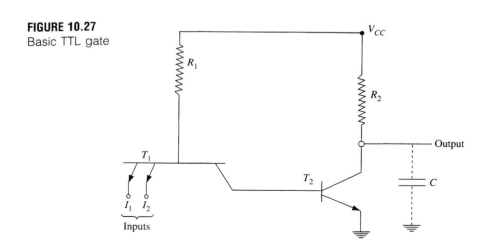

FIGURE 10.27
Basic TTL gate

up—in contrast to the output circuit of Figure 10.27, which is called passive pull-up. Figure 10.28 shows a two-input NAND gate incorporating an active pull-up output circuit. The arrangement of three output components T_3, D_3, and T_4 is referred to as a **totem pole**. The operation of the circuit is explained as follows:

Case i. If any input A or B is low, T_1 is saturated. Since the collector voltage of T_1 is low, T_2 is turned OFF. This results in the grounding of the base of T_4, so that T_4 is also turned OFF. On the other hand when T_2 is OFF, the voltage at the base of T_3 rises almost to V_{CC}; hence, T_3 is turned ON. The output capacitance is now charged through T_3, D_3, and R_4 (100 Ω). Since the forward resistance of D_3 and the saturation resistance of T_3 are very small, the capacitance is very rapidly charged to about 3.6 V. Thus the output voltage is high.

Case ii. When both the inputs are high, the base collector junction of T_1 forms a forward-biased diode, which permits current flow from the supply through R_1 to the base of T_2, turning it ON. When T_2 is ON, the drop across R_3 is sufficient to forward-bias the emitter–base junction T_4, turning it ON. As a result, the voltage output capacitance is quickly discharged and the output is rapidly pulled down from high to low. The totem-pole output does have one disadvantage, however. As the output changes state, both T_3 and T_4 are ON at the same time. This causes a high current spike between supply and ground in spite of the presence of R_4.

The basic circuit of Figure 10.28 can be modified to provide higher speed and drive capability. The modified circuit is shown in Figure 10.29. In this circuit the output transistor T_3 and the diode D_3 are replaced by a **Darlington circuit** consisting of two transistors, T_a and T_b, and a resistor R. The combination of transistors T_a and T_b functions as a single transistor with an overall current gain of approximately the product of the individual current gains. The higher current gives the Darlington circuit a lower output resis-

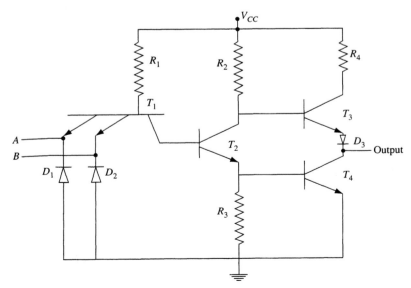

FIGURE 10.28
TTL NAND gate with totem pole

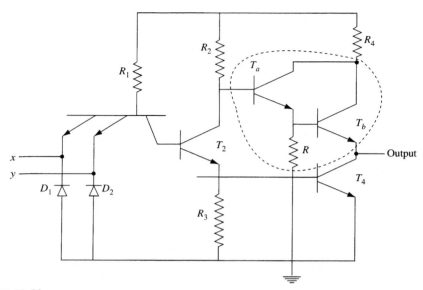

FIGURE 10.29
TTL NAND gate with Darlington totem-pole output

tance and permits the output capacitance to change more quickly to the high level than it can in the basic totem-pole output circuit.

The flexibility of the basic TTL gate can be increased by making the collector remain in an **open** state (i.e., not connected to V_{CC} or ground). Figure 10.30a represents an open collector TTL NAND gate. It is the same as a totem-pole NAND gate (Fig. 10.28)

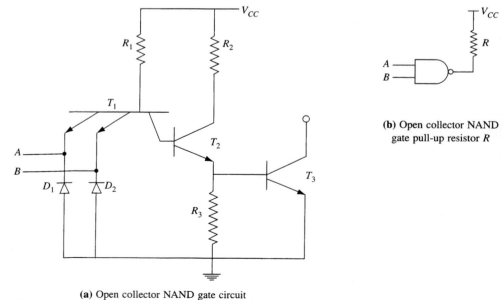

(a) Open collector NAND gate circuit

(b) Open collector NAND gate pull-up resistor R

FIGURE 10.30

10.5 BIPOLAR LOGIC FAMILIES 365

except that transistor T_3, resistor R_4, and diode D_3 have been omitted. A load must be provided at the output in order to make this gate operate; this load is usually a resistor and is connected to the supply voltage to provide pull-up (Fig. 10.30b).

The major advantage of open-collector devices is that their outputs can be connected together to implement **wired-logic** functions. Wired logic often results in reducing the amount of hardware required to implement a function. Figure 10.31a shows a circuit with three open-collector NAND gates connected to perform wired logic; the external pull-up resistor R acts as the load resistor for each of the NAND gates. The output Z of the circuit is 1 only when all the NAND gate outputs are in the 1 state. If the output of any one NAND gate is 0, transistor T_3 in that gate will be ON, and it will pull all other gate outputs low. Clearly the circuit performs the AND operation, $Z = \overline{AB} \cdot \overline{CD} \cdot \overline{EF}$. Thus, the circuit configuration obtained by tying the gate outputs together is said to perform **wired AND** operation. The symbol for the wired AND operation is shown in Figure 10.31b. If the same logic function is to be obtained with conventional NAND gates without using wired AND capability, two additional gates are required (Fig. 10.31c). This is because tying outputs together is not allowed in the case of TTL gates with totem-pole outputs. If the outputs are connected together as illustrated in Figure 10.32, the output in the 0 state would have to absorb current from all the other gates that are in 1 state. Since the ON transistor of gate 2 has a very low resistance, it will draw an extremely high current, which will damage it. Therefore, totem-pole outputs are not used to implement the wired AND function.

One of the more important applications of wired logic is for transferring data onto a **bus**. In a bussed data system, the outputs of several devices are connected to a set of

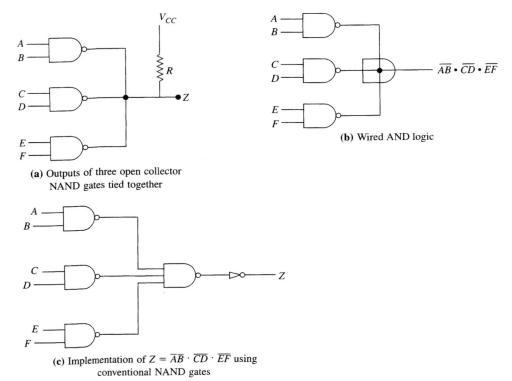

(a) Outputs of three open collector NAND gates tied together

(b) Wired AND logic

(c) Implementation of $Z = \overline{AB} \cdot \overline{CD} \cdot \overline{EF}$ using conventional NAND gates

FIGURE 10.31

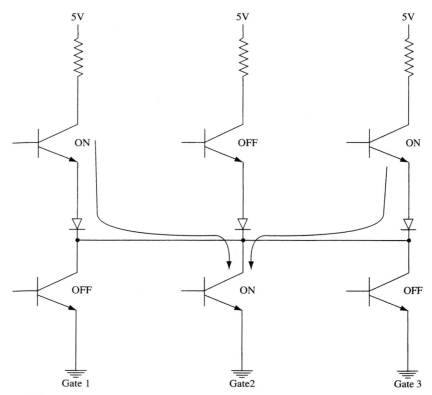

FIGURE 10.32
Totem-pole outputs connected together

common lines, called a bus; only the output of the selected device appears on the bus, and all the other devices are disabled. Such a system must have an enable function, which inhibits unwanted data from appearing on the bus. Figure 10.33 illustrates the formation of a bus by tying together the outputs of open-collector NAND gates.

The disadvantage of open-collector gates is that any load capacitance is charged through the pull-up resistor. This results in a slower rise time if the capacitance is large.

Tristate Devices

As discussed previously, the totem-pole output structure results in a lower resistance for both the 0 and 1 states, thereby providing good high-speed drive capability at reduced power dissipation. One important limitation of gates with totem-pole outputs is that they cannot be used in a bussing structure (for reasons considered in the previous section). However, a modified totem-pole configuration known as a tristate structure can be used in logic systems where the outputs of several devices are connected to a common signal line. A typical tristate TTL NAND gate is illustrated in Figure 10.34.

When the control input is low, the output of T_5 is high; as a result, diode D_2 is reverse-biased and the emitter of T_1 that is connected to the collector of T_5 is at 1. In this condition the NAND gate functions normally.

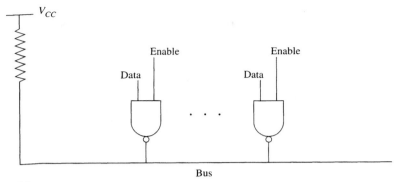

FIGURE 10.33
Bus formation using open-collector gates

When the control input is high, the T_5 output goes low, forward-biasing D_2 and transistor T_1. Thus T_1 is ON, which in turn will switch off both T_2 and T_4. Since D_2 is forward-biased, the base of T_3 will be at 0; consequently, T_3 will also be OFF. The OFF-OFF state configuration of T_3 and T_4 represents the disabled or third state of the gate. In essence the output of the tristate gate is an open circuit when the control input is high.

Figure 10.35 illustrates a situation in which the outputs of three tristate devices are connected to perform wired logic. When a particular tristate device is selected by the control signal, all other devices are in the high impedance state. Thus, only the selected device will transfer its output to the common output node. For example, in Figure 10.35 the common output node corresponds to the output of device 2.

FIGURE 10.34
Tristate NAND gate

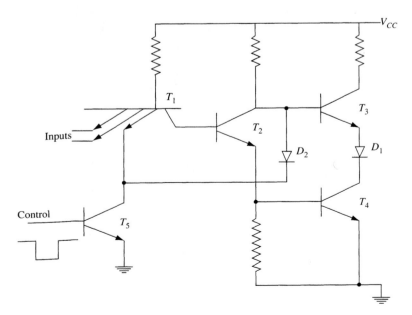

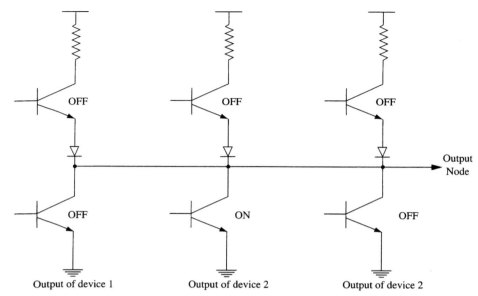

FIGURE 10.35

TTL Subfamilies

The TTL logic circuit shown in Figure 10.27 is known as **standard TTL** and is used in off-the-shelf small and medium scale integration. Several other variations on standard TTL are also available. If all the resistor values in the basic circuit of Figure 10.27 are reduced and transistor T_4 is replaced with a Darlington stage, the circuit speed is significantly increased. This modification is the basis for **high-speed TTL**. However, the increased speed is achieved at the expense of higher power dissipation.

For applications where lower power consumption is more critical than speed, the resistor values in the standard TTL circuit are increased. This variation of the standard TTL configuration is called **low-power TTL**.

The main disadvantage of TTL arises from the fact that the transistors are operated in a saturation mode. When a transistor is switched from the ON state to the OFF state,

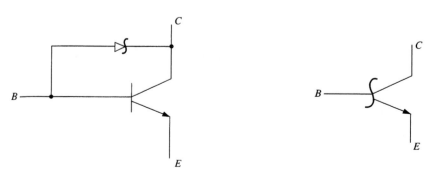

(a) Transistor with Schottky diode (b) Combined diode/transistor symbol

FIGURE 10.36

the stored charge in the transistor slows the rise of the collector voltage to logic 1 state. By preventing the transistor from going into saturation, the transistor switching speed can be significantly improved. One way of preventing transistor saturation is to connect a Schottky diode across the base–collector junction of every transistor, which would otherwise saturate, as shown in Figure 10.36a. The forward voltage drop of the Schottky diode is about 0.3 V compared to 0.5 V for the collector–base junction diode. When the collector voltage of the transistor drops below that of the base, the Schottky diode begins to conduct, thereby preventing the collector–base junction of the transistor from becoming forward-biased by more than 0.3 V. Therefore, the transistor can never saturate and consequently does not suffer from charge storage delay. The symbol for a Schottky-damped transistor, commonly known as Schottky transistor, is shown in Figure 10.36b.

Schottky TTL uses Schottky transistors to achieve at least twice the speed of high-power TTL without increasing the power consumption. By increasing the values of resistors, the Schottky TTL circuits can be converted to **low-power Schottky** TTL, a family that is slightly faster than standard TTL but requires only one fifth of the power. Currently, low-power Schottky is the most widely used TTL variant.

Recent additions to the TTL range are the advanced Schottky and the advanced low-power Schottky, developed by Texas Instruments, and the Fairchild advanced Schottky (FAST). Table 10.1 shows the propagation delay and power consumption of all variations of TTL.

Emitter-Coupled Logic (ECL)

One major limitation of the bipolar logic families we discussed so far is that they all operate in the saturation mode. As we noted in Section 10.5, when a saturated transistor is switched from the ON state to the OFF state, the charge storage effect delays the turn-off time. This results in increased gate propagation delay. In ECL the transistors operate only in the cutoff or in the active region, never in the saturation region. This eliminates the problem of charge storage; consequently, there is a major improvement in speed.

Figure 10.37 shows the basic circuit configuration of an ECL gate. It consists of two transistors T_1 and T_2, and three resistors R_1, R_2, and R_3. The base of T_2 is at a fixed

TABLE 10.1
Comparison of TTL series characteristics

Series	Average Propagation Delay (ns)	Average Power Consumption (mW)
Standard TTL	10	10
Low-power TTL	35	1
High-speed TTL	6	22.5
Schottky TTL	3	20
Low-power Schottky TTL	9	2
Advanced Schottky TTL	1.5	20
Advanced low-power Schottky TTL	4	1
FAST	2	4

FIGURE 10.37
Basic ECL gate

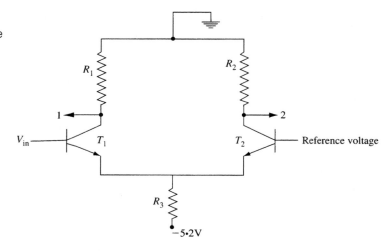

reference voltage, and the input (i.e., logic 1 or 0) is applied to the base of T_1. When input V_{in} is greater than the reference voltage, T_1 conducts and T_2 is OFF. T_2 conducts and T_1 is OFF when V_{in} is less than the reference voltage. The voltage across R_3 tends to remain fairly constant irrespective of which transistor is conducting. Moreover, neither T_1 nor T_2 goes into the saturation stage.

One problem with the circuit of Figure 10.37 is that the output voltage levels are not the same as the input logic levels. This can be taken care of by adding emitter-follower stages (T_3 and T_4) as shown in Figure 10.38. In addition, the emitter followers provide low-output impedance driving capabilities.

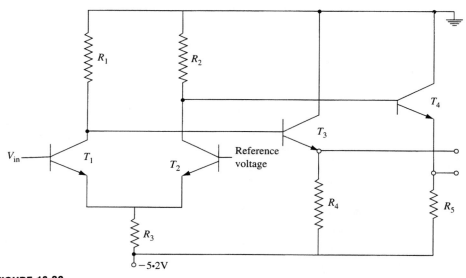

FIGURE 10.38
ECL gate with emitter-follower outputs

The basic circuit of Figure 10.37 can be expanded to more than one input by placing transistors in parallel as shown in Figure 10.39. Here either T_1 or T_3 can switch the current out of T_2. That is, if either input A or B is high, T_1 or T_3 is ON and T_2 is OFF; consequently $V_{out\,2}$ is high, and $V_{out\,1}$ is low. Hence, it is apparent that $V_{out\,1}$ realizes the NOR of inputs A and B, and $V_{out\,2}$ is the complement.

The principal advantage of the ECL family is very fast switching speed. Another advantage is that there are no power rail spikes, since approximately constant current is taken from the supply. The major disadvantage of ECL is its relatively high power dissipation per gate.

10.6 BiCMOS LOGIC

As was mentioned previously, bipolar devices are faster than CMOS devices and allow greater drive capability. CMOS technology, on the other hand, consumes less power and lends itself to very large scale integration (VLSI). The merger of bipolar and CMOS, known as **BiCMOS technology**, retains the attractive characteristics of low power dissipation and high current drive capability. A typical BiCMOS circuit has a basic CMOS gate structure buffered by an *npn* totem pole to provide the additional drive capability. Figure 10.40 shows the BiCMOS inverter circuit. Let us assume first a logic 0 at the input. The *p*-transistor M_1 is then ON, and base current is supplied to transistor T_1. Since the *n*-transistor M_3 is OFF, the output capacitor C_1 will be charged to V_{dd}, thereby changing the output to a logic 1. When the input changes from logic 0 to logic 1, M_2 is turned ON and removes the base charge from T_1. Since M_3 is also turned ON by a logic 1 at the

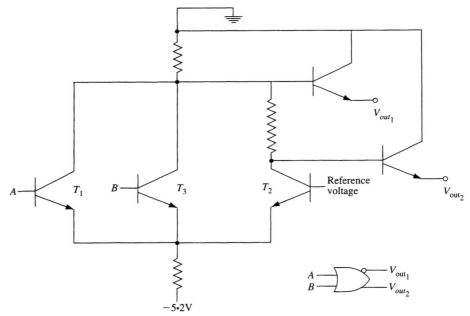

FIGURE 10.39
Two-input ECL OR/NOR circuit

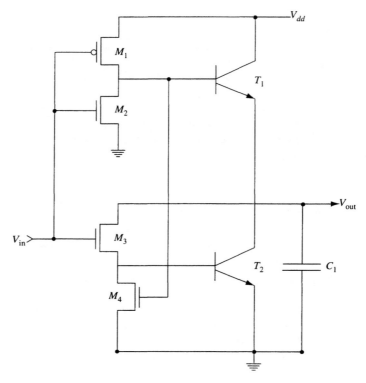

FIGURE 10.40
BiCMOS Inverter [Ref. 7]

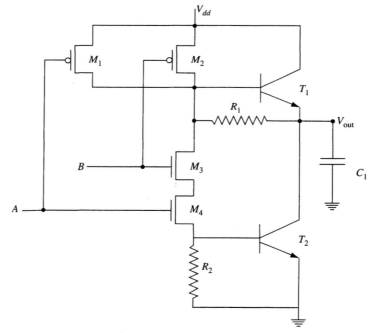

FIGURE 10.41
BiCMOS NAND gate

input, it supplies base current to transistor T_2 to turn it ON. Thus, if V_{out} is at a logic 1, the output capacitance C_I is discharged via T_2 and M_3 to make the output a logic 0. Therefore, unless C_I is already charged to the value of V_{dd}, an input of logic 1 will not change the output state (i.e., the output will remain at logic 0).

Another BiCMOS gate, a two-input NAND, is shown in Figure 10.41. The basic operation of the BiCMOS NAND gate is similar to that of an inverter circuit. When either input A or input B is at logic 0, M_1 or M_2 is ON, and M_3 or M_4 is OFF. Since M_3 and M_4 are in series, no current is supplied to the base T_2. On the other hand, since M_1 or M_2 is ON, base current is supplied to T_1 from $V_{dd,}$ turning it ON. Therefore, output capacitor C_I is charged to a logic 1. When both inputs are at logic 1, M_1 and M_2 are OFF and M_3 and M_4 are ON. Thus, base current is supplied to T_2 turning it ON, which in turn results in the discharging of C_I.

In addition to low power consumption and high drive capabilities, BiCMOS provides lower clock skews compared to conventional CMOS [6]. Despite its obvious advantages, BiCMOS has several drawbacks that have prevented its wide adoption in digital applications. One major problem is that a bipolar/CMOS combination requires more mask levels and more processing steps, which increase the fabrication time. The longer fabrication time results in lower yields, thereby increasing cost. Unless there is significant improvement in performance, the higher cost is unlikely to be acceptable.

EXERCISES

1. The logic diagram of a combinational circuit is shown here. Design an equivalent CMOS circuit for the function.

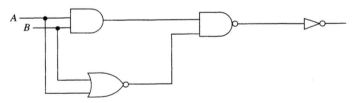

2. A CMOS circuit is shown here. Draw an equivalent gate-level diagram.

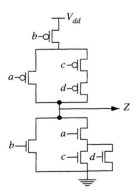

3. Derive an expression for the output Z of the NMOS circuit shown.

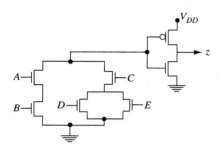

4. Implement the EX-OR function using NMOS transistors so that the propagation delay is of two stages only.
5. The AND-OR-INVERT (AOI) function is very useful in realizing multiplexer functions. The NMOS implementation of such a circuit is shown here. [9]

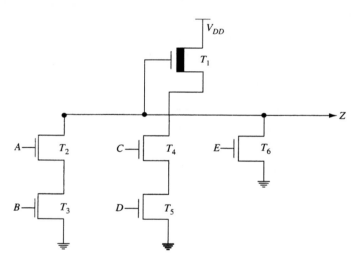

Generate a function table for the circuit with the appropriate status of each transistor (i.e., ON or OFF), and the output of the circuit for eight input patterns using the following format

$$A \quad B \quad C \quad D \quad E \quad T_1 \quad T_2 \quad T_3 \quad T_4 \quad T_5 \quad T_6 \quad Z$$

6. Implement a four-input OR-AND-INVERT CMOS structure. Derive the function table for this structure, in the format used in Exercise 5.
7. Design an NMOS circuit for realizing the EX-NOR function; use a minimal number of transistors.
8. Implement an NMOS set-reset latch using a pair of 2-input AND-OR-INVERTERs.
9. The **Thevenin equivalent** of a TTL NAND gate with an output of logic 1 has the structure shown on p. 377. Find the voltage at node Z when R is

   a. infinite   b. 4K   c. 40K

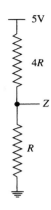

10. Complete the truth table for the CMOS circuit shown. What function does the circuit represent?

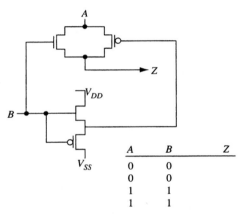

11. A gate-level representation of the EX-OR function is shown. Sketch the CMOS implementation of the circuit.

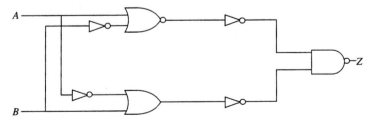

12. Draw a two-stage CMOS domino circuit to implement the function

$$f(w,x,y,z) = \bar{y}(w + x)(y + z)$$

13. For the TTL circuit shown on p. 376, fill in the following table with the status of the transistors (ON or OFF) and the circuit output. What is the function of the circuit?

376 □ CHAPTER 10 / DIGITAL INTEGRATED CIRCUITS

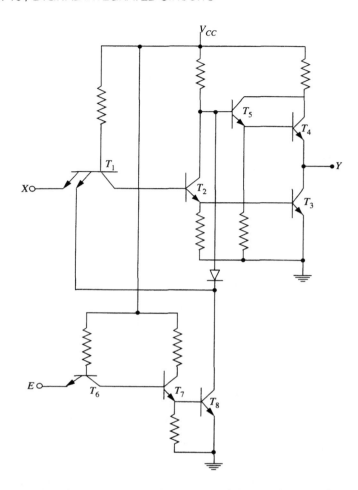

E	X	T_1	T_2	T_3	T_4	T_5	T_6	T_7	T_8	Y
0	0									
0	1									
1	0									
1	1									

14. Derive a functional block diagram of the TTL circuit shown on p. 377.

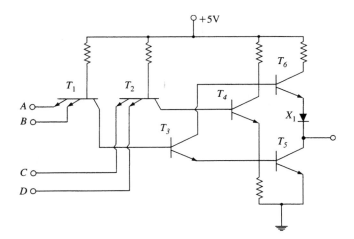

15. The ECL circuit for a 2-input OR/NOR gate is shown below. Assuming two such gates are available, derive the logic expressions for the following cases:
    a. The OR outputs of both gates are tied together, and the NOR outputs are also tied together.
    b. The OR output of one gate is tied to the NOR output of the other. [8]

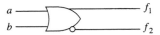

16. Design a BiCMOS tristate inverter. (Hint: Three p-type, seven n-type, and two bipolar transistors are sufficient.)

References

1. S. Muroga, *VLSI System Design*, Wiley, 1982.

2. W. C. Penney and L. Lau, *MOS Integrated Circuits*, Van Nostrand Reinhold, 1972.

3. N. Weste and K. Eshraghian, *Principles of CMOS VLSI Design*, Addison Wesley, 1992.

4. A. Mukherjee, *Introduction to nMOS and CMOS VLSI Systems Design*, Prentice Hall, 1986.

5. M. I. Elmasry, *Digital Bipolar Integrated Circuits*, Wiley, 1983.

6. T. Wang, "BiCMOS making inroads to CMOS, ECL markets," *ASIC Technology and News*, Vol. 1, No. 4, August 1989, p. 29.

7. C. H. Diaz, S. Kang, and Y. Leblebici, "An accurate delay model for BiCMOS driver circuits," *IEEE Trans. on Computer-Aided Design*, Vol. 10, No. 5, May 1991, pp. 577–588.

8. V. Hamacher, Z. Vranesic, and S. Zaki, *Computer Organization*, McGraw-Hill, 1990.

9. T. Dillinger, *VLSI Engineering*, Prentice Hall, 1988.

11
Testing and Testability

The goal of testing a digital circuit is to determine whether there are faults in the circuit. **Fault** refers to a physical defect in a circuit (e.g., a short between signal lines, breaks in signal lines, defective components, etc.). The term **error** is used to describe the incorrect operation of a circuit. An error is usually the manifestation of a fault in a circuit, but a fault does not always cause an error.

A fault is characterized by its nature, value, extent, and duration [1]. The **nature** of a fault can be classified as logical or nonlogical. A logical fault causes the logic value at a point in a circuit to become opposite to the specified value. Nonlogical faults include such things as malfunction of the clock signal, power failure, etc. The **value** of a logical fault at a point in the circuit indicates whether the fault creates a fixed or varying erroneous logical values. The **extent** of a fault specifies whether the effect of the fault is localized or distributed. A local fault affects only a single variable, whereas a distributed fault affects more than one. A logical fault, for example, is a local fault, whereas the malfunction of a clock is a distributed fault. The **duration** of a fault refers to whether the fault is **permanent** or **temporary.** Although a major portion of faults in digital systems are caused by temporary faults, currently there are no reliable techniques for detecting and isolating such faults. In this chapter we shall be concerned only with permanent faults. A permanent fault is often referred to as a **hard fault** and a temporary fault as a **soft fault.**

11.1 FAULT MODELS

The effect of a fault in a logic circuit is represented by a **model**; the usefulness of a model is determined by the accuracy with which it represents the effect of a fault. Several models are in use today for representing faults in logic circuits:

1. Stuck-at fault
2. Bridging fault
3. Stuck-open fault

Single Stuck-at Fault

The most common model used for logical faults is the **single stuck-at fault.** It assumes that a fault in a logic gate results in one of its inputs or the output being fixed to either a logic 0 (**stuck-at-0**) or at logic 1 (**stuck-at-1**). Stuck-at-0 and stuck-at-1 faults are often abbreviated to s-a-0 and s-a-1 respectively.

In an m-input gate, there can be $2(m + 1)$ stuck-at faults. However, a stuck-at fault on an input may be indistinguishable from a stuck-at fault at the output. For example, in a NAND gate any input s-a-0 fault is indistinguishable from the output s-a-1; similarly, in a NOR gate an input s-a-1 fault is indistinguishable from the output s-a-0. Such faults are said to be **equivalent.** Thus, an m-input gate can have a total of $(m + 2)$ logically distinct faults. For example, the two-input NAND gate of Figure 11.1 can have four logical faults: A s-a-1, B s-a-1, Z s-a-0, and Z s-a-1.

Let us assume that the A input of the NAND gate is s-a-1. The NAND gate perceives the A input as a logic 1 irrespective of the logic value placed on the input. The output of the NAND gate is 0 for the input pattern shown in Figure 11.1, when the s-a-1 fault is present. In the absence of the fault, the output will be 1. Thus, $AB = 01$ can be considered as the **test** for the A input s-a-1, since there is a difference between the output of the fault-free and faulty gate.

The single stuck-at fault model is often referred to as the **classical fault model** and offers a good representation for the most common types of defects (e.g., shorts and opens in many technologies). Figure 11.2 illustrates the NMOS realization of the two-input gates. Numbers 1 and 3 indicate possible opens; 2 and 4 indicate possible shorts.

FIGURE 11.1
NAND gate with input s-a-1

FIGURE 11.2
NMOS NAND gate

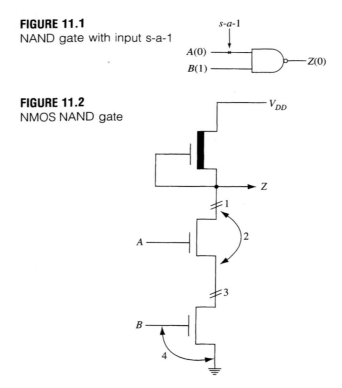

Fault 1 will disconnect the two driver transistors from the circuit and will have the same effect as if both the transistors are off. Thus, the fault can be interpreted as the output s-a-1.

Fault 2 has the effect as if the corresponding transistor is saturated, and hence the input A appears to be s-a-1.

Fault 3 will have the same effect as fault 1, and can be interpreted as either A s-a-0 or B s-a-0.

Fault 4 may be interpreted as B s-a-0 and hence has the same effect as fault 3.

The single stuck-at fault model is also used to represent multiple faults in logic circuits. In a **multiple stuck-at fault** it is assumed that more than one signal line in the circuit are stuck at logic 1 or logic 0; in other words a group of stuck-at faults exist in the circuit at the same time. A variation of the multiple fault is the **unidirectional fault.** A multiple fault is unidirectional if all its constituent faults are either s-a-0 or s-a-1 but not both simultaneously.

Bridging Faults

Bridging faults form an important class of permanent faults that cannot be modeled as stuck-at faults. A bridging fault is said to have occurred when two or more signal lines in a circuit are accidentally connected together. Earlier study of bridging faults concentrated only on the shorting of signal lines in gate-level circuits. It was shown that the shorting of lines resulted in wired logic at the connection.

Bridging faults at the gate level has been classified into two types: **input bridging** and **feedback bridging.** An input bridging fault corresponds to the shorting of a certain number of primary input lines. A feedback bridging fault results if there is shorting of an output and an input line(s). The presence of a feedback bridging fault may cause a circuit to oscillate or convert it to a sequential circuit.

Bridging faults in a transistor-level circuit may occur between the terminals of a transistor or, as in the case or a gate-level circuit, between two or more signal lines. Figure 11.3 shows the NMOS logic realization of the Boolean function

$$Z = \overline{AB + CD}$$

A short between two lines, as indicated by the dotted line in the diagram, modifies the function to

$$Z = \overline{(A + C)(B + D)}$$

The effect of bridging among the terminals of transistors is technology dependent. For example, in CMOS circuits such faults manifest as either stuck-at or **stuck-open** faults, depending on the physical location and the value of the bridging resistance.

Stuck-Open Faults

Stuck-open faults are peculiar to CMOS technology. The major difference between the stuck-at and the stuck-open faults is that a stuck-at fault leaves the faulty gate as a combinational circuit, whereas a stuck-open fault converts the gate into a sequential circuit.

Figure 11.4a shows a two-input CMOS NOR gate. A stuck-open fault causes the output of the NOR gate to be connected neither to the ground nor to V_{DD}. If, for exam-

FIGURE 11.3
An NMOS circuit with a short

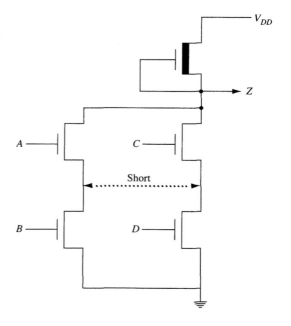

ple, transistor T_2 is open-circuited, then both the p-transistors will be disconnected from the output. Thus, for the input $AB = 00$ there will be no change in the output; in fact, the output will retain its previous logic state. The length of time the state is retained is determined by the leakage current at the output node.

Figure 11.4b shows the truth table for NOR gate. The fault-free output is shown in column Z. The other three columns represent the outputs in the presence of the three stuck-open faults (#1, #2, and #3) shown in Figure 11.4a. Z_t indicates that the gate output is a function of the previous input value rather than the current one. In other words, the NOR gate, which is a combinational element, behaves like a sequential circuit in the presence of a stuck-open fault.

A fully complementary CMOS circuit with stuck-open faults can be converted into an equivalent gate-level circuit with stuck-at faults [2]. This is accomplished by replacing each CMOS gate by a network consisting of a **B-block** (Fig. 11.5a). The inputs S_0 and S_1 of a B-block are driven by two logic circuits constructed from AND, OR, and inverters. The truth table of the B-block shown in Figure 11.5b records the output of the CMOS gate in the presence of a stuck-open fault, as well as the fault-free operation. The logic circuit driving the S_1 input of the B-block is obtained by replacing the series connections of p-transistors by AND gates, and parallel connections of p-transistors by an OR gate. The inputs to the p-transistors are applied to the AND gates via inverters. The logic circuit driving the S_0 input of the B-block is derived by replacing parallel connections of n-transistors by OR gates and the series connections of n-transistors by an AND gate. The inputs to the n-transistors are directly connected to the OR gates.

FIGURE 11.4

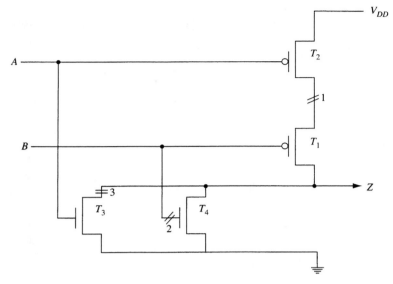

(a) CMOS NOR circuit

A	B	Z	Z(#1)	Z(#2)	Z(#3)
0	0	1	Z_t	1	1
0	1	0	0	Z_t	0
1	0	0	0	0	Z_t
1	1	0	0	0	0

(b) Truth table for the NOR gate

FIGURE 11.5

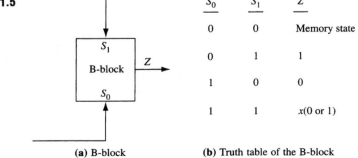

S_0	S_1	Z
0	0	Memory state
0	1	1
1	0	0
1	1	x(0 or 1)

(a) B-block   **(b)** Truth table of the B-block

EXAMPLE 11.1

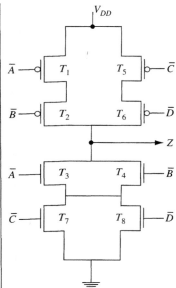

FIGURE 11.6
A CMOS transistor circuit

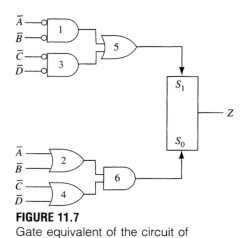

FIGURE 11.7
Gate equivalent of the circuit of Figure 10.6

We convert the CMOS circuit shown in Figure 11.6 into an equivalent gate-level circuit shown in Figure 11.7. If an input combination cannot create a conducting path from V_{DD} or ground to the output due to an open fault, then $S_0 = S_1 = 0$ and the output is in memory state, which indicates that the output retains its value due to the previous input. The open and short faults at the circuit level (Fig. 11.6) can be mapped into equivalent faults at the gate-level circuit (Fig. 11.7). Figure 11.8 shows the corresponding mapping. This procedure for modeling circuit-level faults is applicable only to the series-parallel type of CMOS circuits. The problem with this procedure is that in general the gate equivalent circuit becomes much larger than the original circuit.

Fault in CMOS Circuit	*Equivalent Fault in Gate-Level Circuit*
$\bar{A}$ s-a-0/1	$\bar{A}$ s-a-0/1
$\bar{B}$ s-a-0/1	$\bar{B}$ s-a-0/1
$\bar{C}$ s-a-0/1	$\bar{C}$ s-a-0/1
$\bar{D}$ s-a-0/1	$\bar{D}$ s-a-0/1
T_1 ON (source and drain shorted)	$\bar{A}$ (Gate 1) s-a-0
T_2 ON	$\bar{B}$ (Gate 1) s-a-0
T_3 ON	$\bar{A}$ (Gate 2) s-a-1
T_4 ON	$\bar{B}$ (Gate 2) s-a-1
T_5 ON	$\bar{C}$ (Gate 3) s-a-0
T_6 ON	$\bar{D}$ (Gate 3) s-a-0

FIGURE 11.8
Mapping of faults from circuit to gate level

T_7 ON	$\overline{C}$ (Gate 4) s-a-1
T_8 ON	$\overline{D}$ (Gate 4) s-a-1
T_1 Open	$\overline{A}$ (Gate 1) s-a-1
T_2 Open	$\overline{B}$ (Gate 2) s-a-1
T_3 Open	$\overline{A}$ (Gate 1) s-a-0
T_4 Open	$\overline{B}$ (Gate 2) s-a-0
T_5 Open	$\overline{C}$ (Gate 3) s-a-1
T_6 Open	$\overline{D}$ (Gate 3) s-a-1
T_7 Open	$\overline{C}$ (Gate 4) s-a-0
T_8 Open	$\overline{D}$ (Gate 4) s-a-0

FIGURE 11.8
(Continued)

11.2 FAULT DETECTION IN LOGIC CIRCUITS

Fault detection in a logic circuit is carried out by applying a sequence of tests and observing the resulting outputs. A **test** specifies the expected response that a fault-free circuit should produce. If the observed response is different from the expected response, a fault is present in the circuit.

The aim of testing at the gate level is to verify that each logic gate in the circuit is functioning properly and the interconnections are good. If we assume only a single stuck-at fault can be present in the circuit under test, then the problem is to construct a test set that will detect the fault by utilizing only the inputs and the outputs of the circuit.

In order to detect a fault, the circuit must first be **excited** (i.e., a certain input combination must be applied so that the logic value appearing at the fault must be **sensitized,** that is, the effect of the fault must be propagated through the circuit to an observable output). For example, in Figure 11.9 the input combination $abc = 111$ must be applied for the excitation of the fault and $d = 1$ for sensitizing the fault to the output Z. Thus, the test for the s-a-1 fault is $abcd = 1111$. This input combination is also a test for other faults (e.g., gate 1 s-a-0, gate 3 s-a-1, and input a s-a-0, etc.).

One of the main objectives in testing is to minimize the number of test patterns. If the function of a circuit in the presence of a fault is different from its normal function (i.e., the circuit is **non-redundant**), then an n-input combinational circuit can be completely tested by applying all 2^n combinations to it; however, 2^n increases very rapidly as n increases. For a sequential circuit with n inputs and m flip-flops, the total number of input combinations necessary to exhaustively test the circuit is $2^n \times 2^m = 2^{m+n}$. If, for example, $n = 20$ and $m = 40$, there would be 2^{60} tests. At a rate of 10,000 tests per second, the total test time for the circuit would be about 3.65 million years! Fortunately, a complete truth table exercise of the logic circuit is not necessary—only the input combinations that detect most of the faults in the circuit are required.

FIGURE 11.9
Circuit with a single stuck-at fault

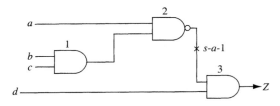

The efficiency of a test set is measured by a figure of merit called **fault coverage**. The term *coverage* refers to the percentage of the possible single stuck-at faults that a test set will detect. The computation time needed to generate tests for combinational circuits is proportional to the square of the number of gates in the circuit. For example, the test generation time for a 100,000-gate circuit is 100 times that for a 10,000-gate circuit. The task is even more complicated for sequential circuits, because the number of internal states is an exponential function of the number of memory elements. Thus, sequential circuits have to be designed so that the fault detection in such circuits becomes easier. We discuss some of these techniques in Section 11.4.

11.3 TEST GENERATION FOR COMBINATIONAL CIRCUITS

Several distinct test generation methods have been developed over the years for combinational circuits. These methods are based on the assumptions that a circuit is nonredundant and only a single stuck-at fault is present at any time.

Truth Table and Fault Matrix

The most straightforward method for generating tests for a particular fault is to compare the responses of the fault-free and the faulty circuit to all possible input combinations. Any input combination for which the output responses do not match is a test for the given fault.

Let the inputs to a combinational circuit be $x_1, x_2, \ldots, x_n$ and let z be the output of the circuit. Let z_α be the output of the circuit in the presence of the fault α. The test generation method starts with the construction of the truth tables of z and z_α. Then for each row of the truth table, $z \oplus z_\alpha$ is computed; if the result is 1, the input combination corresponding to the row is a test for the fault.

■ **EXAMPLE 11.2**

Let us consider the circuit shown in Figure 11.10a, and assume that tests for faults α s-a-0 and β s-a-1 have to be derived. The truth for the circuit is shown in Figure 11.10b, where column z denotes the fault-free output, and z_α and z_β correspond to the circuit output in presence of faults α s-a-0 and β s-a-1 respectively. The tests for the faults are indicated as 1's in the columns corresponding to $z \oplus z_\alpha$ and $z \oplus z_\beta$. Thus, the test for α s-a-0 is $x_1x_2x_3 = 110$, and the test for β s-a-1 is $x_1x_2x_3 = 001$. For all other input combinations the output of the fault-free circuit is the same as the output in the presence of the fault; consequently, they are not tests for α s-a-0 and β s-a-1. The minimum number

FIGURE 11.10

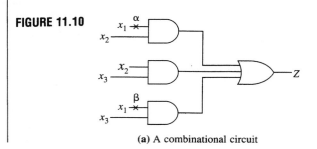

(a) A combinational circuit

**FIGURE 11.10
(Continued)**

x_1	x_2	x_3	z	z_α	z_β	$z + z_\alpha$	$z + z_\beta$
0	0	0	0	0	0	0	0
0	0	1	0	0	1	0	1
0	1	0	0	0	0	0	0
0	1	1	1	1	1	0	0
1	0	0	0	0	0	0	0
1	0	1	1	1	1	0	0
1	1	0	1	0	1	1	0
1	1	1	1	1	1	0	0

(b) Truth table for the fault-free and the faulty circuit

of tests required to detect a set of faults in a combinational circuit can be obtained from a **fault matrix**.

The columns in a fault matrix list the possible faults, and the rows indicate the tests. A fault matrix for the circuit of Figure 11.11a is shown in Figure 11.11b. A 1 at the intersection of the ith row and the jth column indicates that the fault corresponding to the jth column can be detected by the ith test. As can be seen from Figure 11.11b, a fault matrix is identical to a prime implicant chart discussed in Chapter 4. Thus, the

FIGURE 11.11

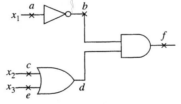

(a) Circuit under test

x_1	x_2	x_3	a s-a-0	b s-a-0	c s-a-0	d s-a-1	e s-a-0	f s-a-1
0	0	0				1		1
1	0	0						1
0	1	0		1	1			
1	1	0	1					1
0	0	1		1			1	
1	0	1	1					1
0	1	1		1				
1	1	1	1					1

(b) Fault matrix

FIGURE 11.11 (Continued)

x_2	x_3	a s-a-0	b s-a-0	c s-a-0	d s-a-1	e s-a-0	f s-a-1
0	0				1		1
1	0		1	1			
1	0	1					1
0	1			1		1	

(c) Minimal test set

problem of finding the minimum number of tests is the same as the problem of finding the minimum number of prime implicants (i.e., rows) so that every column has a 1 in at least one row. In Figure 11.11b rows 110, 101, and 111 are equivalent (i.e., each test detects the same faults as the other two), hence 101 and 111 can be omitted. Furthermore, row 000 covers row 100 and row 001 covers row 011; thus, rows 100 and 011 can be omitted. Elimination of rows 100, 101, 011, and 111 yields the minimal test set as shown in Figure 11.11c. These four tests detect all of the six faults under consideration. ∎

It is obvious from this example that the truth table approach to test generation is not practicable when the number of input variables is large. We now discuss some alternative techniques developed to solve test generation problems.

Boolean Difference

The Boolean difference of a function $F(x_1, x_2, \ldots, x_n) = F(X)$ with respect to one of its inputs x_i is defined as

$$\frac{dF(x_1, \ldots, x_i, \ldots, x_n)}{dx_i} = \frac{dF(X)}{dx_i}$$
$$= F(x_1, \ldots, x_i, \ldots, x_n) \oplus F(x_1, \ldots \bar{x}_i, \ldots, x_n)$$

It is easy to see that $dF(X)/dx_i = 1$ implies that $F(x_1, \ldots, x_i, \ldots, x_n) \neq F(x_1, \ldots, \bar{x}_i, \ldots, x_n)$. Also, $dF(X)/dx_i = 1$ indicates that the function $F(X)$ is dependent on the value x_i alone. Similarly, $dF(X)/dx_i = 0$ when $F(x_1, \ldots, x_i, \ldots, x_n) = F(x_1, \ldots \bar{x}_i, \ldots, x_n)$; that is, the function $F(X)$ is not sensitive to variable x_i.

In order to find a test for a fault on x_i, it is necessary to find an input combination so that $F(x_1, \ldots, x_i, \ldots, x_n)$ will be different from $F(x_1, \ldots, \bar{x}_i, \ldots, x_n)$. However, before finding a test or tests for a fault on x_i, it is necessary to assign values to other variables in the function so that the function is only sensitive to x_i. In other words, the objective is to find input combinations that will make $dF(X)/dx_i = 1$ for a fault occurring on x_i. Thus, the tests for a fault on x_i are as follows:

$$x_i \frac{dF(X)}{dx_i} \quad \text{for } x_i \text{ s-a-0}$$

$$\bar{x}_i \frac{dF(X)}{dx_i} \quad \text{for } x_i \text{ s-a-1}$$

EXAMPLE 11.3

Let us derive the test(s) for fault a s-a-1 in the circuit shown in Figure 11.12. The circuit implements the function

$$F(x_1, x_2, x_3) = x_1x_2 + x_2x_3 + x_1x_3$$

The tests that will detect the fault a s-a-1 are defined by the Boolean expression $\bar{x}_3 dF/dx_3$, where

$$\frac{dF}{dx_3} = \frac{d(x_1x_2 + x_2x_3 + x_1x_3)}{dx_3}$$
$$= (x_1x_2 + x_2 \cdot 0 + x_1 \cdot 0) \oplus (x_1x_2 + x_2 \cdot 1 + x_1 \cdot 1)$$
$$= \bar{x}_1 x_2 + x_1 \bar{x}_2$$

FIGURE 11.12
Circuit with the fault α s-a-1

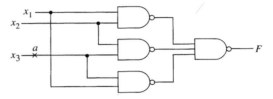

Thus, the tests that detect the fault a s-a-1 are given by

$$\bar{x}_3(\bar{x}_1 x_2 + x_1 \bar{x}_2) = \bar{x}_1 x_2 \bar{x}_3 + x_1 \bar{x}_2 \bar{x}_3$$

In other words, the fault a s-a-1 will cause an output error only if either $x_1 = 0$, $x_2 = 1$ or $x_1 = 1$, $x_2 = 0$, and x_3 equal to 0. This can be verified by inspection of Figure 11.12. ∎

Path Sensitization and the D-Algorithm

The basic principle of the path sensitization method is to choose some path from the origin of the fault to the circuit output. As mentioned earlier, a path is sensitized if the inputs to the gates along the path are assigned values such that the effect of the fault can be propagated to the output.

EXAMPLE 11.4

Let us consider the circuit shown in Figure 11.13 and assume that line α is s-a-1. To test for α, both G_3 and C must be set at 1. In addition, D and G_6 must be set at 1 so that $G_7 = 1$ if the fault is absent. To propagate the fault from G_7 to the circuit output f via G_8 requires the output of G_4 to be 1. This is because if $G_4 = 0$, the output f will be forced to be 1 independent of the value of gate G_7. The process of propagating the effect of the fault from its originational location to the circuit output is known as the **forward trace.**

The next phase of the method is the **backward trace,** in which the necessary signal values at the gate outputs specified in the forward trace phase are established. For ex-

FIGURE 11.13
A combinational circuit with α s-a-1

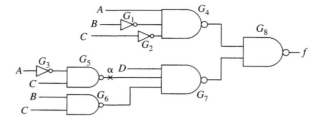

ample, to set G_3 at 1, A must be set at 0, which also sets $G_4 = 1$. In order for G_6 to be at 1, B must be set at 0; note that G_6 cannot be set at 1 by making $C = 0$ because this is inconsistent with the assignment of C in the forward-trace phase. Therefore, the test $ABCD = 0011$ detects the fault α s-a-1, since the output f will be 0 for the fault-free circuit and 1 in the presence of the fault. ∎

In general, a test pattern generated by the path-sensitization method may not be unique. For example, the fault α s-a-0 in the circuit of Figure 11.14 can be detected by $ABC = 01-$ or $0-0$. In the first test C is unspecified, and B is unspecified in the second test. An unspecified value in a test indicates that the test is independent of the corresponding input.

The drawback of the path sensitization method is that only one path is sensitized at a time. This does not guarantee that a test will be found for a fault even if one exists.

■ **EXAMPLE 11.5**

Let us derive a test for the fault α s-a-0 in Figure 11.15. To propagate the effect of the fault along the path G_2–G_6–G_8 requires that B, C, and D should be set at 0. In order to propagate the fault through G_8, it is necessary to make $G_4 = G_5 = G_7 = 0$. Since B and D have already been set to 0, G_3 is 1, which makes $G_7 = 0$. To set $G_5 = 0$, A must be set to 1; as a result, $G_1 = 0$, which with $B = 0$ will make $G_4 = 1$. Therefore, it is not possible to propagate the fault through G_8. Similarly, it is not possible to sensitize the single path G_2–G_5–G_8. However, $A = 0$ sensitizes the two paths simultaneously and also makes $G_4 = 0$. Thus, two inputs to G_8 change from 0 to 1 as a result of the fault α s-a-0, while the remaining two inputs remain fixed at 0. Consequently, $ABCD = 0000$ causes the output of the circuit to change from 1 to 0 in the presence of α s-a-0 and is the test for the fault. ∎

This example shows the necessity of sensitizing more than one path in deriving tests for certain faults and is the principal idea behind the D-Algorithm.

FIGURE 11.14
Circuit with α s-a-0

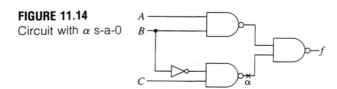

FIGURE 11.15
Three-level NOR logic circuit

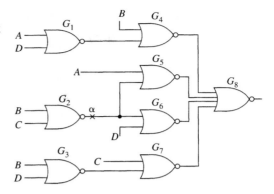

The D-Algorithm is guaranteed to find a test if one exists for detecting a fault. It uses a **cubical algebra** for automatic generation of tests. Three types of cubes are considered:

1. Singular cube
2. Propagation D-cube
3. Primitive D-cube of a fault

Singular Cube

A singular cube corresponds to a prime implicant of a function. Figure 11.16 shows the singular cubes for the 2-input NOR function; x's or blanks are used to denote that the position may be either 0 or 1.

Propagation D-Cube

D-cubes represent the input/output behavior of the good and the faulty circuit. The symbol D may assume 0 or 1. $\overline{D}$ takes on the value opposite to D (i.e., if $D = 1$, $\overline{D} = 0$ and if $D = 0$, $\overline{D} = 1$). The definitions of D and $\overline{D}$ could be interchanged, but they should be consistent throughout the circuit. Thus, all D's in a circuit imply the same value (0 or 1) and all $\overline{D}$'s will have the opposite value.

The propagation D-cubes of a gate are those that cause the output of the gate to depend only on one or more of its specified inputs. Thus, a fault on a specified input is propagated to the output. The propagation D-cubes for a two-input NAND gate are

a	b	f
1	D	$\overline{D}$
D	1	$\overline{D}$
D	D	$\overline{D}$

FIGURE 11.16
Singular cubes for the 2-input NOR function

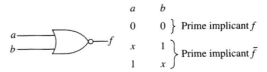

The propagation D-cubes $1D\overline{D}$ and $D1\overline{D}$ indicate that if one of the inputs of the NAND gate is 1, the output is the complement of the other. $DD\overline{D}$ propagates multiple input changes through the NAND gate.

Propagation D-cubes of a gate can be constructed by intersecting its singular cubes having different output values. The intersection rules are as follows:

$$0 \cap 0 = 0 \cap x = x \cap 0 = 0$$
$$1 \cap 1 = 1 \cap x = x \cap 1 = 1$$
$$x \cap x = x$$
$$1 \cap 0 = D$$
$$0 \cap 1 = \overline{D}$$

For example, the propagation D-tube of a three-input NOR gate can be formed as shown in Figure 11.17.

	a	b	c	f		a	b	c	f
c_1	0	0	0	1	$c_2 \cap c_1$	0	0	D	$\overline{D}$
c_2	x	x	1	0	$c_3 \cap c_1$	0	D	0	$\overline{D}$
c_3	x	1	x	0	$c_4 \cap c_1$	D	0	0	$\overline{D}$
c_4	1	x	x	0					

(a) Singular covers of the NOR gate

(b) Propagation D-cube of the NOR gate

FIGURE 11.17

Primitive D-Cube of a Fault

The primitive D-cube of a fault (pdcf) is used to specify the existence of a given fault. It consists of an input pattern which shows the effect of a fault on the output of the gate. For example, if the output of the NOR gate shown in Figure 11.16 is s-a-0, the corresponding pdcf is

a	b	f
0	0	D

Here D is interpreted as being 1 if the circuit is fault-free and is 0 if the fault is present. The pdcfs for the NOR gate output s-a-1 are

a	b	f
1	x	$\overline{D}$
x	1	$\overline{D}$

The pdcfs corresponding to an output s-a-0 fault in a gate can be obtained by intersecting each singular cube having output 1 in the fault-free gate with each singular cube having output 0 in the faulty gate. Similarly, the pdcfs corresponding to an output s-a-1 fault can be obtained by intersecting each singular cube with output 0 in the fault-free gate with each singular cube having output 1 in the faulty gate. The intersection rules are similar to these used for propagation D-cubes.

EXAMPLE 11.6

Let us consider a three-input NAND gate with input lines a, b, and c, and output line f. The singular cubes for the fault-free NAND gate are

	a	b	c	f
c_1	0	x	x	1
c_2	x	0	x	1
c_3	x	x	0	1
c_4	1	1	1	0

Assuming the input line b is s-a-1, the singular cubes for the faulty NAND gate are

	a	b	c	f
c'_1	0	x	x	1
c'_2	x	x	0	1
c'_3	1	x	1	0

Hence,

$$c_1 \cap c'_3 = \overline{D}x1D \quad c_4 \cap c'_1 = D11\overline{D}$$
$$c_2 \cap c'_3 = 101D \quad c_4 \cap c'_2 = 11D\overline{D}$$
$$c_3 \cap c'_3 = 1x\overline{D}D$$

Therefore, the primitive D-cube of the b s-a-1 fault is $101D$. The pdcfs for all single stuck-at faults for the three-input NAND gate are

a	b	c	f	Fault
0	x	x	D	f s-a-0
x	0	x	D	f s-a-0
x	x	0	D	f s-a-0
1	1	1	$\overline{D}$	f s-a-1
0	1	1	D	a s-a-1
1	0	1	D	b s-a-1
1	1	0	D	c s-a-1

Let us now consider how the various cubes described are used in the D-Algorithm method to generate a test for a given fault. The test generation process consists of three steps:

Step 1: Select a pdcf for the given fault.

Step 2: Drive the D (or $\overline{D}$) forward from the output of the gate under test to an output of the circuit, by successively intersecting the current **test cube** with the propagation D-cubes of successor gates. A test cube represents the signal values at various lines in the circuit during each step of the test generation process. The intersection of a test cube with the propagation D-cube of a successor gate results in a test cube.

Step 3: Justify the internal line values by driving back toward the inputs of the circuit, assigning input values to the gates so that a consistent set of circuit input values may be obtained.

EXAMPLE 11.7

Let us demonstrate the application of the D-Algorithm by deriving a test for detecting the α s-a-1 fault in Figure 11.18a. The test generation process is explained in Figure 11.18b. As can be seen in Figure 11.18b, the consistency operation at step 4 terminates unsuccessfully because the output of G_3 has to be set to 1. This can be done only by making input $B = 0$; however, B has already been assigned 1 in step 1. A similar problem will arise if D is propagated to the output via G_3 instead of G_2. The only way the consistency problem can be resolved is if the $\overline{D}$ output of G_1 is propagated to the output of the circuit via both G_2 and G_3 as shown in Figure 11.18c. No consistency operation is needed in this case, and the test for the given fault is $AB = 11$. This test also detects the output of G_2 s-a-0, the output of G_3 s-a-0, and the output of G_4 s-a-1.

FIGURE 11.18
Derivation of test for α s-a-1

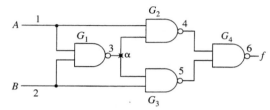

(a) Circuit example for D-Algorithm application

	1	2	3	4	5	6
Step 1: Select pdcf for α s-a-1.	1	1	$\overline{D}$	x	x	x
Step 2: Intersect the test cube with an appropriate propagation D-cube of G_2 (e.g., $1\overline{D}D$). (N.B., the parity of the D-cube is inverted.)	1	1	$\overline{D}$	D	x	x
Step 3: Intersect the test cube with the propagation D-cube $D1\overline{D}$ of G_4.	1	1	$\overline{D}$	D	1	$\overline{D}$
Step 4: Check that line 5 is at 1 from G_3 singular cubes.	1	#	$\overline{D}$	D	1	D

(b)

	1	2	3	4	5	6
Step 1: Select pdcf for α s-a-1.	1	1	$\overline{D}$	x	x	x
Step 2: Intersect the test cube with the propagation cube $1\overline{D}D$ of G_2.	1	1	$\overline{D}$	D	x	x
Step 3: Intersect the test cube with the propagation cube $\overline{D}1D$ of G_3.	1	1	$\overline{D}$	D	D	x
Step 4: Intersect the test cube with the propagation cube $DD\overline{D}$ of G_1.	1	1	$\overline{D}$	D	D	$\overline{D}$

(c)

EXAMPLE 11.8

As a further example of the application of the D-Algorithm let us derive a test for the s-a-0 fault at the output of gate G_2 in the circuit shown in Figure 11.19a. The test derivation is shown in Figure 11.19b. The test is $ABC = 011$.

FIGURE 11.19
Derivation of test for line 5 s-a-0

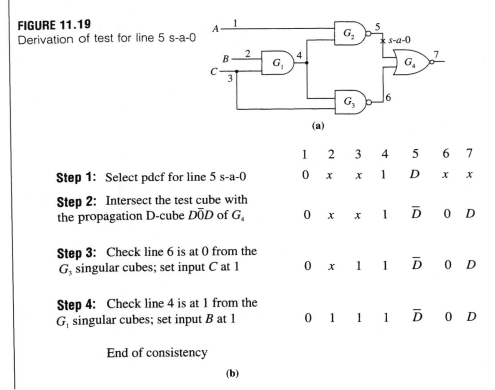

	1	2	3	4	5	6	7
Step 1: Select pdcf for line 5 s-a-0	0	x	x	1	D	x	x
Step 2: Intersect the test cube with the propagation D-cube $D\bar{0}D$ of G_4	0	x	x	1	$\bar{D}$	0	D
Step 3: Check line 6 is at 0 from the G_3 singular cubes; set input C at 1	0	x	1	1	$\bar{D}$	0	D
Step 4: Check line 4 is at 1 from the G_1 singular cubes; set input B at 1	0	1	1	1	$\bar{D}$	0	D

End of consistency

(b)

11.4 DESIGN FOR TESTABILITY

The phrase "design for testability" refers to how a circuit is either designed or modified so that the testing of the circuit is simplified. Several techniques have been developed over the years for improving the testability of logic circuits. These can be divided into two categories: ad hoc and structured. The ad hoc approaches simplify the testing problem for a specific design and cannot be generalized to all designs. On the other hand, the structured techniques are generally applicable to all designs.

Ad Hoc Techniques

One of the simplest ways of improving the testability of a circuit is to provide more test and control points. Test points are in general used to observe the response at a node inside the circuit, whereas control points are utilized to control the value of an internal node to any desired value, 0 or 1. For example in the circuit shown in Figure 11.20a, the fault α s-a-0 is undetectable at the circuit output. By incorporating a test point at node α as shown in Figure 11.20b, the input combination 010 or 011 can be applied to detect the fault.

The usefulness of adding a control point can be appreciated from the circuit shown in Figure 11.21a. If the output of the EX-NOR gate in the circuit is always 1, indicating that both the outputs of the logic block are the same, it is not possible to say whether the EX-NOR gate is operating correctly or not. If a control point is added to the circuit, as shown in Figure 11.21b, the input of the EX-NOR gate, and hence the operation of the circuit, can be controlled via the added point. During the normal operation of the circuit, the control point is set at logic 1. To test for an s-a-1 fault at the output of the EX-NOR gate, the control point is set at logic 0 and an input combination that produces logic 1 at the outputs has to be applied.

Another way of improving the testability of a particular circuit is to insert multiplexers in order to increase the number of internal nodes that can be controlled or observed from the external points. For example, the fault α s-a-0 in Figure 11.20a can also be detected by incorporating a 2-to-1 multiplexer as shown in Figure 11.22. When the test input (i.e., the select input of the multiplexer) is at logic 1, the output of the circuit is transferred to the output of the multiplexer. On the other hand, if the control input is at logic 0 and the input combination 010 or 011 is applied to the circuit, the state of node α can be observed at the multiplexer output.

A different way of accessing internal nodes is to use tristate drivers as shown in Figure 11.23. A test mode signal could be used to put the driver into the high impedance state. In this mode the input of the OR gate can be set to logic 0 or logic 1 from an external point. When the driver is enabled, the same external point becomes a test point.

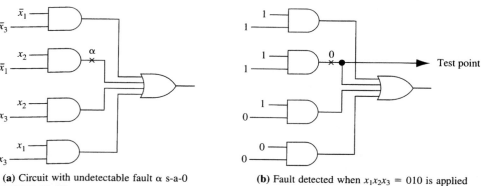

(a) Circuit with undetectable fault α s-a-0

(b) Fault detected when $x_1 x_2 x_3 = 010$ is applied

FIGURE 11.20

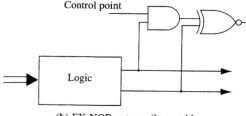

(a) EX-NOR gate not testable

(b) EX-NOR gate easily testable

FIGURE 11.21

FIGURE 11.22
Use of multiplexer to enhance testability

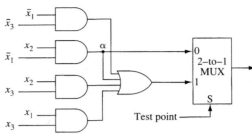

FIGURE 11.23
Use of tristate drivers to improve testability

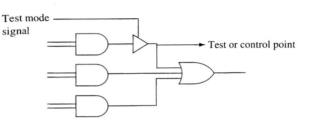

The test mode signals required by the added components such as multiplexers, tristate drivers, and so on cannot always be applied via external points, because it is often not practicable to have many such points. To reduce the number of external points, a **test state register** may be included in the circuit. This could in fact be a shift register that is loaded and controlled by just a few pins. The testability hardware in the circuit can then be controlled by the parallel outputs of the shift register.

Frequently flip-flops, counters, shift registers, and other memory elements assume unpredictable states when power is applied, and they must be set to known states before testing can begin. Ideally, all memory elements should be reset from external points (Fig. 11.24a). Alternatively, a power-up reset may be added to provide internal initialization (Fig. 11.24b).

A long counter chain presents another test problem. For example, the counter chain in Figure 11.25 requires thousands of clock pulses to go through all the states. One way

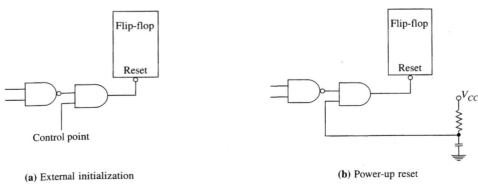

(a) External initialization (b) Power-up reset

FIGURE 11.24

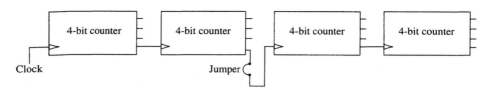

FIGURE 11.25
Breaking up a counter chain

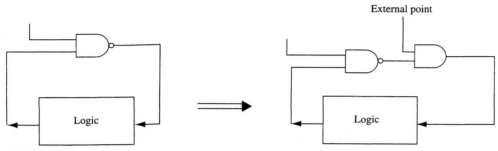

FIGURE 11.26
Breaking a feedback loop by using an extra gate

to avoid this problem is to break up the long chains into smaller chains by filling jumpers to them; the jumpers can be removed during testing.

A feedback loop is also difficult to test because it hides the source of the fault. The source can be located by breaking the loop and bringing both lines to external points that are shown during normal operation. When not shorted, the separate lines provide a control point and a test point. An alternative way of breaking a feedback loop is to add to the feedback path a gate that can be interrupted by a signal from a control point (Fig. 11.26).

On-circuit clock oscillators should be disconnected during test and replaced with an external clock. The external clock can be single-stepped to check the logic values at various nodes in the circuit during the fault diagnosis phase. Figure 11.27 shows how the on-board clock can be replaced by an external one.

FIGURE 11.27
Replacement of on-circuit clock

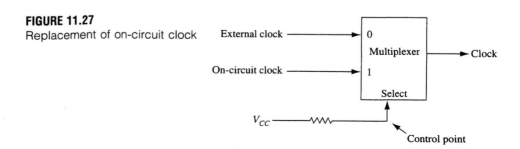

Structured Techniques

The ad hoc techniques for improving testability are incorporated after the logic design has been completed. Although they provide some help in simplifying testing, they are not by themselves sufficient to alleviate the problem of testing high-density chips. Research in this area indicates that testability has to be taken into account at the logic design level. Structured techniques are used to design logic circuits so that the testing will be significantly less complicated. These techniques are based on certain design rules and can be grouped into three categories [3]:

1. Scan path
2. Built-in self-test (BIST)
3. Autonomous self-testing

Scan Path

As mentioned earlier test generation for sequential circuits is significantly more complex than for combinational circuits. This is because of the difficulties in setting and checking the states of the flip-flops. As can be seen in Figure 11.28, there is access only to the primary inputs and primary outputs of a sequential circuit. The inputs of the flip-flops have to be controlled and their outputs observed through the combinational logic section of the circuit.

This problem can be overcome if the state information can be loaded into the flip-flops and their outputs observed directly. Thus, the task of testing sequential circuits simplifies to that of testing purely combinational circuits. The testing strategy based on this concept is generically referred to as the scan-path technique.

The basic idea of the scan path technique is to add an extra input into the flip-flop excitation logic in order to control the mode of operation of the circuit. In Figure 11.29 when $m = 0$ the circuit operates in its normal mode, but when $m = 1$ the circuit enters into a mode in which all the flip-flops are connected together to form a shift register. A 2-to-1 multiplexer is inserted at the input(s) of every flip-flop. The select inputs of all the multiplexers are tied together, and are driven by the signal on input m (known as the **scan select input**).

During normal operation, the inputs of the multiplexers are obtained from the combinational logic section. In shift-register mode, the first flip-flop receives its input from a primary input (known as the **scan-in input**) and the output of the last flip-flop is observed

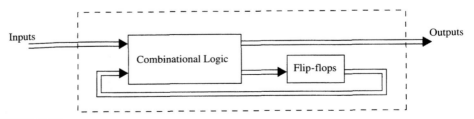

FIGURE 11.28
Sequential circuit

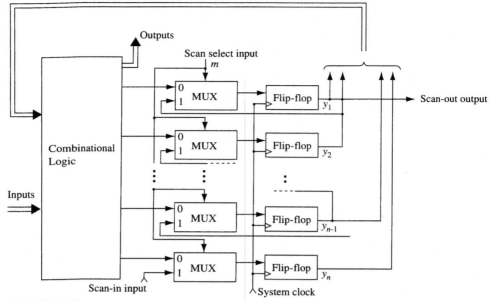

FIGURE 11.29
Sequential circuit with scan capability

on a primary output (known as the **scan-out output**). With a scan-in and a scan-out point accessible externally, the states of the flip-flops can be set and read directly by shifting data in and out.

■ **EXAMPLE 11.9**

Let us demonstrate the application of the scan-path technique by designing the circuit specified by the state table of Figure 11.30. Since there are five states, we need three flip-flops to implement the sequential circuit. We use D flip-flops and choose the following state assignment:

	y_1	y_2	y_3
A	0	0	0
B	0	0	1
C	0	1	0
D	0	1	1
E	1	0	0

The resulting excitation equations for the D flip-flops and the output equation are

$$D_1 = \bar{x}y_1 + xy_2y_3$$
$$D_2 = \bar{x}y_2 + x(\bar{y}_2y_3 + y_2\bar{y}_3)$$
$$D_3 = \bar{x}y_3 + x\bar{y}_1\bar{y}_3$$
$$z = xy_2y_3$$

FIGURE 11.30
A state table

State	Input	
	x = 0	x = 1
A	A,0	B,0
B	B,0	C,0
C	C,0	D,0
D	D,0	E,1
E	E,0	A,0

Next State/Output

The corresponding testable circuit, based on the scan-path design concept, is shown in Figure 11.31.

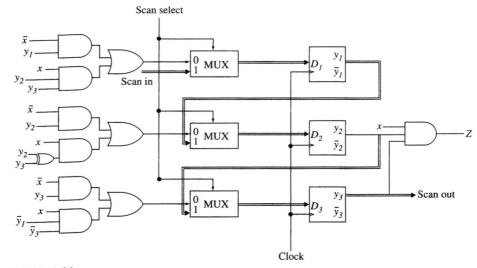

FIGURE 11.31
Testable sequential circuit

Sequential circuits designed using the scan-path technique can be tested by using the following procedure:

Step 1: Set the circuit to shift register mode by applying logic 1 to the scan select input.

Step 2: Check the operation of the shift register by using scan-in input, scan-out output, and the clock.

Step 3: Set the initial state of the shift register.

Step 4: Apply logic 0 to the scan select input to return to normal mode.

Step 5: Apply the test pattern to the combinational logic.

Step 6: Apply logic 1 to the scan select input to return to shift-register mode.

Step 7: Shift out the final state while shifting in the next starting state for testing the circuit.

Step 8: Go to step 4.

This procedure is applied to the circuit of Figure 11.31 to derive a diagnostic test sequence for the circuit, which is shown in Table 11.1. Any fault in the circuit will either produce a wrong output or show an erroneous state transition. Thus, by making the state variables externally controllable and observable, the testing of the circuit is considerably simplified.

TABLE 11.1
Diagnostic test sequence for the circuit of Figure 11.31

Step 1: Reset all the flip flops. Set scan select = 1 and apply 0101 at the scan-in input. The scan-out output should be 0101000; the first three bits of output e.g. 000 correspond to the original states of the flip-flops.

Step 2: Set scan select = 1, and scan in $y_1y_2y_3$ = 000. Set scan select = 0 and $x = 0$, which will make $z = 0$. Apply the clock pulse. Next, set scan select = 1 and scan out $y_1y_2y_3$; the scanned-out bits should be 000.

Step 3: Repeat step 2 with $x = 1$; in this case z should be 0 and the scanned out bits will be 001.

Step 4: Set scan select = 1, and scan in $y_1y_2y_3$ = 010.

Step 5: Set scan select = 0 and $x = 1$; this should produce $z = 0$. Apply the clock pulse, and scan out $y_1y_2y_3$ after making scan select = 1; the scanned out bits should be 011.

Step 6: Repeat step 4, and step 5 with $x = 0$; this should produce $z = 0$, and the scanned out bits should be 010.

Step 7: Set scan select = 1, and scan in $y_1y_2y_3$ = 011.

Step 8: Set scan select = 0 and $x = 1$; this should produce $z = 1$. Apply the clock pulse, and scan out $y_1y_2y_3$ after making scan select = 1; the scanned out bits should be 100.

Step 9: Repeat step 7, and step 8 with $x = 0$; this should produce $z = 0$, and the scanned out bits should be 011.

Step 10: Set scan select = 1, and scan in $y_1y_2y_3$ = 100.

Step 11: Set scan select = 0 and $x = 0$, which should result in $z = 0$. Apply the clock pulse, and scan out $y_1y_2y_3$ after making scan select = 1; the scanned out bits should be 100.

Step 12: Repeat step 9, and step 11 with $x = 1$; this should produce $z = 0$, and the scanned out bits should be 000.

Step 13: Set scan select = 1, and scan in $y_1y_2y_3$ = 001.

Step 14: Set scan select = 0 and $x = 0$, which should result in $z = 0$. Apply the clock pulse, and scan out $y_1y_2y_3$ after making scan select = 1; the scanned out bits should be 001.

Step 15: Repeat step 13, and step 14 with $x = 1$; this should produce $z = 0$, and the scanned out bits should be 010.

There are some disadvantages associated with the use of the scan-path technique. For example, setting a state requires a number of clock pulses equal to the length of the shift register. This can be reduced somewhat by simultaneously shifting in a new state while the current state is being shifted out. Alternatively, several smaller shift registers may be used rather than using a single long one. The main advantage of the scan-path technique is that a sequential circuit can be considered as a combinational circuit, the flip-flops providing some additional inputs and outputs. A number of techniques based on the concept of scan path are used in practice; the discussion of these can be found in Reference 4.

11.5 BUILT-IN SELF-TEST (BIST)

Over the past few years a considerable amount of work has been done in the area of built-in self-testing of logic circuits. This surge of interest in BIST arises from the fact that conventional test generation techniques cannot cope with high-density integrated-circuit chips. The basic philosophy of the BIST approach is to build the test pattern generator and output response checker directly on the chip (Fig. 11.32).

Currently the most popular approach used for built-in test pattern generation is using pseudo-random number generators. As discussed in Section 8.10, pseudo-random numbers are generated by an LFSR implementing a primitive polynomial of degree n whose period is $2^n - 1$. The output patterns produced by the pseudo-random number generator are applied to the circuit under test. The resulting response of the circuit is then compressed into a unique code; the purpose of the output data compression is to reduce the number of data bits that would have to be examined. The idea of compressing output data into a unique code was first proposed by Hewlett Packard and is known as **signature analysis.** In the signature analysis technique the output data from the circuit under test is fed into an n-bit LFSR. Selected taps of the LFSR are fed back to the input via EX-OR gates. After the data stream has been clocked through, a residue is left in the LFSR. This residue is unique to the data stream and is known as its signature. To form the signature of a data stream, the LFSR must first be initialized to a known state. Then the data stream is shifted in. Normally the all 0's state is chosen as the initial state. Figure 11.33 shows a simplified 16-bit signature generator.

FIGURE 11.32
BIST scheme

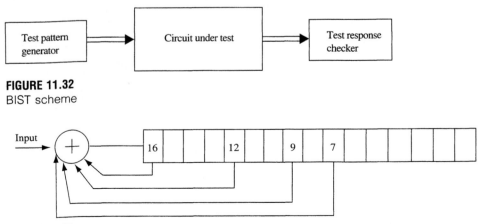

FIGURE 11.33
A 16-bit signature generator

EXAMPLE 11.10

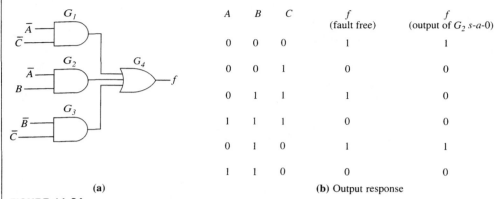

FIGURE 11.34

Let us apply the signature analysis technique for fault detection in the circuit of Figure 11.34a. The input patterns applied at A, B, and C, the correct circuit response, and the response of the circuit in the presence of the fault (output G_2 s-a-0) are shown in Figure 11.34b. The output of the circuit of Figure 11.34a is connected to the input of the signature generator circuit of Figure 11.35a. Each output bit produced by the circuit under test in response to one of the input patterns is clocked into the signature generator circuit before the next input pattern is applied to the circuit. Figure 11.35b shows the contents of the register in Figure 11.35a after each output bit produced at f is clocked into it. As can be seen from Figure 11.35b, the contents of the signature register after all six output bits

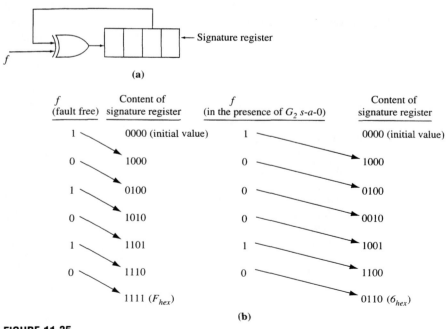

FIGURE 11.35
Generation of signatures for the fault-free and faulty circuit

have been clocked into it is 1111 (i.e., F_{hex}), which is the signature corresponding to the fault-free response. If the assumed fault (i.e., G_2 s-a-0) is present in the circuit under test, the corresponding faulty response (shown in Figure 11.34a) will be loaded into the signature register. The contents of the signature register after each of the faulty response bits is clocked into it is also shown in Figure 11.35b. It can be seen that the signature produced by the circuit in the presence of the fault (i.e., 6_{hex}) is different from that produced by the fault-free circuit, indicating the presence of the fault. ∎

An n-stage signature generator can generate 2^n signatures. However, many data streams can map into one signature. In general, if there are m symbols in a data stream, then in an n-stage signature generator, 2^{m-n} data sequences map into each signature. This mapping may give rise to **aliasing** (i.e., the signature generated from the faulty output response of a circuit may be identical to the signature obtained from the fault-free response), which masks the effect of the fault. The probability P that the output sequence of a circuit has deteriorated into another that has the same signature as itself is

$$P = \frac{2^{m-n} - 1}{2^m - 1}$$

P is calculated on the assumption that any of the possible output sequences of a given length may be good or faulty. For $m \gg n$, the probability P reduces to

$$P = \frac{1}{2^n}$$

Hence, the probability of fault-masking will be low if the signature generator has many stages.

The single-input signature generator has a drawback in that for a multi-output circuit, the input pattern necessary to produce an output signature has to be repeated for each output. This results in an undesirable overhead if the input sequence is very long. This problem can be overcome by using a multiple-input signature generator (**MISR**). In an MISR, an EX-OR gate is added between every two consecutive flip-flops (Fig. 11.36). Each flip-flop is fed by the output of the previous flip-flop and the output of the circuit under test. If the MISR is long enough to accommodate all the outputs of the circuit under test, then the input sequence has to be applied only once to produce a unique signature.

A major drawback of the BIST approach just discussed is that it does not provide any fault-coverage figure. Moreover, additional circuitry is needed to implement test pattern generators and signature analyzers.

Another approach to built-in self-testing, known as the **BILBO** scheme, combines the scan-path technique with the BIST approach described earlier. Figure 11.37 demonstrates how the scheme works. The first BILBO (Built-In Logic Block Observer) generates the pseudo-random patterns that are applied to the circuit under test. The second BILBO functions as an MISR.

The details of a 4-bit BILBO are shown in Figure 11.37. It is essentially a multimode shift register with facilities for EX-OR feedback. The two control inputs c_1 and c_2 are used to select one of the four modes of operation:

Mode 1: $c_1 = 0$, $c_2 = 1$. All the flip-flops are reset.

Mode 2: $c_1 = 1$, $c_2 = 1$. The input data $x_1 \ldots x_4$ can be parallel-loaded into the flip-flops and read on the flip-flop outputs.

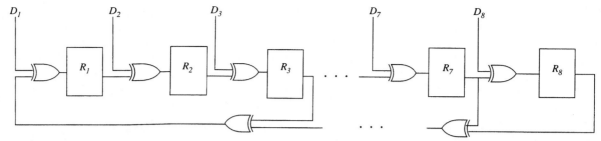

FIGURE 11.36
A multiple-input 8-bit signature register

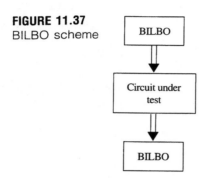

FIGURE 11.37
BILBO scheme

Mode 3: $c_1 = 0$, $c_2 = 0$. The BILBO acts as a serial shift register. The test inputs can be scanned in via the serial input, or the test outputs via the serial output.

Mode 4: $c_1 = 1$, $c_2 = 0$. The BILBO functions as an LFSR, which can be used as an MISR or as a test pattern generator by setting its inputs to a given value.

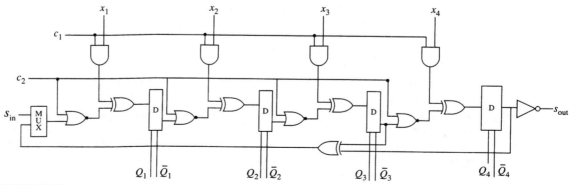

FIGURE 11.38
A 4-bit BILBO

The merit of this technique lies in the fact that a specific sequence of test inputs can be scanned in to maximize fault coverage. It is specially suitable for circuits which can be partitioned into independent modules. The flip-flops in a module can be configured to function as a test pattern generator or as a signature analyzer instead of using additional circuitry to implement these.

11.6 AUTONOMOUS SELF-TESTING

The structured design for testability techniques provides substantial improvements over designs that do not consider testability. However, there are a number of problems associated with these techniques:

i. Test generation is still necessary for combinational section of sequential circuits.
ii. Test time is substantially increased because test patterns have to be shifted in and the corresponding responses have to be shifted out.
iii. Tests are generated on the assumption that only the stuck-at types of faults occur in a circuit, which is not true in VLSI environments.

The autonomous self-testing approach, also known as **verification testing,** does not require test generation [5]. It consists of two steps:

Step 1: Partition a circuit into subcircuits. Each subcircuit should have few enough inputs so that all possible input combinations can be applied to it.

Step 2: Incorporate additional circuitry or reconfigure existing circuits to provide the input combinations needed to test all subcircuits exhaustively and to verify their responses.

A circuit designed in this way can operate either in the normal mode or in the test mode. In the normal mode the circuit performs its normal function, whereas in the test mode each subcircuit is tested exhaustively.

■ EXAMPLE 11.11

Let us modify the circuit of Figure 11.39a by incorporating two multiplexers. The modified circuit is shown in Figure 11.39b. The operation of the circuit can be controlled by setting appropriate values at the control inputs c_1 and c_2.

When $c_1 = 0$ and $c_2 = 1$, subcircuit 1 is disabled. By making $x_1 = 0$, subcircuit 2 can be exhaustively tested by applying the input combinations at x_6, x_7, and x_8. Similarly, when $c_1 = 1$ and $c_2 = 0$, subcircuit 1 can be tested by setting $x_8 = 0$ and applying all possible input combinations at x_1, x_2, x_3, x_4, and x_5. When $c_1 = c_2 = 0$, the NOR gate from which the output of the circuit is derived can be exhaustively tested via inputs x_1 and x_8. Finally, when $c_1 = c_2 = 1$, the circuit functions as the unmodified circuit, except for the added delay due to the multiplexers.

Notice that $2^8 = 256$ test patterns are needed if the original circuit of Figure 11.39a is to be exhaustively tested, whereas fewer than 50 tests are needed to exhaustively test all the subcircuits of Figure 11.39b.

FIGURE 11.39
Partitioning of a circuit for autonomous testing

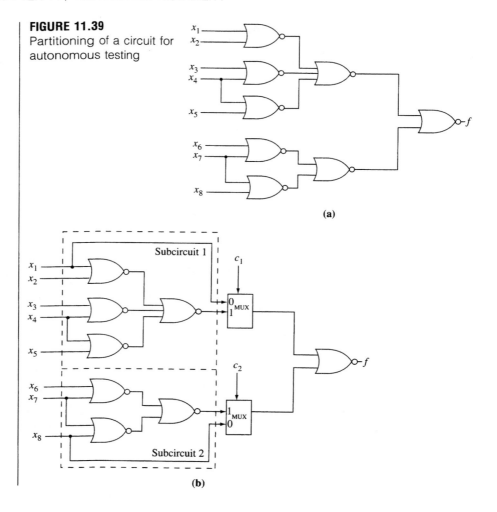

EXERCISES

1. Use path sensitization to find a test(s) for α being stuck at 1 in the following circuit. What other (if any) stuck-at faults will be detected by the test pattern(s)?

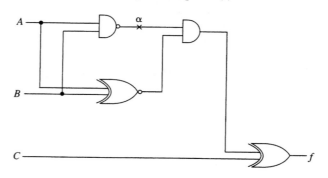

2. For the circuit shown, use the path sensitization method to detect the multiple fault (α s-a-0, β s-a-1).

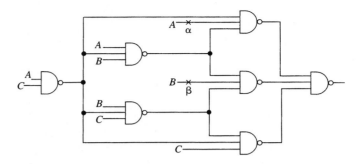

3. Find the minimal test that detects all single stuck-at faults in a 3-input NOR gate.
4. Find the minimal test set to detect all single stuck-at faults in the following circuit:

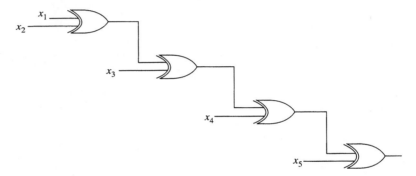

5. Show that all single stuck-at faults in the following circuit can be detected by using only four tests.

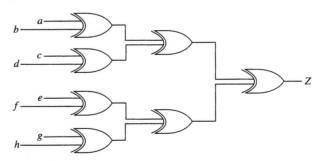

6. Find the Boolean difference of

$$F(x_1, x_2, x_3, x_4) = \bar{x}_1 \cdot \bar{x}_3 + x_2 \cdot x_4 + \bar{x}_2 \cdot \bar{x}_3 + x_3 \cdot \bar{x}_4 + x_1 \cdot x_4$$

using a Karnaugh map.

7. Find $\dfrac{dF}{dx_1}, \dfrac{dF}{dx_2}, \dfrac{dF}{dx_3}$ for the following functions:

   a. $F(x_1, x_2, x_3) = x_1 x_2 + x_1 x_3 + x_2 x_3$
   b. $F(x_1, x_2, x_3) = x_1 \bar{x}_2 + x_1 x_2 \bar{x}_3 + \bar{x}_1 \bar{x}_3$
   c. $F(x_1, x_2, x_3) = x_1(x_2 + x_3) + \bar{x}_2 \bar{x}_3$

8. Prove that the Boolean difference of a function F with respect to x_i is 0 if F is independent of x_i.
9. Determine the test(s) to detect the fault α s-a-0 in the following circuit using the Boolean difference method.

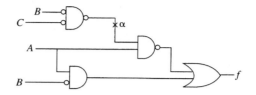

10. Find the Boolean expression for a 2-out-of-3 majority voting circuit, assuming the inputs to the circuit are a, b, and c. Derive the Boolean difference of the function with respect to ab and explain the results.
11. Find dF/db for the following function using a Karnaugh map.

$$F(a,b,c,d) = ab \cdot (\bar{c} + \bar{d}) + cd \cdot (\bar{a} + \bar{b})$$

12. For the circuit shown on p. 413,
    a. Determine the primitive D-cubes.
    b. Determine all propagation D-cubes for faults on input lines.
    c. Determine all primitive D-cubes for single faults g s-a-1 and h s-a-0.

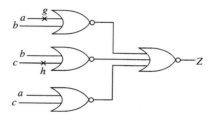

13. For the circuit shown, determine a test for detecting the fault
    a. β s-a-0   b. α s-a-1
    using the D-Algorithm

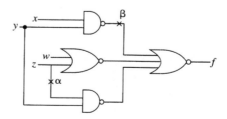

14. For the circuit shown, which, if any, of the following input patterns detect the fault α stuck-at-0?
    a. 100   b. 010   c. 101   d. 110

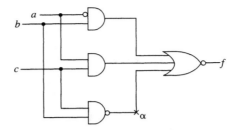

15. For the circuit shown, derive tests to distinguish the following pairs of faults.
   a. a stuck-at-0 and b stuck-at-0
   b. a stuck-at-1 and b stuck-at-1

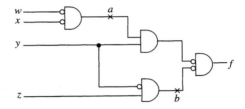

16. Determine a minimal set of tests that will detect any single stuck-at fault on an input or output line of a 4-to-1 multiplexer device.

17. Find a test pattern to detect the fault a stuck-at-1 in the circuit shown on p. 414. Is the fault b s-a-1 detectable? If not, why not?

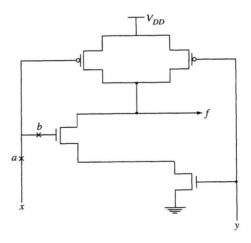

18. A possible realization of the logic function is shown below

$$f = \overline{A}\overline{B}\overline{C} + ABD + ACD$$

The faults α stuck-at-1 and β stuck-at-1 are indistinguishable in this circuit. Suggest an alternative realization of the circuit so that either fault is locatable to one gate instead of two.

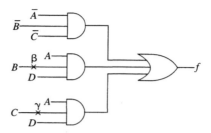

19. Implement the following state table with a control input α so that the flip-flops in the resulting circuit are connected as a shift register (assume D flip-flops), when α is set to logic 1:

Present State	Input x = 0	x = 1
A	B,0	C,0
B	C,1	D,0
C	D,0	E,0
D	E,0	F,1
E	F,1	A,0
F	A,0	B,1

20. Show how the following circuit can be partitioned so that the autonomous self-test approach can be used to reduce the number of test patterns.

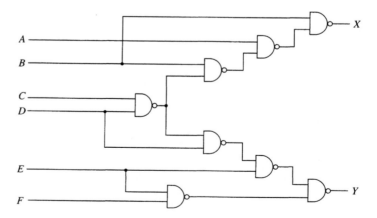

References

1. A. Avizienis, "Fault-tolerant systems," *IEEE Trans. on Computer,* December 1976, pp. 1304–1311.

2. S. K. Jain and V. D. Agrawal, "Test generation for MOS circuits using the D-Algorithm," *Proc. 20th Design Automation Conf.,* 1983, pp. 1304–1311.

3. E. J. McCluskey, *Logic Design Principles,* Prentice Hall, 1986.

4. M. Abramovici, M. A. Breuer, and A. D. Friedman, *Digital System Testing and Testable Design,* Computer Science Press, 1990.

5. E. J. McCluskey and S. Bozorgui-Nesbat, "Design for autonomous test," *IEEE Trans. on Computer,* November 1981, pp. 866–875.

Index

1's complement, 14
2's complement, 17
a-numbers, 323, 324
ABEL, 131, 201
   fuse map, 131
   JEDEC file, 131
Absorption law, 41, 48
Access time, 309
Acyclic, 46
Addition, 12
Adjacent, 46
Algebraic division, 99
Algebraic product, 99
Algebraic substitution, 98
Algorithmic state machine (ASM), 198
Aliasing, 405
AND gates, 59
Antisymmetric relation, 43
Applicable input sequence, 195
Arcs, 46
ASM, 198
Associative law, 41
Asynchronous operation, 141, 142
   fundamental mode, 237
Asynchronous up-down counters, 272
ATV750, 220
B-block, 382
Backward trace, 389
Base, 359
Binary Arithmetic, 5
   borrow, 6
   division, 7
   minuend, 6
   multiplication, 7
   subtrahend, 6
BCD, 23
BCD adders, 325
BCD subtractors, 330
Berger codes, 31
   self-checking circuits, 31
BiCMOS logic, 371
Bijection, 45
BILBO, 405
Binary encoding, 23
   non-weighted codes, 23, 25
      cyclic code, 26, 27
      excess-3 code, 25
      Gray code, 27
      reflected code, 27
   weighted codes, 23
      BCD, 23
      self-complementing, 24
      weight, 23
Binary numbers, 2
Binary relation, 42
Bipolar, 343
Bipolar logic families, 358
   base, 359
   collector, 359
   current transfer ratio, 359
   d.c. current gain, 360
   forward-bias, 359
   reverse-bias, 359
Biquinary codes, 30
BIST, 403
Black box, 166
BLIF, 192
Boolean algebra, 47
   absorption law, 48
   closure property, 47
   complement, 47
   consensus, 50
   DeMorgan's law, 49
   duality, 47
   idempotent law, 48
   involution law, 49
Boolean difference, 388
Boolean division, 99
Boolean functions, 52, 56
   canonical product of sums, 56
   canonical sum of products, 56
   complementation, 53
   literals, 56
   maxterm, 56
   maxterm list form, 57
   minterm, 56
   minterm list form, 57
   product, 52
   product of sums, 55
   sum, 53
   sum of products, 55
   truth table, 53
Boolean product, 99
Boolean substitution, 98
Borrow, 6
Branches, 46

Bridging faults, 381
   feedback bridging, 381
   input bridging, 381
Buried cells, 213
Bus, 365
Bytes, 309
Canonical product of sums, 56
Canonical sum of products, 56
Capacity, 309
Carry-anticipation, 317
Carry-generate, 319
Carry-lookahead, 317
Carry-lookahead adders, 317
   carry-anticipation, 317
   carry-generate, 319
   carry-propogate, 319
Carry save multiplication, 335
Carry-propogate, 319
Carry-save addition, 319
Cartesian product, 42
Chain connected, 45
Channel, 343
Chips, 108
   LSI chips, 109
   MSI chips, 108
   SSI chips, 108, 109
   TTL, 109, 110
   VLSI chips, 109
Circulating ring counters, 296
Classical fault model, 380
Clock, 141
Clocked CMOS circuits, 356
   evaluate phase, 356
   precharge phase, 356
Clock inhibit, 301
Clock skew, 154
Clocked latch, 144
Closed, 197
Closure, 197
Closure property, 47
Coincidence gate, 63
Cokernel, 100
Cokernel cube matrix, 101
Collapsing, 98
Collector, 359
Combinational logic circuits, 69
Commutative law, 41
Comparators, 338
Compatibility class, 196
Compatible, 195
Complement, 14, 47
Complement set, 41
Complementary approach, 78
Complementation, 53
Connection matrix, 46
Consensus, 50
Controlled inverter, 104
Cover, 197, 247
Covering, 197
Critical race, 243
Crosspoints, 122
Cubical algebra, 391
   test cube, 393
Current transfer ratio, 359
Cutoff, 360
Cycle, 46, 245
Cycle time, 309
Cyclic, 85

Cyclic code, 26, 27
D-Algorithm, 389
   primitive D-cube of fault, 391
   propagation D-cube, 391
   singular cube, 391
D flip flop, 146
D.C. current gain, 360
Darlington circuit, 363
Decade counter, 270
Decimal numbers, 1
Decoders, 119, 120
Decomposition, 97
Decomposition charts, 104
DeMorgan's law, 41, 49
Demultiplexers, 117
Depletion mode, 344
Design equations, 160, 174
   state assignment, 174
Diminished radix complement, 14
   1's complement, 14
   end-around carry, 15
   negative zero, 16
   positive zero, 16
Digraph, 46
Directed graph, 46
Distributive law, 41
Divide-by-n counter, 269
Division, 7, 336
   restored, 337
Domain, 45
Domino logic, 358
Don't care conditions, 77, 87
Double-rail inputs, 94
Drain, 343
Drivers, 346
Duality, 47
Duration, 379
Duty cycle, 142
Dynamic hazard, 256, 259
Dynamic logic, 349
   dynamic NMOS logic, 350
   static NMOS logic, 350
Dynamic NMOS logic, 350
Dynamic RAM, 309
Edges, 42, 46
EEPLDs, 217, 222
Emitter, 359
Empty set, 39
End-around carry, 15
Enhancement mode, 344
EPLDs, 217
Equivalence gate, 63
Equivalence classes, 44
Equivalence partition, 171
Equivalence relation, 43
Equivalent, 170
Error correcting codes, 32
   Hamming code, 34
   Hamming distance, 32
Error detecting codes, 28
ESPRESSO, 88
   off-set, 88
   on-set, 88
   prime cube, 88
Essential hazard, 256, 261
Essential prime implicant, 82
Evaluate phase, 356
EX-OR gates, 62, 103–108

Excess-3 code, 25
Excitation, 385
Excitation equations, 166
Excitation variables, 141
Extent, 379
Extraction, 97
Factoring, 97
Fall delay, 152
Fall time, 362
Falling edge, 142
Fault, 379
    duration, 379
    extent, 379
    hard fault, 379
    nature, 379
    permanent, 379
    soft fault, 379
    temporary, 379
Fault coverage, 386
Fault detection, 385
    excitation, 385
    fault coverage, 386
    non-redundant, 385
    sensitization, 385
Fault matrix, 387
Feedback bridging, 381
Finite state machine, 141
Flattening, 98
Flip-flops, 145
    hold time, 146
    metastable state, 145
    setup time, 146
Floating-point numbers, 19
    mantissa, 19
    normalization, 20
Flow table, 238
Forward trace, 389
Forward-bias, 359
FPLA, 124
Full adders, 313
Full-simplify, 91
Full subtractors, 326
Function hazard, 256
Functions, 45
    bijection, 45
    domain, 45
    injection, 45
    mapping, 45
    one-to-one, 45
    onto, 45
    range, 45
    transformation, 45
Fundamental mode, 237
Fuse map, 131
GAL16V8, 222
Gated latch, 144
Gates, 39, 59, 344
    AND, 59
    coincidence, 63
    equivalence, 63
    NOT, 61
    OR, 60
Glitch, 256
Graphs, 46
    acyclic, 46
    adjacent, 46
    arcs, 46
    branches, 46
    connection matrix, 46
    cycle, 46
    digraph, 46
    directed, 46
    edges, 46
    in-degree, 46
    leaves, 46
    nodes, 46
    non-directed, 46
    out-degree, 46
    path, 46
    root, 46
    tree, 46
    vertices, 46
Gray code, 27
Gray code counters, 276
Greatest lower bound, 44
Half-adders, 313
Half-subtractors, 325
Hamming code, 34
Hamming distance, 32
Hard fault, 379
Hazard, 255
    dynamic, 256, 259
    essential, 256, 261
    function, 256
    glitch, 254
    logic, 256, 257
    static, 256, 257
    static-0, 256
    static-1, 256
Hexadecimal numbers, 12
    addition, 12
    subtraction, 13
High-speed TTL, 368
Hold-time, 146
Idempotent law, 41, 48
Implication table, 172
    implications, 172
    incompatibles, 173
Implications, 172
Incompatibles, 173, 195
Incompletely specified functions, 77
Incompletely specified sequential circuits, 195
Incompletely specified state table, 195
    applicable input sequence, 195
    closed, 197
    closure, 197
    compatibility class, 196
    compatible, 195
    covering, 197
    covers, 197
    incompatible, 195
    maximal compatible, 196
In-degree, 46
Injection, 45
Input bridging, 381
Integrated circuit counters, 278
Inverts, 361
Involution law, 49
Irreducible polynomial, 303
Irreflexive relation, 42
Is-included-in, 40
Jam inputs, 302
JEDEC file, 131
JEDI, 191–193

418 □ INDEX

JK Flip-flop, 148
Johnson counters, 299
    clock inhibit, 302
    jam inputs, 302
    preset enable, 302
    pseudo Johnson, 299
    twisted ring, 299
Karnaugh maps, 73
Kernels, 99
    cokernel, 99–100
    cokernel cube matrix, 101
    level-0 kernel, 100
    rectangular cover, 101–102
KISS2 format, 190
Latches, 143
    clocked latch, 144
    gated latch, 144
    reset, 143
    set, 143
    transparent latches, 144
Leaves, 46
Level-0 kernel, 100
Linear feedback shift register (LFSR), 303
Literals, 56
Load, 346
Logic circuits, 69
    combinational, 69
    sequential, 69
Logic hazard, 256, 257
L/W ratio, 346
Low-power Schottky TTL, 369
Low-power TTL, 368
Lowest upper bound, 45
LSI chips, 109
M-out-of-*N* codes, 29
Macrocells, 210, 213
Majority function, 323
Mantissa, 19
Mapping, 45
Maximal compatible, 196
Maximal length sequence, 304
Maxterm, 56
Maxterm list form, 57
Mealy model, 157
Memory, 141
Metastable state, 145
Minimization, 70
    full-simplify, 91
    multiple-output functions, 90
    shared product terms, 90
Minterm, 56
Minterm list form, 57
Minterm table, 116
Minuend, 6
MISR, 405
Mod (modulo) number, 269
Moore model, 157
MOS, 343
MSI chips, 108
Multilevel logic design, 97
    collapsing, 97, 98
    decomposition, 97
    extraction, 97
    factoring, 97
    flattening, 98
    substitution, 97
        algebraic, 98

Boolean, 98
Multiple-output functions, 90
Multiplexers, 109
Multiplication, 7, 331
    carry-save multiplication, 335
NAND-NAND logic, 92, 93
    double-rail inputs, 94
    single-rail inputs, 94
Nature, 379
Negative edge, 142
Negative zero, 16
Next state, 141
NMOS inverters, 344
    drivers, 346
    load, 346
    L/W ratio, 346
    ratioed inverter, 346
NMOS transistors, 343
    channel, 343
    depletion-mode, 344
    drain, 343
    enhancement-mode, 344
    gate, 344
    source, 343
    threshold voltage, 344
Nodes, 46
Non-binary counter, 288
Non-critical race, 243
Non-directed graph, 46
Non-redundant circuit, 385
Non-redundant fault detection, 385
Non-saturated switch, 361
Non-weighted codes, 23, 25
NOR-NOR logic, 92, 95
Normalization, 20
NOT gates, 61
NOVA, 190–193
Null partition, 44
Null set, 39
Observability product term, 213
Octal numbers, 8
Off-set, 88
Off-time, 142
On-set, 88
On-time, 141
One-shot assignment, 248
One-to-one function, 45
Onto function, 45
Open state, 364
OR gates, 60
Ordered pair, 41
Out-degree, 46
Overflow, 14, 18
PAL, 128
PAL22V10, 210
    macrocell, 210
PAL23S8, 213
    buried cells, 213
    macrocells, 213
    observability product term, 213
PAL29M16, 228
Parallel in/parallel out, 287
Parallel in/serial out, 287
Parity checked codes, 29
Partially symmetric function, 323
Partitions, 43
    chain connected, 45

INDEX

equivalence classes, 44
equivalence partition, 171
greatest lower bound, 44
lowest upper bound, 45
null partition, 44
product, 44
sum, 45
unity partition, 44
PAL16R4, 208
PAL22V10, 208, 210
PAL23S8, 208, 213
Parity function, 323
Partitioning, 170, 186
Pass transistors, 350
Path, 46
PEEL18CV8, 225
Permanent fault, 379
PLDs, 121–136, 201–229
PLS155, 202
Positive edge, 142
Positive zero, 16
Power set, 40
Precharge phase, 356
Present state, 141
Preset enable, 302
Primary input signals, 141
Prime cube, 88
Prime implicant, 82
    essential prime implicant, 82
    prime implicant chart, 85
    cyclic, 85
Prime implicant chart, 85
Primitive D-cube of a fault, 391
Primitive flow table, 241
Primitive polynomial, 303
Product, 44, 52
Product of sums, 55
PROM, 122
Proper subset, 40
Propogation D-cube, 391
Pseudo Johnson, 299
Pseudo-random sequence, 303
    linear feedback shift register (LFSR), 303
    maximal length sequence, 304
    Quine-McCluskey method, 82
Quine-McCluskey method, 82
Radix complement, 17
    2's complement, 17
    overflow, 18
    sign-extended, 19
Random access memory, 309
    access time, 309
    bytes, 309
    capacity, 309
    cycle time, 309
    dynamic RAM, 309
    read, 309
    records, 309
    sequential access memory, 309
    static RAM, 309
    write, 309
Range function, 45
Ratioed inverter, 346
Read, 309
Records, 309
Rectangular cover, 101, 102
Redundant state, 169
Reed-Muller canonical form, 108

Reflected code, 27
Reflexive relation, 42
Relations, 41
    antisymmetric, 43
    binary relation, 42
    Cartesian product, 42
    edges, 42
    equivalence relation, 43
    irreflexive, 42
    ordered pair, 41
    reflexive, 42
    symmetric, 42
    transitive, 43
    vertices, 42
Reset, 143
Residue codes, 30
Residue function, 114
Restored, 337
Reverse-bias, 359
Ring counters, 296
    circulating, 296
    self-decoding, 297
Ripple adder, 317
Ripple (asynchronous) counters, 267
    decade counter, 270
    divide-by-n, 269
    mod (modulo) number, 269
Rise delay, 152
Rise time, 362
Rising edge, 142
Root, 46
Saturated switch, 361
Saturation, 360
Scan path, 399
    scan-in input, 399
    scan-out output, 400
    scan-select input, 399
Scan-select input, 399
Scan-in input, 399
Scan-out output, 400
Schottky TTL, 369
Secondary input signals, 143
Self-checking circuits, 31
Self-complementing code, 24
Self-starting counter, 276
Sensitization, 385
Sequential machine, 141
Sequential access memory, 309
Sequential logic circuits, 69, 141
Serial and parallel adders, 316
    ripple adder, 317
Serial in/parallel out, 286
Serial in/serial out, 286
Set, 143
Set theory, 39
    absorption law, 41, 48
    associative law, 41
    commutative law, 41
    complement set, 41
    DeMorgan's law, 41, 49
    distributive law, 41
    empty set, 39
    idempotent law, 41, 48
    is included in, 40
    null set, 39
    power set, 40
    proper subset, 40
    singleton, 39

Set-up time, 146
Shared product terms, 90
Shift registers, 284
    parallel in/parallel out, 287
    parallel in/serial out, 287
    serial in/parallel out, 286
    serial in/serial out, 286
    shift-left register, 285
    shift-right register, 285
Shift register counters, 287
Shift-left register, 285
Shift-right register, 285
Sign-extended, 19
Sign-magnitude representation, 14
Signature analysis, 403
Signed numbers, 14
    complement, 14
    overflow, 14, 18
Single transition time assignments, 248
Single-rail inputs, 94
Singleton, 39
Singular cube, 391
SIS, 89, 190
Soft fault, 379
Source, 343
SSI chips, 109, 110
Stable state, 237
Standard TTL, 368
State assignment, 166, 174, 182
    partitioning, 186
    substitution property, 186
State diagrams, 154, 155, 160
State minimization, 166
State reduction, 169
    redundant, 169
    equivalent, 170
State tables, 154, 155
State transition graph, 156
State variables, 141
Static, 256, 257
Static NMOS logic, 350
Static RAM, 309
Static-0, 256
Static-1, 256
Stuck-at fault, 380
Stuck-open fault, 381
Substitution, 97
Substitution property, 186
Subtraction, 13
Subtrahend, 6
Sum, 45, 53
Sum of products, 55
Switches, 343
Symmetric relation, 42
Synchronous circuits, 141
    clock, 141
    duty cycle, 142
    falling edge, 142
    negative edge, 142
    off-time, 142
    on-time, 141
    positive edge, 142
    rising edge, 142
Synchronous counters, 272

self-starting counter, 276
T Flip-flop, 151
    toggle, 151
    trigger, 151
Temporary fault, 379
Test cube, 393
Test state register, 397
Threshold voltage, 344
Timing, 151
    clock skew, 154
    fall delay, 152
    rise delay, 152
Toggle, 151
Totally symmetric function, 321
    a-numbers, 323, 324
    majority function, 323
    parity function, 323
Totem pole, 363
Transformation function, 45
Transistor switch, 360
    cutoff, 360
    inverts, 361
    non-saturated switch, 361
    saturated switch, 361
    saturation, 360
Transition diagram, 245
Transition table, 160
Transitive relation, 43
Transmission gates, 353
Transparent latches, 144
Tree graph, 46
Trigger, 151
Tristate device, 354
Truth table, 53
TTL, 109, 110, 361
    bus, 365
    Darlington circuit, 363
    fall time, 362
    open state, 364
    rise time, 362
    totem pole, 363
    wired AND, 365
    wired logic, 365
TTL subfamilies, 368
    high-speed TTL, 368
    low-power Schottky TTL, 369
    low-power TTL, 368
    Schottky TTL, 369
    standard TTL, 368
Twisted ring, 299
Two's complement subtractors, 329
Unidirectional fault, 381
Unity partition, 44
Universal shift register, 287
Unstable state, 237
Verification testing, 407
Vertices, 42, 46
VLSI chips, 109
Weight, 23
Weighted codes, 23
Wired AND, 365
Wired logic, 365
Write, 309